Landscape Ecolog

Concepts, Methods and App

D0084362

Landscape Ecology
Concepts, Methods and Applications

Françoise Burel
Director of the ECOBIO research Unit
Centre National de la Recherche Scientifique (CNRS)
Rennes, France

Jacques Baudry
Director of the SAD-Armorique research Unit
Institut National de la Recherche Agronomique (INRA)
Rennes, France

Illustrations
Yannic Le Flem

Photographs
Jacques Baudry

Science Publishers, Inc.
Enfield (NH), USA   Plymouth, UK

SCIENCE PUBLISHERS, INC.
Post Office Box 699
Enfield, New Hampshire 03748
United States of America

Internet site: *http://www.scipub.net*

sales@scipub.net (marketing department)
editor@scipub.net (editorial department)
info@scipub.net (for all other enquiries)

Library of Congress Cataloging-in-Publication Data

Burel, Françoise.
    [Écologie du paysage, English]
    Landscape ecology: concepts, methods, and applications/
    Françoise Burel, Jacques Baudry; illustrations, Yannie
    Le Flem; Photographs, Jacques Baudry.
      p. cm.
    Includes bibliographical references (p.).
    ISBN 1-57808-214-5
      1. Landscape ecology. I. Baudry, Jacques. II. Title.

    QH541.15.L35 B8713 2003
    577.5'5--dc21   2002042731

ISBN 1-57808-214-5

Published by arrangement with Technique & Documentation, Paris

© 2003, copyright reserved

Reprint 2004

Translation of: *Écologie du paysage concepts, méthodes et applications*,
                   Technique & Documentation, Paris, 1999.
French edition: © Technique & Documentation, Paris, 1999
                   ISBN 2-7430-0305-7

Published by Science Publishers, Inc. Enfield, NH, USA
Printed in India.

We are grateful to the research and economic affairs service of the Ministry in charge of the Environment for its support of landscape ecology research since 1985 and for its encouragement of the publication of this work. Finally, we are most grateful to the service for having ensured a wide dissemination of this work.

Acknowledgements

This book is one step in our research agenda, an agenda marked by encounters with many colleagues. Those colleagues have posed incisive questions: Jean-Claude Lefeuvre, for example, asked us how to describe a territory from the ecological point of view. Meetings with Richard Forman, Michel Phipps, and Gray Merriam, just as we started work on *Landscape Ecology*, were decisive in the orientation of our studies. We had the good fortune of their diverging opinions and their warm welcome, as well as that of Barbara, Jenny and Aileen. Our work was also warmly received by Jean-Pierre Deffontaines (and Colette), who contributed to its heterogeneity by adding an agronomic point of view. With Jean-Pierre, we were able to measure the richness and complexity of the commonly used word "landscape". These meetings allowed us to be part of the networks and international connections and to discover many landscapes, and they would not have been possible without the motivation of George Cancela de Fonseca and Alain Ruellan. To participate in the birth of landscape ecology has been an exciting adventure. It has given us many occasions for collaboration and the benefit of encouragement, since there were moments of doubt. Frank Golley, Bob Bunce, Zev Naveh, Jon Marshall, Gary Fry, Hubert Gulink, Maurizio Paoletti, Chantal Blanc-Pamart and others were important supporters.

These studies were completed with the collaboration of our students, who shared a part of our life in our laboratories, shook up our certainties, and pushed us towards new roads. Some became our colleagues, including Mark McDonnell, Thierry Tatoni, Sandrine Petit, Claudine Thenail and Didier Le Coeur. We are grateful to all those who had confidence in us and helped to make the 1993 IALE colloquium in Rennes a success.

We are grateful to our laboratory colleagues, to Yann Rantier for the preparation of figures, and to Yann and Maryvonne Chevallereau for statistical help.

Thanks to Lou and Sandrine for the commas and the weeding out of excessively long phrases and many errors, and to Yannic Le Flem, the nearly permanent artist of *Landscape Ecology*, for the illustrations.

Contents

Part I

Introduction

1

Definition of a Discipline

1. EMERGENCE OF LANDSCAPE ECOLOGY IN THE HISTORY OF ECOLOGY

1.1. History of ecology from its origin to the 1970s

The term *ecology*, proposed in 1866 by the German biologist Haeckel, denotes the study of relations of living things with their environment. Etymologically, it combines the Greek words *oikos* and *logos* and signifies the science of the habitat. In fact, ecology aims to establish the laws regulating not only relationships between living things and their physicochemical environments, but also relationships that develop between organisms.

Over time, the objects of ecological study have become progressively more complex, from the individual to the landscape, in relation with the development of sciences in general and, more recently, of technologies. Ecology has gradually changed its scientific objects and widened its field of investigation from the study of species considered in their relationships with the surrounding physical environment (autecology) to analysis of groups of species (settlements, communities) in "natural" environments, and finally consideration of complex systems integrating humans and their activities (Sheail, 1987; Acot, 1988; Barnaud and Lefeuvre, 1992).

Di Castri (1981) distinguished five stages in this progression, chronologically and conceptually: autecology, synecology, the ecosystem, the biosphere, and man in the biosphere. In reality, the weight of naturalist disciplines, where many researchers come from, influences the evolution of ecology: from the persistence of principles from botany and zoology rose a marked interest in the levels of organization corresponding to the individual (autecology), the population, and groups of species (synecology).

1.1.1. Autecology

Autecology is the study of relationships between individuals and the environment in which they live. These relationships involve individuals in themselves and the population they constitute. Autecology thus addresses physiological problems (ecophysiology) as well as demographic problems. In this perspective, the environment is conceived as a set of factors of two types: abiotic factors, linked to the physical and chemical environment, and biotic factors, linked to the living things present in the ecosystem under study. The major research issues addressed at this level have treated the functioning of the individual organism in the framework of the constraints the environment imposes on it, in order to understand its adaptation to these constraints and to determine its capacity to survive when they change. In essence, ecophysiological studies can be grouped according to two major axes: the establishment of individual energy budgets and the study of adaptation to the physicochemical environment. Demographic studies aim to define the role of factors, biotic as well as abiotic, that determine the mode in which populations are renewed and the variation of their total numbers. This causal approach constitutes population dynamics.

1.1.2. Synecology

Synecology (sometimes called biocoenotics) considers essentially the structure and functioning of communities. The most developed approach at this level is description and understanding of trophic structures of ecological systems. Apart from these trophic relations (prey-predator, plant-phytophage, etc.), a certain number of interspecific relationships control the functioning of communities. These include the exploitation of one population by another, competition for resources, and symbiosis or mutualism, in which each partner benefits from the activity of the other.

1.1.3. Ecology of ecosystems

The notion of ecosystem was introduced by Tansley in 1935: "These ecosystems, as we may call them, are of the most various kinds and sizes. They form one category of the multitudinous physical systems of the universe, which range from the universe as a whole to the atom." Tansley thus defined the concept of ecosystem as an element in the hierarchy of physical systems from the universe to the atom, a basic system of ecology, and a composite of all living things and the physical environment. From that definition to the present, the ecosystem has remained a key concept of ecology.

Golley (1993) traced the history of ecosystem research, emphasizing the importance of scientific personalities involved, parallel advances of theory and application of research, interactions and competition with other scientific disciplines, influences of institutions and the public, the role of political events such as wars, depressions, and revolutions, and the role of cultural paradigms, which structure our thoughts on the nature and objectives of the research.

The history of the ecosystem has been mostly American. Even though the field of study emerged partly in Europe, scientists there abandoned it after World War II, during which the systemic theories were used by the Nazis to justify the connection between regionalism and racism. In the United States, on the contrary, the concept of ecosystem was developed and modernized. It drew on systemics, information theory, data processing, and modelling. The major research conducted in the 1950s, following the use of nuclear arms, concerned the effect of radiation on organisms and the transfer of radioactive elements along the food chain. In the early 1960s, the warnings of Rachel Carson in *Silent Spring* (Carson, 1962) triggered the environment movement. The ecosystem, even the ecology, was seen to be in danger. For managers and industrialists, ecosystem studies seemed to be a means of managing the complexity of natural systems. It was in this context that major research programmes on ecosystems were developed, especially the International Biological Programme (IBP). The ecosystem having been conceived as a machine, represented by a computer model, the multitude of complex interactions between elements of the ecosystem was reduced to a few flows of energy and matter (Fig. 1). Traditional systems studies often consisted of analyses of energy and matter flows across units considered as black boxes for which entry-exit budgets were calculated (Schlesinger, 1989). This prolific stream of research was long dominated by the influence of E.P. Odum (1969, 1971). The ecosystems studied were natural or semi-natural

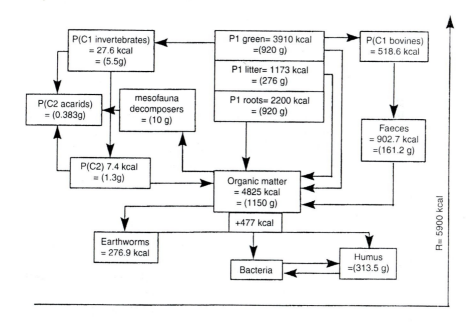

Fig. 1. Energy budget in various compartments of a Normandy grassland (Ricou, 1978), in g dry matter and kcal/m^2/year (in cases). Transfers of matter are indicated by arrows. C1 and C2, primary and secondary consumers. R, respiration.

environments. In France, for example, the studies of the IBP involved the grasslands of northwest France, the natural beech grove of Fontainebleau, the Mediterranean forest of green oak, the Sahelian savannah of Fete-Ole in Senegal, the pre-forest grasslands of Lamto in Ivory Coast, and the evergreen forest of lower Ivory Coast.

The ecosystem has long been defined as a homogeneous and aspatial entity. Duvigneaud (1980) defined it as a homogeneous biocoenosis that developed in a homogeneous environment. Soukatchev designated as *biogeocoenosis* any part of the earth's surface in which, over a certain extent, the biocoenosis remains uniform, as well as the corresponding parts of the atmosphere, lithosphere, hydrosphere, and pedosphere, and in which, consequently, the interaction of all these parts, interdependent and forming a unique complex, also remains uniform (Soukatchev, 1954 in Guinochet, 1973).

1.2.   The emergence of landscape ecology

Troll introduced the term *landscape ecology* in 1939, three years after Tansley introduced the term *ecosystem* (Troll, 1939; Vink, 1983). This German biogeographer aimed to combine the two disciplines of geography and ecology, i.e., to link spatial structures, which are the object of geography, to ecological processes. In this context the landscape was seen as the spatial expression of the ecosystem (Richard, 1975).

1.2.1.   The first developments of landscape ecology: ecological mapping

From 1939 to the 1970s, landscape ecology, very strongly dominated by a "geographical" component, was developed mainly in Eastern Europe, Canada, and Australia and was used in the study of ecological potential of vast territories (ecological mapping in Canada, geosystems in the USSR). In Eastern Europe, geographers played a driving role in the formulation of landscape science in close relation with problems of natural resource management (Preobrazensky, 1984).

Cartography is the basic tool for representing the landscape. It involves identifying, on a given territory, ecological and spatial units the ontology of which derives from some homogeneity relative to one or several attributes of the territory (biotic communities, nature of soil, forms of terrain, drainage, etc.). This notion of homogeneity is fundamental and rests on the concept of ecotope, or elementary ecological unit defined spatially. According to Phipps and Berdoulay (1985), the notion of system is present at two levels. Within the ecotope the various attributes of landscape are linked in the local ecosystem and, therefore, the various ecotopes are expressed in relation to one another in a spatial system of integration. This practice of successive frames raises various methodological questions, especially that of grouping procedures. The process relies on a postulate of spatial organization between the biotic and abiotic components of the landscape. It has given rise to many theoretical reflections and practical applications such as the Regional Land Survey in Australia in the 1950s (Christian, 1952), maps of ecological potential in the USSR, and the inventory of natural resources in Canada (Jurdant et al., 1972).

1.2.1.1. Example of ecological mapping: Ducruc's inventory of natural resources

From 1970 to 1985, over 600,000 km^2 of Quebec territory was mapped and classified in an ecological inventory programme (Ducruc, 1985). The programme involved regions of the James Bay and the North Coast, located north of Quebec, in territories that were not affected or little affected by human occupation or activities. In both cases it was the potential for generating hydroelectricity that triggered the inventories, which had three major objectives. The first was to obtain knowledge of the ecological determination of territories for which biological data were occasional and disparate. The second was to provide data for evaluation of the environmental impact of hydroelectric projects. The third was to establish an ecological basis for planning and for the integrated management of resources. The inventory of natural resources aimed to give managers and developers, among other things, a geographic framework in which were inscribed the ecological characteristics, aptitudes, potentials, and risks of land degradation.

The basic unit of this work was the ecological system (Ducruc, 1980), that is, a portion of territory characterized by a particular combination of geology, relief, and nature and form of geological materials of the land and water bodies. Each combination induced a computerized distribution of soil series and plant chronosequences. The mapping of ecological systems corresponded to a geographic unit. There was no hierarchy of variables that determined the layout of units. It was, however, possible to state certain basic principles around which the mapping strategy was constructed: to respect the complexity of the natural environment; to ensure a continuity in documents produced and rely on criteria such as permanence and stability (on the human scale); and to use criteria that could be identified in aerial photographs.

The geographic framework was defined by means of photo-interpretation (Fig. 2). Permanent elements of the environment (geology, relief, surface deposits, water bodies, and plant cover) were taken into account to delimit ecological systems, then quantify the elementary portions of the landscape (zones homogeneous with respect to one criterion or another). These elementary portions were identified mainly on the basis of morphological and topographical criteria: form of terrain and/or the slope, position on slope, steepness of slope, and length of the active slope.

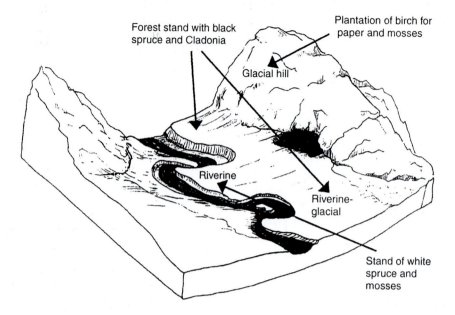

Fig. 2. Example of an ecosystem along the river Aguanus, northern coast (51°16'N, 62°30'W) (Ducruc, 1985)

The second phase was an ecological classification arising from statistical analysis of sampled variables. Major ecological traits of the territory were identified. Elementary portions such as the combination of a plant chronosequence and a soil series were defined. This definition situated the elementary landscape unit (or ecological type) in a higher bioclimatic framework: the framework of ecological regions.

The mapping of ecological systems demands a complete and relatively detailed ecological inventory. Even if some ecological types (elementary landscape units) do not figure directly in the map, they must be analysed and classified. Knowledge of them is essential to add a biological dimension to the map. By its overall geographic dimension, mapping of ecological systems allows us to describe the landscapes of a territory on a given scale and constitutes a frame of reference for its ecological determination. It also sheds light on problems of land management and ecological planning.

1.2.1.2. The geosystem
Bertrand (1978) introduced the notion of geosystem in France (Beroutchachvili and Radvanyi, 1978). The geosystem is characterized by a morphology and function. The term denotes a natural, homogeneous geographic system associated with a territory. It is characterized by a morphology, that is, by vertical spatial structures (geohorizons) and

horizontal spatial structures (geofacies). It has a function that encompasses all the transformations associated with solar or gravitational energy, water cycles, biogeocycles, as well as movements of aerial masses and processes of geomorphogenesis. It has a specific behaviour in that changes of state intervene in the geosystem for a given time sequence.

These authors compare the geosystem to the ecosystem by way of analogous concepts: there is a biocentric and metabolic approach in the ecosystem, while in the geosystem there is no preferential approach. Biotic and abiotic processes are understood overall. It is the natural hierarchy of elements as they appear in the quantitative analysis of concrete space-time that determines the analysis priorities.

This geographic concept aims to designate a spatial unit that is well delimited and analysed on a given scale. It has been implemented in mapping studies, notably by Alet (1986) in a mapping assay of birds in the Gresine massif (Tarn).

During the same period, Long (1974, 1975), a phytoecologist, defined the bases of a phytoecological diagnostic of land. Vegetation was presented as a good integrator of environmental conditions. Long's study and cartography made it possible to understand the physical factors and land potential for various uses. Large-scale studies have been done in Sologne (France) and in Margeride (Daget, 1975).

1.2.2. Environmental questions related to landscape transformation

1.2.2.1. Consequences of forest fragmentation in the United States

During the 1970s, a high degree of environment consciousness emerged in the developed countries of Western Europe and in the United States. The work of Rachel Carlson (1962) highlighted the harmful environmental consequences of transformation of practices linked to technological development in our society. Carlson emphasized the complexity of responses of ecosystems to disturbances and particularly to pollution from various sources. At the same time, a certain number of species emblematic of the American West became rare, and some came to the brink of extinction. The change in the status of these species was associated with mining in the northwestern American forests. Foresters, for the most part private agencies, preferred techniques of clear cutting that left the forest cover fragmented (Plate 1). The wooded parcels thus became smaller and increasingly isolated from one another, no longer constituting habitats favourable to species associated with large forests, such as the large spotted owl.

1.2.2.2. Changes in land use in Europe

In Europe, the repercussions of World War II led to an important revolution in agricultural production and technology. The need to increase production to "feed Europe" led to an agricultural politics centred on intensification and concentration. In Brittany, this very rapid development, further accelerated around 1970, led to major upheavals in the economy as well as in the regional landscapes. Systems of production were transformed from the cultivation and raising of a variety of species that had characterized Breton agriculture to more specialized systems. New crops were introduced, notably maize. Livestock rearing was concentrated. The raw materials needed to feed livestock were no longer produced on the spot but imported from all over the world.

The property structures in place in 1950 were entirely unsuitable in the face of these transformations. As land was inherited over the centuries it had been parcelled, and a single farmer sometimes farmed parcels that were very far from each other. In hedged farmlands or "bocages", the network of hedgerows, which had reached its greatest density since the beginning of the 20th century, was an obstacle to increasing parcel size. Between 1960 and 1970, farmers saw advantages in removing hedgerows on an individual basis. At the same time, many holdings were consolidated. The aim was to reorganize the parcels belonging to each farmer and to clear the vegetation on each new consolidated parcel. The connecting works accompanying these consolidations involved the removal of many hedgerows and embankments located within the new larger parcels or along farm lanes. The transformations of the rural landscape, with their most visible facet, the eradication of a large number of hedgerows (Fig. 3), were rapidly associated in the public mind with environmental problems that occurred subsequently, such as floods and wind damage. Similar developments took place in many regions that came under intensive agriculture in Europe. The decline of hedgerows was massive in England (Rackham, 1986). In Denmark,

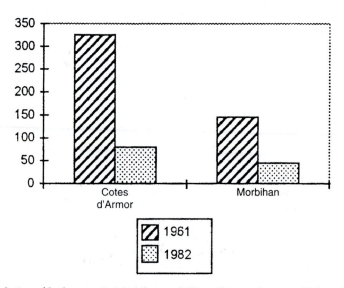

Fig. 3. Evolution of hedgerows in Morbihan and Côtes d'Armor between 1961 and 1982

non-cultivated areas, such as groves and ponds, progressively disappeared from the landscape (Agger and Brandt, 1988).

1.2.2.3. *Response of governments and the scientific community*

In the United States, as in Western Europe, the rise of environment consciousness led government and public opinion to question the scientific community. The question posed was: how can one evaluate the environmental effects of human activities, whether exploitation of the forest in the western United States or intensification of agriculture in Western Europe? In 1970, a multidisciplinary research programme on the functioning of farm hedges was financed jointly by the ministries of environment and agriculture in France (INRA et al., 1976). The objective of this concerted action, conducted by researchers of the University of Rennes, CNRS, and INRA, was to understand the functioning of these agrarian landscapes in order to manage them and ensure good working conditions for farmers, while maintaining an environment of "good quality".

The promulgation by the French President of an environment protection law (10 July 1976) marked a decisive turn in the relationship of society and environment in France. This law stipulated that the protection of natural spaces and landscapes and the maintenance of biological equilibrium should be declared a matter of general interest. From this declaration of intention, the legislature in the judicial framework of the law considered the possibility of being able to control various development works and projects that by their size or their impact on the natural environment may lead to its degradation.

This was the beginning of impact studies that allowed identification, description of the organization, and evaluation of the physical, chemical, biological, ecological, aesthetic, social, and cultural effects of a facility or decision (technical, economic, or political) (Definition de l'Atelier Central de l'Environnement). Impact studies were defined as follows by the law.

The contents of the impact study must be appropriate to the size of the works and developments projected and their foreseeable impact on the environment. The study presents:

1. An analysis of the initial state of the site and its environment, covering especially the natural wealth and agricultural, forest, and leisure spaces and water bodies that will be affected by the proposed development or works.
2. An analysis of effects on the environment, particularly on the sites and landscapes, fauna and flora, natural environments, and biological equilibrium and, if necessary, on the comfort of the neighbourhood (noise, vibrations, odours, light emissions) or on public health and hygiene.
3. The reasons for which, out of all the options considered, the project presented has been proposed, especially from the perspective of environmental concerns.

4. The measures envisaged by the manager of the facility or the petitioner to suppress, reduce, and, if possible, compensate for the damaging consequences of the project on the environment as well as estimation of corresponding expenses.

One year after the decree of the application of the law, it was unanimously judged that impact studies did not work well (Lefeuvre, 1979). The causes were not only administrative at the level of evaluation of studies and the low efficiency of public enquiry, but also scientific. In effect, the knowledge and fundamental research of ecology were not yet ready to provide the relevant tools and data for impact studies.

1.2.2.4. The emergence of landscape ecology

In order to be able to answer questions about environment protection measures addressing the effects of forest fragmentation on the decline of animal populations, and about ecological consequences of transformations in agricultural areas, the scientific community had to evolve and change the object of study. The study of ecosystems gave way to the study of more complex systems for which multidisciplinary approaches proved to be essential. To understand the functioning of an agricultural or forest land, to evaluate the impact of clear cutting, forestation, or removal of hedgerows on the fauna, flora, and flow of water and nutrients, it was necessary to:

- explicitly take the spatial dimension into account,
- recognize man as an integral part of the ecological system, and
- recognize the spatial and temporal heterogeneity of the environments studied.

Ecologists then came together with biogeographers on the basis of the ideas of Troll and there was a desire—a need—to share concepts, tools, and methods. From these exchanges arose landscape ecology as it has presently developed. In 1982, the International Association for Landscape Ecology (IALE) was created. It comprised geographers, mostly from Eastern Europe, who had developed a system of ecological cartography, ecologists working on populations and communities, and practitioners—landscape developers and architects. IALE published a journal from 1987 onward on landscape ecology and organized itself into regions. Each regional association has its representative and organizes seminars. Among these regional associations, that of the United States rapidly achieved a wide coverage. In Europe, the associations of the southern and eastern countries have been very active. A cultural and disciplinary specificity can be found in each region. The United States has mostly favoured modelling of populations or nutrient flow in fragmented landscapes. The countries of southern Europe have emphasized links between population dynamics and fragmentation of non-cultivated parts of the landscape.

2. RECOGNITION OF HETEROGENEITY IN ECOLOGICAL SYSTEMS

Until 1970, most major research programmes in ecology worked within the framework of the ecosystem defined as a homogeneous biocoenosis developing in a homogeneous environment (Duvigneaud, 1980). The main subjects of research were "natural" systems such as the forest or the savannah (Lamotte, 1978).

For the study of an agricultural land with cultivated fields or an exploited forest with regenerating parcels of different ages, the need to explicitly take into account the heterogeneity of the system studied is obvious. But this need had already emerged during multidisciplinary studies of ecosystems.

Lefeuvre and Barnaud (1988), tracing the evolution of their research team between 1972 and 1987, indicated the progressive transition from assumption of a structural homogeneity of systems studied to the recognition of a functional heterogeneity. In the framework of a research programme on the Armorican moors (Lefeuvre, 1980), they used the definition of homogeneous ecosystems as a point of departure. According to these authors, the moor of Cap Frehel, dominated like all formations of this type by some species of gorse and heather, and thus easily differentiated from neighbouring formations, may appear to be homogeneous. However, it is in fact made up of 16 units of vegetation. The notion of homogeneous moor rapidly became blurred as the team probed further. Clement (1978) distinguished as many as 24 units of vegetation in the moors of the Arre mountains. More recently, Morvan et al. (1995) indicated the role of this heterogeneity in the spread of fires.

The realization that a system is not homogeneous, which is evident to the eye of a walker or researcher, poses a dilemma that will recur throughout this work: the choice of spatial scale and elementary units. When we study an ecological system, when we study an animal or plant population, when we map a site to be managed, how do we choose the appropriate scale of analysis and representation?

2.1. Heterogeneity depends on the nature of elements and scale on which the system is represented

Heterogeneity is the property of being formed of dissimilar, disparate, often contrasting elements (Larousse, 1979). If heterogeneity is to be taken into account, the elements that form the mosaic of the territory considered must be identified, and their spatial arrangement must be determined (Baudry and Baudry-Burel, 1982).

Plate 2 and Fig. 4 illustrate two aspects of the process leading up to the representation of a study area.

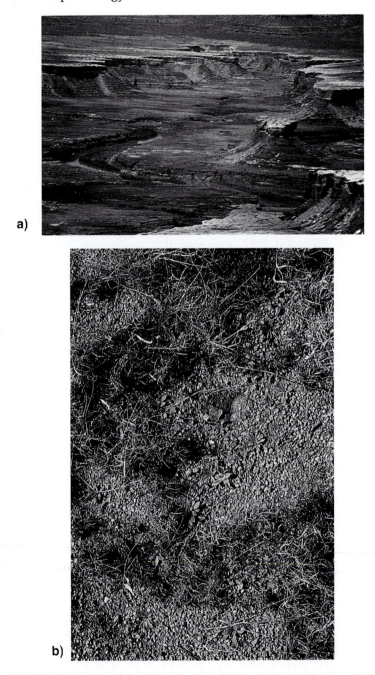

Fig. 4. Aerial photographs of the Colorado desert (USA). (a) Two landscape units are clearly seen: canyons with a tortuous relief and plateaux that appear homogeneous at this scale. (b) On a more refined scale, the vegetation of a plateau appears heterogeneous. Here two plant species grow in tufts on a coarse soil.

In the first case, the size and form of basic elements of the landscape mosaic are defined by the researcher's determination of their nature. The ecologist interested in the continuity of the plant cover for its role in the possible control of physical flows such as water or wind classifies the land into wood (areas of higher and long-term vegetation), grassland (zone with permanent grassy cover), and cultivated fields, sown each year and left bare for some time after each harvest. The agronomist, on the other hand, must take into account the diversity of production within the exploited zones in order to understand the organization of the landscape. He or she differentiates between a thicket and a poplar grove, between a permanent grassland and a short-rotation grassland, between cereals and various cash crops. Depending on the objective of the research, and the perspective of the researcher, a single landscape will be represented differently, and the spatial heterogeneity of this representation will vary with the diversity of elements present. The differentiation of representations depends on the scale of perception.

In the second case, the same landscape is seen at different distances. Seen from an aeroplane, the contrast between the plateau and canyons is striking and shows us two distinct units. The canyon zone appears heterogeneous even from a high altitude, with a succession of crests and valleys with sinuous contours, but the plateau seems uniform and homogeneous. Seen from the ground, that uniformity becomes a mosaic of tufts of creosotes growing more or less randomly on a stony soil. The aerial view in this case allows us to differentiate geomorphologically contrasting zones, while on the finer scale, on the plateau, a heterogeneity is created by the spatial distribution of the vegetation.

In Chapter 3, devoted to analysis of landscape structures, methods of measuring heterogeneity are described that are used to quantify the variations of heterogeneity created by these differences of representation. These two aspects must be kept in mind so that we can compare landscapes and identify the scales of description of space that are relevant in order to associate landscape structure and ecological processes.

2.2. Heterogeneity is a factor of organization of ecological systems

The consequences of heterogeneity on the organization of communities have been recognized in ecology since the works of Cowles (1899) on plant successions along the coasts of the Great Lakes. The tolerance of different species with respect to inundation, burial, and competition creates gradients of spatial distribution of the vegetation, which are correlated to the distance from the bank. Thus, each site, each station, comprises a subset of the flora present on the landscape level. The composition of this subset is defined as a function of responses of each species to biotic or abiotic conditions present and past (Milne, 1991).

2.3. Heterogeneity is both spatial and temporal

Ecological systems, no matter what the intensity of human activities they support, are dynamic. The factors of the dynamics, often called disturbances, are highly diverse. They may be abiotic, as for example hurricanes, tree falls, floods, fires, or volcanic eruptions. They may be disturbances of biotic origin, such as an epidemic that may eradicate a species or an attack of processionary caterpillars, which could be limited to some parcels of land. Depending on the intensity of these "natural catastrophes", the effects may be more or less long-lasting, even irreversible. On the other hand, in many places human activities such as agriculture, forest exploitation, and urbanization become the main driving force of landscape dynamics.

The heterogeneity perceived at a given moment, in a given place, is the result of the spatio-temporal heterogeneity of environmental constraints, ecological processes, and man-made or natural disturbances (Fig. 5).

2.4. New methods to account for heterogeneity

The recognition of heterogeneity as a factor of organization of ecological systems has profited from theoretical progresses in the analysis of complex systems. The theory of fractal geometry (Mandelbrot, 1984; Peitgen and Saupe, 1988), allowing simultaneous consideration of several spatial scales, is a powerful tool for determining the range of scales for which a given ecological factor is relevant in explaining an ecological process. Associated with fractal geometry, the theory of percolation (Grassberger, 1991; de Gennes and Guillon, 1977-1989) allows us to study flows in a heterogeneous space, whether they are propagated by disturbances or individual movements of animals in landscapes (Gardner et al., 1989). The characteristic of percolation is that it involves a critical phenomenon: below a threshold, the probability of passage of flows considered is nul, and above it, it is equal to 100%. In the case of landscapes, the threshold corresponds to a particular spatial representation (percentage of the total area covered by a type of landscape element) of fragments of each type of element (Milne, 1992). Landscape ecology has profited from developments in theoretical physics for the description of heterogeneous structures (Leduc et al., 1994; Chapter 3 of the present work), the comprehension of processes of organization of landscape structures (Rex and Malanson, 1990), the modelling of individual movements (Turchin, 1996), and the modelling of choice of habitats (Falardeau and Desgranges, 1991). All these points are elaborated in detail in the following chapters.

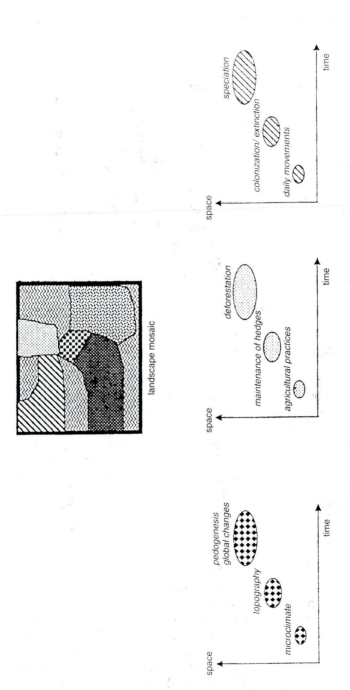

Fig. 5. Heterogeneity of landscape mosaic (Urban et al., 1987). The landscape mosaic is the result of environmental constraints, disturbances, and biological processes, each operating at the appropriate spatio-temporal scale.

3.   TAKING HUMAN ACTIVITIES INTO ACCOUNT IN ECOLOGICAL SYSTEMS

When landscape ecology emerged in the 1980s, the objective of ecologists was to differentiate the ecosystem not only by its heterogeneity, but also by the role of human activities in its dynamics and in the emergence of environmental problems. The landscape is the result of a continuous confrontation between a society and its environment.

Crumley and Marquardt (1987), in their analysis of the landscape dynamics of Bourgogne, define the relationships between society and landscape as follows: "People make their territories, houses, living spaces and work spaces their own by consciously modifying them in terms of their effects on the senses, their utility, and their economic value. The landscape is the spatial manifestation of the relations between humans and their environment." This balance between environment and society in the genesis of anthropized landscapes is recognized by many authors (Birks et al., 1988), the relative importance of one or the other varying according to the site and the time period (Olsson, 1988; Crumley and Marquardt, 1987).

3.1.   Genesis of agrarian landscapes: example of hedged farms in western France

This section contains a quick overview of the long-term dynamics of hedged landscapes as an example of ancient human occupation of an environment. Landscapes have been established over several millennia in regions with an old agrarian civilization, as in Europe. The history of their dynamics is reconstructed by geographers (Meynier, 1970), archaeologists (Chapelot, 1980), palynologists (Groupe d'histoire des forêts françaises, 1986), botanists (Hooper, 1976), and historians (Rackham, 1986). This multidisciplinary approach is essential because of the difficulty of collating historical documents and combining the spatial and temporal scales appropriate to each discipline. Research advances and progress in analytical techniques, especially in archaeology, often raise questions about existing information. In this section is presented a synopsis of the history of the landscape of Brittany as it is understood today.

The Breton landscape is a mosaic of woods, moors, and agricultural land partitioned by a network of living hedgerows. It is a bocage zone.

The first cattle-rearing societies appeared at the end of the Mesolithic era, and the agriculture-livestock association dates from the 4th millennium B.C. (Giot et al., 1979; Gebhardt, 1988). Along with this agricultural settlement there was a corresponding clearing of primeval forest that involved more than 80% of the area (Lefeuvre, 1986). The first hedgerow networks were also very likely of prehistoric origin. In fact, traces of certain hedgerows running into the sea suggest that they were submerged by the last oceanic transgression. That suggestion has, however, been disputed (Pitte, 1983) because the number of such traces is small, and the hypothesis of a medieval transgression cannot be excluded.

The Gallo-Roman period left little documentary evidence on the countryside. Only the findings of coins within the embankments indicate the origin of those embankments. Archaeologists nevertheless sketch out a description of the landscape of this era: the presence of a dense, evenly distributed settlement, without a real central forest, and predominance of open land comprising enclosed parcels.

In the Middle Ages, technological innovations, particularly the appearance of the wheel plough, led to a wave of clearing during the 11th and 12th centuries. The area cultivated extended considerably and at the same time fields were enclosed so that they would be protected and grazing conflicts would be avoided. After an optimum period of settlement in the 13th century, a demographic depression settled over Brittany until 1470. During this period a large part of the region and perhaps more than half of the land consisted of moors, without farm fields or hedges.

Until the 19th century, the dominant trait of the Breton landscape was the presence of isolated pockets of cultivation scattered in the middle of the moors. Within these cultivated zones there were alternating open and closed fields. The enclosures were often temporary. Tenants of a leased parcel had to destroy the fences when they finished using the field in order to allow animals to graze (Meyer, 1972).

In the 18th century, the upper class, supported by the parliament of Brittany, attempted to clear the moors, but it faced resistance from the peasants. In the middle of the 19th century the demand for arable land increased because of several transformations in cultivation practices. Siliceous soils could now be amended, and potato production improved the food supply and also favoured the development of pig-farming. The cultivation of moors thus met with a favourable response from all social groups. The commons were distributed among the peasants. The new land-owners marked the borders of their property with hedgerows and embankments. These protected crops against the movement of cattle that still grazed on the surrounding moors. This period of massive construction of hedgerows, begun in 1840-1880, continued until 1914.

After World War II, the rapid development of agricultural technology militated against the small size of hedged fields. The hedges, considered obstacles to modernization (Deniel, 1965), were targeted for clearing on an individual as well on a collective basis during the process of land development.

In conclusion, this rapid survey of the evolution of hedged structures in Brittany indicates a great instability on this time scale. Periods of important modifications of the landscape, building or removal, have corresponded to revolutions in agricultural technology. These streams of progress were significant enough to spread widely and induce the same effects throughout a region. We can thus speak of Breton hedgerow farming at this level.

There are still many gaps in this history of landscapes, because historical sources are rare or contain few descriptions of the landscape, and the archaeological data are spatially scattered and often do not help us reconstitute the landscapes. Only through multidisciplinary approaches, such as those developed in the CNRS programme on environment and hominids, can we refine, or perhaps again challenge, our present understanding.

In any region with an ancient civilization, we can find relationships similar to those that have been stated above between the history of societies and the establishment of landscapes. Overviews have been written in Sweden (Berglund, 1991) and in England, where *The Making of the English Landscape* by Hoskins (1955) has been a best-seller since 1955.

3.2. The existing landscape structure is the result of past dynamics

The shapes of fields and the organization of the landscape mosaic are the result of this long history between societies and the environments in which they live. In the Armorican Massif, for example, the division of land can be related to the history of land property, agricultural technology, and environmental heterogeneity.

Figure 6 presents four portions of territory, of the same area (64 ha), that are representative of various parts of the Armorican Massif.

Example 1 (Fig. 6a)

In the northeastern limit of the Morbihan department, Concoret is enclosed in the middle of the Ille-et-Vilaine commons. Purple schist and sandstone constitute the subsoil of the perimeter of the commons, and there are hard rocks that form heights on which the thin soil is not favourable for agriculture. In the centre, on green schist, the relief is less marked and the deeper soils are suitable for cultivation. The woods, which are continuous with the Paimpont forest, cover 108 ha, or 6.8% of the community land. Forage cultivation and permanent grasslands occupy more than 60% of the useful agricultural area, the rest being used to cultivate cereals. The plant production is essentially meant to feed cattle, which constitute the main animal production. The hedged sector represented in the figure is located on the purple schist. The farms are gathered into large villages surrounded by very small fields. Beyond that, cultivated land coexists with remaining scraps of moors or woods. The hedgerows are continuous with the uncultivated zones. Towards the northern part of the zone, the parcels of moor are larger and more numerous. The hedged fields are enclosed in a matrix of uncultivated land. The spatial intricacy of the hedged fields and

a) Concoret

b) Lalleu

c) Campeneac

d) Quetreville

Fig. 6. Form of land parcels and representation of uncultivated zones

woods and moors is linked here to the constraints of the physical environment. The difficulties of exploiting these thin and irregular soils has been an obstacle to the collective appropriation of moor land. Each parcel is cultivated or left as it is depending on its agronomic qualities, heterogeneity, and constraints of substrate, which leads to heterogeneity of the land use.

Example 2 (Fig. 6b)

Lalleu is located 40 km south of Rennes. The subsoil is formed of alternating bands of Brioverian schist and sandstone, oriented east-west. The sandstone crests shelter the landscape, and the schist zones are softly undulating. The uncultivated zones occupy less than 1% of the territory. The main agricultural enterprise is milk production. Crops are chiefly meant to be used to feed cattle, and forage crops and grassy pastures cover 75% of the total area. The farm structures are efficient. The sector presented here is located at the limit of schist and sandstone. The farm buildings, which form hamlets, are found in the north, on the schist. The fields are small and irregular in shape. There are many roads. As soon as the land changes to sandstone, the field size changes entirely. The average size of fields is larger, and their more geometric shape is determined by a network of straight roads. This is a typical example of the recent clearing of the moors by a process of collective appropriation. The shared space is flattened and all the parcels are cultivated simultaneously. The geometric character of this landscape is typical of zones that have been collectively exploited, in a relatively short time. It is found in Roman centuriations (Pitte, 1983), German knicks (Luhning, 1984), the Canadian range, or American townships. In Brittany, hedges were established more or less spontaneously on the borders of the fields thus created.

In Lalleu and Concoret, there are similar conditions of environment and agriculture, a result of landscapes that were similar up to the end of the 19th century. However, during the 20th century, the progress of agricultural techniques in Lalleu led to the cultivation of soils on the Armorican sandstone, which are hydromorphous but thicker than those of Concoret, which are on purple sandstone.

Example 3 (Fig. 6c)

Campeneac is located in the southwest of the Morbihan department. Its subsoil is formed of 95% Brioverian schist, which creates a smooth relief. In the north, purple schist makes the ground significantly uneven over a small area. This zone is largely covered with heather. Agriculture here is essentially oriented to meet the needs of cattle. Forage crops are grown and there is permanent grassland.

The sector represented in the figure is characterized by the presence of a set of long parcels arranged in strips and not separated by hedges. This landscape of open field, with very fragmented parcels isolated within a hedged region, is called *méjou*. There are farmed areas on the protruding parts of the relief. These open islands are surrounded by a bocage with narrow parcels, in which there are clusters of farm buildings. In the present case, it is delimited by alluvial axes, in which a dense network of hedges on embankments encloses the permanent grasslands. This type of organization—cultivation on strips of open fields and permanent grasslands associated with a dense hedging—is characteristic of the landscape of the municipality. There is differentiation in land use and mode of appropriation of land as a function of the relief and in correlation with the degree of hydromorphy.

Meynier (1966) attributes the establishment of these agrarian structures to the collective use of *méjous* that would be the result of relatively recent clearing in comparison to the surrounding hedged fields.

Example 4 (Fig. 6d)

Quettreville is located in the central west of the department of Manche. The subsoil is formed of sedimentary rocks of Brioverien origin covered in places with deep silt soils. The relief is shallow overall; the water courses create only small depressions in a plateau sloped towards the principal river. Out of the 1515 ha of agricultural area used, 1333 is always grassy, 144 ha planted with forage crops, and the rest planted with cereals. The agriculture is almost exclusively focused on milk and meat farming through the exploitation of permanent grasslands. In relation to the Breton departments, there is transformation from a largely farmed land to a land largely covered by permanent grasslands. The grasslands are small and cut up fairly regularly by a network of roads. Such regularity has been favoured by the homogeneity of environmental conditions. This grassland landscape is relatively recent; we will see (Chapter 4) that it did not exist at the end of the 19th century. The farms are grouped in large villages linked by many roads.

The hedgerow network is characterized by its density and by the regularity of the parcels in shape and size. This homogeneity is related to the uniformity of the substrate and the land use. The permanent grassland used for grazing is also a form of exploitation highly favourable to the establishment and maintenance of hedgerows.

Example 5

Another network of hedgerows and scattered groves is found on the eastern coast of the United States, in New Jersey, on red schist with a slightly undulating relief, except along the water course, where the rock is cut and the banks are wooded. When European colonists arrived in the 18th century, they left some woods uncleared. The land was divided up. Until the middle of the 20th century, the enclosures were retained. Subsequently, the farmers allowed those trees that grew from seed disseminated by birds to grow. At present, there are spontaneous linear woods, which may be up to 10 to 15 m wide, abundantly bordered with eglantine. There is thus a hedgerow network that has an origin and a form that is very different from the Armorican hedgerow networks described above (Fig. 7).

3.3. Human activities are the main factor of evolution of landscapes on the global level

We have seen in the preceding paragraphs that in countries with older civilizations the present landscape is closely dependent on the history of societies and technology. Societies have always evolved in an unstable equilibrium with their environment. Certain civilizations such as the Sumerian collapsed after poor management of water resources (Ponting, 1991). However, Pfister and Brimblecomte (1990) show that societies have often reacted effectively to modifications and destruction of their environment. Since the 16th century, measurements were taken at Hambourg to fight air and water pollution caused by paper factories, tanneries, and dyeing plants. Human ecology and historians can provide answers to questions on the genesis of present environmental problems and on conditions that allowed certain ancient societies to maintain an equilibrium with their environment. At present, humans have increasing control over their environment, owing to demographic growth, industrialization, and technological developments. Their actions influence all landscapes and ecosystems on the planet directly through exploitation of resources and occupation of space for agriculture and urbanization, as well as indirectly through global climatic changes or pollution caused by industrial development.

4.   EXPLICIT ACCOUNTING FOR SPACE AND TIME

We saw earlier that the ecosystem as it was studied until the 1970s was essentially aspatial and often considered "at an equilibrium". To understand the functioning of a dynamic system such as the landscape, we must again place it in its spatial context by studying especially the exchanges between neighbouring systems, and in its temporal context because we recognize a close interdependence of stages of a system in the course of its dynamics and more particularly their dependence on the initial state.

4.1.   Spatially explicit representation of ecological systems

Taking spatial heterogeneity into account is the only means of expressing the control of ecological processes at the landscape level. For example, studies conducted before 1980 on the flora of hedges (Roze, 1978; Delelis-Dussollier, 1973) linked plant groups of hedges and embankments to local parameters. They took into account the width of the hedge, the quality of the substrate, and the climatic data; part of the variability of the floristic composition of hedges could thus be explained. It was only in recognizing the spatial distribution of hedges in the landscape, and more particularly within a hedgerow network, that the importance of processes of colonization could be demonstrated.

   Baudry (1985) studied two continuous networks of hedges in New Jersey (Fig. 7). They were spontaneous hedges, with tree species whose seeds are generally dispersed by birds (*Prunus serotina* and *P. avium*). The seeds were

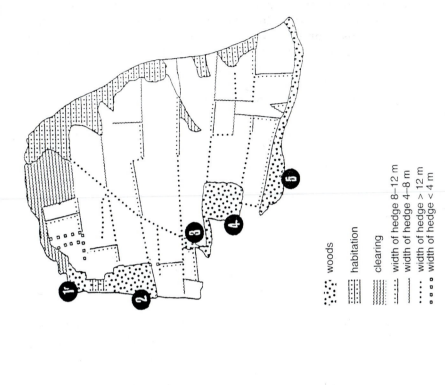

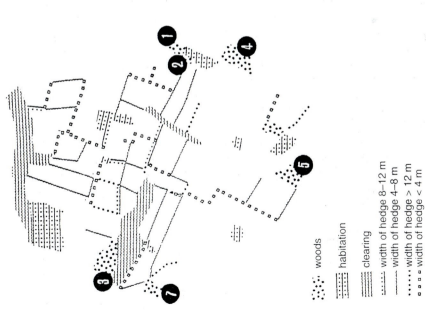

woods

habitation

clearing

width of hedge 8–12 m
width of hedge 4–8 m
width of hedge > 12 m
width of hedge < 4 m

woods

habitation

clearing

width of hedge 8–12 m
width of hedge 4–8 m
width of hedge > 12 m
width of hedge < 4 m

Fig. 7. Map of two nearby hedgerow networks in New Jersey (USA)

dispersed along the borders of fields, which provided nearby perches. The objective of the research was to test the hypothesis that the remanent forest patches included in these landscape units served as sources of forest species and thus contributed to the organization of the hedge flora. Some 118 hedges of width varying from 1 to 13 m were sampled, as well as 13 woods connected to hedges of these networks. The hedges with the largest number of forest species were the widest and the nearest to the woods (*Podophyllum peltatum, Sanicula* sp., *Polygonatum biflorum, Viburnum acerifolium*), while others were found more than 1000 m away (*Galium aparine, Impatiens capensis, Ranunculus abortivus, Viburnum prunifolium*). These differences in the distance of colonization of species depend on two factors: the dispersal potential and the ecological conditions of the hedge. In all cases, continuity of the network is indispensable for forest species to propagate themselves in the agricultural landscape.

A more refined study was done on spatial distribution of plants near intersections to evaluate the importance of these key elements in the diffusion of forest species. Contrary to what had been observed for trees and shrubs, the species richness of which did not vary when far from the intersection, the number of herbaceous species diminished significantly between 30 and 40 m, then remained practically constant (Fig. 8). When the whole network was divided into sectors 10 m long, it could be demonstrated that the species were present discontinuously. This suggests a colonization by saltatory movements.

From these results taken together, a scenario of dispersal of forest species in this type of landscape can be presented. On the one hand, there is progressive dispersal of forest species along the hedges between connected hedges. On the other, there is a possibility of establishment and development in certain portions of the hedge only (due to conditions of light, humidity, concurrence with other species). The intersections of hedges are favourable to the development of forest species, especially when they are wide.

Here we see organizational levels that are differentiated on the basis of factors of control of the state and dynamics of the vegetation. At the local level the structure of the hedge and the microclimate that it induces (more or less shade) select the species present (species of shade or light). Within landscapes it is the dispersal capacities that control the distribution of species. Finally, the factors of establishment of the landscape, clearing with maintenance of groves, then spontaneous growth of woods from trees and shrubs that are not harvested, constitute the limiting conditions of ecological processes.

These studies illustrate the contribution of spatialization of data to the understanding of ecological processes at the landscape level. We will see in the rest of this work that this method and the development of associated tools are closely linked to the problematics of landscape ecology, whether this be the dynamics of fragmented populations and their applications to

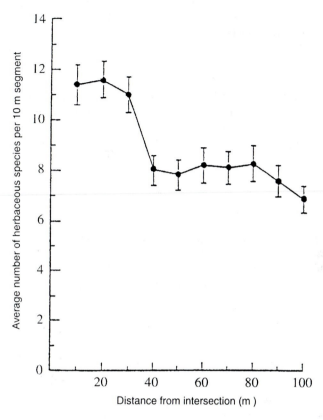

Fig. 8. Effect of distance from intersection on species richness of herbaceous species

the biology of conservation, organization of rural landscapes by agricultural activity, or control of flows of water and nutrients in the watersheds.

4.2. Taking time into account in the analysis of ecological processes

We have seen in section 2 of this chapter that the dynamics of ecological systems, under the action of natural or man-made disturbances, contributes to their spatial and temporal heterogeneity. Time is a key factor for the understanding of ecological processes and evolutionary mechanisms of landscapes. The organization of animal and plant communities and cycles of matter and energy depends not only on the present environmental conditions, but also on recent or ancient history. In studying the organization of natural landscapes (tundras of the Yukon in Canada) or man-made landscapes (agrarian and pre-urban landscapes), Phipps (1981) showed that the part explained by environmental factors is always small, even in "natural" zones that are highly constrained, such as the Yukon. There is no determinism between the landscape patterns and the environmental factors, and the indeterminate part is partly linked to the history, that is, the dynamics and initial conditions of the system after each new disturbance.

4.2.1. Historical information needed to understand evolutionary mechanisms of "natural" systems and their management

Landscapes in which the human influence was slight, such as the forests of western North America before the arrival of colonists, or those in which environmental constraints are strong, such as the large, un-channelled river valleys, are subject to changes that are more or less similar over centuries. Through archaeological techniques such as palynology, malacology, carbon dating, and micro-morphology of soils, we can reconstitute the history of these landscapes and discover the extent and periodicity of major disturbances. Historical methods (analysis of archives combined with the methods mentioned above) and, for a more recent period, the study of aerial photographs supplement the tools with which the history of landscapes is analysed.

Amoros and Bravard (1985), in a synthesis of research done on the Rhone valley, showed how a study on several temporal scales allows us to reconstitute the dynamics of the river hydrosystem. This system is made up of all the aquatic, semi-aquatic, and terrestrial biocoenoses and biotopes that are directly and indirectly related to the river. There is a heterogeneous set of ecosystems that evolve at different rhythms as a function of varied processes of colonization and competition, as well as disturbances linked to the river, such as erosion, silting, and rise or fall of water levels. The scales preferentially used in analysing these different parameters are the geological time scale of the order of a millennium or more, the historical time scale of the order of centuries or decades, and biological time scales involving the pace of physical phenomena, chemical or biological, ranging from a day or even an hour to a season. Over each scale there occur phenomena that influence the present state. On the geological scale, climatic variations and natural evolution of plant cover impose changes in the water regime and the nature and quantity of the materials transported. On the historical scale, the natural development and recovery of a meander may occur, the phenomena of eutrophication or alluvial deposit may be observed, successions of settlements occur, and human activities may modify the dynamics of a water course. It is also on this scale that predictive scenarios can be proposed in response to questions posed by managers. On the biological time scale, the analysis of physical, chemical, and biological processes allows us to define the functional units that, placed in the historic framework, indicate the general evolutionary trends of the landscape.

In the forests of western North America, fire is a major disturbance. A study of forest history was done in Yellowstone National Park, an area that has never been exploited by colonists for agriculture or timber (Romme, 1982). Before anti-fire measures were put into place by park managers, individual trees often caught fire (because of lightning). In rare cases, large areas caught fire, and still more rarely immense forest fires occurred, such as the fire of 1988, which involved an area of 700,000 ha. The extent and intensity of a fire depends on the vegetation, climate, and terrain. Fires often start: 79,131 were recorded at Yellowstone between 1940 and 1995. However, their extent depends on environmental conditions. In 1988, drought and a strong wind favoured the extent and intensity of the fire. From the study of soils and sediments of charred wood in the park, it is clear that such fires have occurred in the past with a periodicity of around 400 years. Following such a disturbance, the forest gradually regenerates itself, until the accumulated organic matter and climatic conditions favour another such fire. At a given moment, the diversity of the landscape mosaic is the result of the interlocking of these disturbances of varying frequency and extent (Romme and Knight, 1982). An understanding of this landscape dynamics in the case of Yellowstone led to a review of the fire management policy by park authorities: According to De Golia (1993), "Fire presents opportunities for life that don't exist until a burn. Each place responds in its own way and its own time. While the forests and grasslands of today are products of earlier fires, they are also setting the stage for fires to come. After a century perfecting our techniques for fighting fire, we are now perfecting our ability to live with it and to use it to achieve important goals in the National Parks."

4.2.2. Present organization may reflect past environmental conditions: ecological systems may slowly adapt to environmental changes

We saw earlier that landscapes, no matter what their degree of anthropization, are dynamic. There is no synchronous relationship *a priori* between the spatial distribution of species and the landscape patterns. Changes in the spatial structure of landscapes and changes in the spatial organization of settlements are not simultaneous. The distribution of species depends on the present landscape and its earlier stages. For example, the composition of bird populations in the Mediterranean zone is explained directly by the history of the vegetation (Blondel, 1986). Reconstitution of paleo-landscapes of Europe shows that in the Mediterranean geographic area there was a juxtaposition of all the conditions of habitat required by all the fauna of Europe. Drifting with the climatic oscillations of the Pleistocene, the spatio-temporal distributions of flora and fauna underwent gigantic contractions and re-expansions in terms of latitude and altitude. During the cold phases the mountains of this region offered ecological conditions for survival that the high latitudes presently offer for Arctic fauna. In association with the diversification of the vegetation mosaic, fauna from throughout Europe came together in a relatively small area. This explains why the bird populations of the Mediterranean zone, notably the forest and water birds, have no particularly "Mediterranean" characteristics.

Over shorter periods of time, as will be seen in the rest of this book, a delayed effect can be traced between the dynamics of the landscape and the dynamics of populations or species assemblages. The delay depends on the taxa, the efficiency with which they adapt to changes in ecological conditions, and the rapidity with which they colonize new habitats or disappear from habitats that have become less favourable.

4.2.3. Knowledge of the initial state is fundamental to predicting the dynamics of a landscape

Landscapes, thus, have a history: at a given moment, the potential evolutions depend on the present state, as well as on the information accumulated in the past. These constitute the initial conditions. The significance of initial conditions is often put forward in physics to explain the impossibility of predicting evolution linked to deterministic processes. In ecology, plant and animal successions, notably in the case of abandoned agricultural land, offer examples of this dependence on initial conditions. In this case, the state of the soil, its physical condition and chemical condition (level of pesticides and minerals), and the sources of plant seeds or of dispersing animals constitute the initial conditions. An abandoned field in the middle of other fields will not be colonized in the same way as an identical field that is hedged. Similarly, for two hedged fields, their land cover at the time they are abandoned (grassy or bare) and past cultivation practices (fertilization, mode of tilling) will play a role in the selection of new species that are able to establish themselves. Clearly, it is important to consider the spatial heterogeneity, to take into account differences between apparently identical landscape elements (two fields) and differences in neighbourhood. Also, the close interweaving of time and space must be noted. The spatial differentiation of processes that occur in time (agricultural practices) creates a heterogeneity that will modify future evolution.

5.   LANDSCAPE ECOLOGY IS BASED ON SCIENTIFIC THEORIES LINKED TO ECOLOGY AND RELATED DISCIPLINES

Every scientific discipline emerges and develops in continuity with earlier theories and methods and draws on them even while refuting or going beyond them. Landscape ecology is no exception to this continuity of the history of sciences. We have seen that it was identified in reaction to the ecology of ecosystems, and that it recognizes the complexity and dynamics of ecological systems. It accompanied the recent development of a unitary thought linking the sciences of matter and the life sciences on the functioning of complex systems and their dynamics. The principal theories on which studies of landscape ecology are based are the hierarchy theory (Allen and Starr, 1982), the theory of chaos (Gleick, 1991) and the associated fractal geometry (Mandelbrot, 1984), and the percolation theory as far as the apprehension of complex systems is concerned. Spatial and temporal analysis of ecological systems is based on the island biogeographical theory (MacArthur and Wilson, 1967) and the disturbance theory (Pickett and White, 1985).

5.1. Hierarchy theory

Landscapes are complex systems in which an entire series of ecological phenomena occur, each having its own spatio-temporal scale. The hierarchy theory (Allen and Starr, 1982; O'Neill et al., 1986; Urban et al., 1987; May, 1989; Baudry et al., 1991) is a conceptual framework appropriated to treat sets of phenomena that occur at several scales of space and time. Its two essential predictions are as follows. (1) There is a correlation between scales of time and space. Phenomena that occur over a large area are slower than those that occur over a small area (Fig. 9). (2) The levels of organization are essentially characterized by rates of functioning of phenomena, and phenomena having very different rates of functioning interact little. For example, daily movements of carabid Coleoptera are hardly influenced by the evolution of landscapes (Fig. 10). At higher levels of the hierarchy,

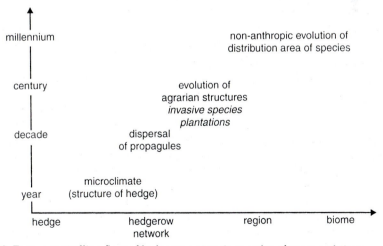

Fig. 9. Factors controlling flora of hedgerows at various scales of space and time

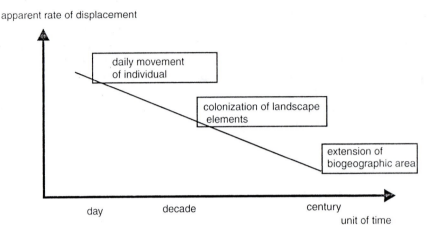

apparent rate of displacement

daily movement
of individual

colonization of landscape
elements

extension of
biogeographic area

day   decade   century

unit of time

Fig. 10. Rates of functioning of ecological processes involved in the spatial distribution of animal species

phenomena occur over long time periods and large spaces, while at the lower levels they occur rapidly and locally (Koestler, 1967).

Within the landscape an entire range of scales can be identified at which processes are related, and they can be grouped into speed classes and rays of similar action. These groups form levels of the hierarchy (O'Neill, 1989). Hierarchy theory predicts that there is no continuum in scales but rather a certain number of distinct levels. This was shown by Krummel et al. (1987) in studying wooded elements of landscapes. Their form, measured by their fractal dimension, is not distributed continuously along a gradient, but oscillates around two values. These two types of form are the result of two different processes. The small woods of simple form are determined by constraints of land ownership, while the larger ones depend on constraints of the physical environment. The processes that created these woods can thus be separated into two distinct levels.

The system can be divided into levels of organization corresponding to scales of space and time appropriate to each phenomenon (Fig. 9). These levels of organization have a quasi-autonomous property, which renders the hierarchic systems decomposable (Auger, 1992). If too many levels are analysed simultaneously, the relevant information of each is blurred by signals from others that are perceived as white noise.

To study a phenomenon, however, several levels of the hierarchy must be taken into account. The higher levels have a constraining role in fixing a framework for the processes. The lower levels impose limiting conditions linked for example to the nature of elements that fit into a hierarchy. The example of flora of hedgerows in New Jersey cited earlier is based on such a process.

One method of analysing landscape is to consider several levels in the spatial hierarchy. At each level, factors that explain the processes studied are identified. Most of the time, two spatial levels are taken into account: one is the landscape and the other consists of the elements of the landscape. Each level is identified by different criteria, a function of the problems posed. At the landscape level, criteria most often retained are the spatio-temporal structure and the factors of organization. At the level of landscape elements, the parameters most frequently identified are related to forms, quality of habitat, and relationships with other landscape elements. There is no universal description of a space or an object. The observer defines it as a function of his or her objective.

The hierarchic organization of ecological systems is most often perceived as a set of interlocking entities of lesser order that make up larger entities (e.g., individual, population, metapopulations, community). But criteria other than structural ones can be used to describe hierarchic organizations. The higher hierarchic levels (1) have a slow dynamics, (2) constitute the framework of functions, (3) are links between looser elements, and (4) impose constraints on lower levels (Allen and Hoekstra, 1984) (Fig. 10). O'Neill (1989) presents the differences in rates of behaviour as a sign of differences between levels of organization. In the field of animal behaviour, Senft et al. (1987) use the number of decisions taken by an animal over a unit of time to distinguish levels of organization. At the lower level (tuft of grass), the number of decisions per day is large; at the higher level (migration route, for example), only a few decisions are taken in a year.

If there is a link between the levels of a hierarchized system, those levels are also characterized by a certain autonomy (O'Neill, 1989) that allows us to understand the systems more easily. We can thus, in relation to a problem, choose a level of organization, then observe how the characteristics of the level above constitute constraints on its behaviour and how the properties of the level studied emerge from the behaviour of elements at a lower level.

5.2. Theories of physics of complex systems: percolation, fractal geometry

5.2.1. Theory of chaos

The recent development of the theory of chaos that has allowed us to approach phenomena that do not seem to obey any law, and are thus impossible to predict, has had an influence on all the sciences treating of complexity. Theoreticians of population dynamics (May, 1976; Molofsky, 1994) have shown by using series of data over a long term that population fluctuations in certain vertebrates can be described by chaotic functions. That is, in phenomena sensitive to initial conditions, the complex dynamics can be the result of simple and deterministic processes.

5.2.2. Percolation theory

The percolation theory was formulated to describe the behaviour of fluids diffusing through heterogeneous and random media. It considers the communication between "sites" that can relay information locally (De Gennes, 1990). These may be porosities in the case of diffusion of fluids, or favourable pockets of resources in the case of movement of living organisms, individuals or populations.

The characteristic of percolation is that it is a critical phenomenon: below a threshold, a piece of information or an individual remains confined in the small island in which it was created. As soon as it passes the threshold, the information "percolates" and is found far from its starting point. This is a phenomenon of all or nothing that links the structure of a medium to possibilities of transfer of propagules, of whatever nature they may be. The measurement of this threshold of percolation (pc) depends on the number of different elements present in the environment. In the simple case of two types of elements, one favourable, the other hostile, this threshold, which represents the relative proportion of these two types, is 0.5928. If the proportion of favourable elements is lower than the value of this threshold, there is no diffusion; if it is higher, there is total diffusion. Figure 11 represents two graphs formed of light and dark grey squares. The dark squares are favourable, the light ones hostile. Each map is formed of elementary pixels arranged randomly. A cluster is defined as a group of favourable cells that have at least one common side on the map. When the number of dark cells

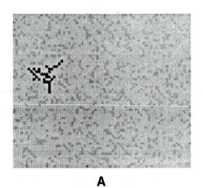
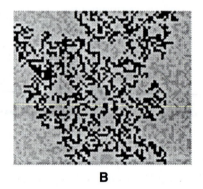

A   **B**

Fig. 11. Percolation maps, 60 × 70 pixels. The percolating element being studied is depicted in black for the largest group of connected pixels. In A, it occupies 19% of the area and is distributed in 291 groups (or clusters), the largest being made up of 33 pixels. In B, the rate of occupation is 40% and the largest of 77 clusters is made up of 1135 pixels.

is progressively increased, the number, size, and shape of the cluster vary. This variability increases greatly when the critical value pc is approached. When the proportion of dark squares is low, the clusters are small and isolated from one another. When the proportion increases, the average size of the clusters also increases. When it becomes higher than the critical value, nearby clusters combine into a vast cluster that connects the edges of the squares, and it is thus possible for an individual for example to move across the map using only favourable sites.

Percolation allows us to describe a large number of physical, biological, or sociological phenomena by means of geometric or statistical concepts. In landscape ecology, many studies on the movement of individuals (Wiens and Milne, 1989) and the propagation of disturbances (Gardner et al., 1987) draw on this theory.

5.2.3. *Fractal geometry*

Fractal geometry is used to measure complex objects and especially to measure the size, shape, and perimeter of clusters such as the one described above in the percolation maps. Fractal geometry can be used to describe complex objects the dimensions of which are in fractions and not integers. A fractal object is characterized by a form that is either extremely irregular or extremely interrupted or fragmented, and remains so no matter what scale it is examined at (Mandelbrot, 1984). It contains distinctive elements of highly varied scale and covering a very wide range. The fractal dimension is a number that serves to quantify the degree of irregularity and fragmentation of a geometric set or a natural object. The fractal dimension is not necessarily an integer. For example, a line with such a complex trajectory that it nearly fills out the plan has a dimension between 1 and 2 (Fig. 12).

The tools of fractal geometry allow us to quantify the structure of the landscape in integrating the complexity on a set of scales (Mandelbrot, 1982; Milne, 1992; Voss, 1988) and to define a self-similarity over a certain range of scales.

To measure the complexity of landscape structures, we can for example measure a fractal dimension using a grid at different scales (Voss, 1988; Milne, 1991). The linearity of the relation between the measurements and the scales is a characteristic of fractal objects, linked to their invariance of scale despite successive increases.

This capacity for defining a space by a single measure over an entire range of scales is very useful in landscape ecology, where the scale of ecological processes is rarely defined *a priori*. The approach has been widely used to study the organization of landscapes and in theoretical research on transfers across scale.

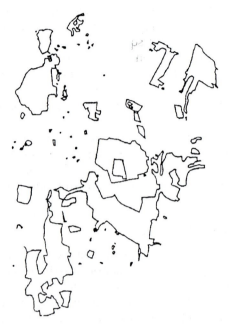

Fig. 12. Borders of forest in southern Sweden, original map at 1/50,000 (Sarlov-Herlin, unpublished). The fractal dimension in this case is 1.29.

5.3. Island biogeography theory

The theory of dynamic equilibrium was developed from islands in the Pacific Ocean (MacArthur and Wilson, 1963; Preston, 1962). It predicts the species richness (number of species) of bird populations on each island as a function of spatial parameters: the area of the island and its distance from the mainland.

The hypothesis that supports the elaboration of this theory (MacArthur and Wilson, 1967) is that the species richness on an island at a given moment is the result of two dynamic processes: the immigration of propagules and the extinction of populations. The rate of immigration of new species decreases to the extent that the number of species established on the island increases and approaches the total number of species on the nearby continent, a source of individuals (pool of potentially colonizing species). The rate of extinction increases with the number of species already present. The more species there are, the greater the risk of extinction linked to demographic or environmental accidents, because of the small size of each population. The number of species present on an island, the species richness, is measured at the intersection of the two curves of immigration and extinction (Fig. 13).

The function of immigration I is expressed as a concave decreasing curve because on average the species that disperse most rapidly are the first

Model of species equilibrium of islands

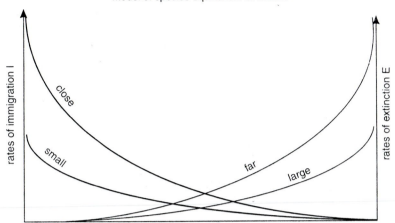

Fig. 13. Rates of colonization and immigration in an island as a function of its size and its distance from the continent (MacArthur and Wilson, 1967). The rates of immigration and extinction depend respectively on the distance from the continent and the area of the island. The species richness at equilibrium is given by the intersection of the curves.

to establish themselves, leading to a rapid fall in the rate of colonization, while the last to arrive, which are slower colonizers, establish themselves over a longer period of time and make the curve diminish more slowly. The curve of rates of extinction E is increasing and has a nearly exponential form. This is explained by the combination of the reduction in population size and increase in competitive interactions of predation, which accelerates the processes of extinction.

These curves I and E vary, respectively, with the distance to the source-continent and with the area of the island. The species richness increases with island size and as the distance to the continent diminishes.

This theory, apart from the pioneering approaches of Watt (1947) on changing mosaics generated by plant communities, is the first approach proposing that the spatial organization of the environment controls ecological processes. It has given rise to controversy and a large number of studies on species richness of fauna and flora, as well as on the organization and dynamics of species assemblages. It has also been the basis of several studies on continental islands, more particularly isolated woods.

5.4. Theory of disturbances

We have seen that the spatio-temporal heterogeneity of a landscape is the result of a set of natural or man-made disturbances. These disturbances are localized and unpredictable events that damage, displace, or kill one or several individuals or communities, creating an opportunity for colonization by new organisms (Blondel, 1995).

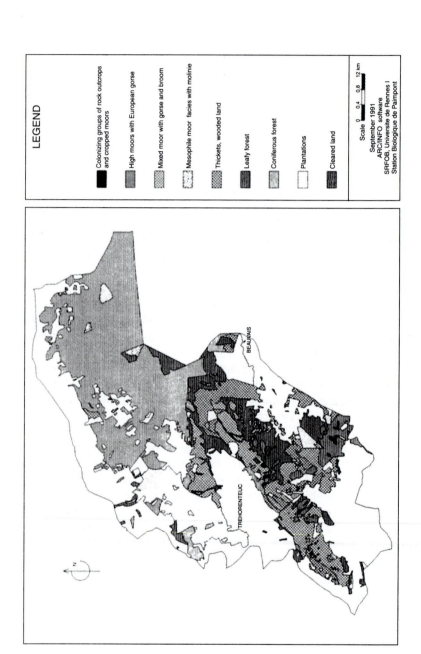

LEGEND

- Colonizing groups of rock outcrops and cropped moors
- High moors with European gorse
- Mixed moor with gorse and broom
- Mesophile moor facies with molinie
- Thickets, wooded land
- Leafy forest
- Coniferous forest
- Plantations
- Cleared land

Scale　0　0.4　0.8　12 km

September 1991
ARC/INFO software
SRFOB, Universite de Rennes I
Station Biologique de Paimpont

BEAUVAIS

TREHORENTEUC

N

Fig. 14. Map of plant groups of the Paimpont Massif forest, western zone, in 1990 (DDAF of Ille-et-Vilaine)

The characteristics of disturbances can be summarized in terms of two concepts. The first relates to the disturbance "regime", which determines the spatial and temporal organization of the creation of patches. It defines the genesis and distribution of patches. The variability of disturbances as well as their effects are vast and can be linked to the initial conditions, as well as the internal heterogeneity at different places. The second relates to patch dynamics (Pickett and White, 1985). Patch dynamics refers to the functioning of landscapes that are essentially subjected to "natural" disturbances. The habitats here are considered discrete, distributed in patches that are individualized and different from one another. The processes of succession scar the disturbed habitat, the structure of which gradually becomes similar to what it was before disturbance. Networks of exchange exist between habitat patches, facilitating the processes of recolonization. In highly anthropized landscapes, the dynamics is more complex and unpredictable (Burel and Baudry, 1990), but the concepts and methods developed in the framework of recent studies on disturbances are used to understand the dynamics of agricultural abandonment, for example (Acx and Baudry, 1993), or the establishment of vegetation in environments with strong constraints, such as wetlands.

The recognition of the universal character of disturbances and heterogeneity in ecological systems has called into question the definition of "climax", which was considered till now a state of equilibrium, the ultimate phase of a successional process. This historical concept was based solely on the temporal dimension and processes of colonization in a given place. The concept of patch dynamics has given rise to the notion of metaclimax: a set of successional sub-systems at different phases from each other, but all equally necessary to the functioning of the system on the landscape scale (Blondel, 1986). Its dynamics is maintained by the regional regime of disturbances, which is a necessary condition to maintain the regional diversity, because the structuration of space that results from it ensures the coexistence of a large number of species on the landscape scale. For example, the successive fires that occurred in the Paimpont Massif (Ille-et-Vilaine) generated a mosaic of plant communities (Fig. 14) the composition of which on a given site depends on the number of fires that have burnt there, their intensity, the conditions before fire, and the place of the site within the landscape (Morvan et al., 1995).

2

Landscape Ecology: Definition of a Multidisciplinary Approach

The ecologist assigns a specific content to the concept of landscape, by which he or she distinguishes an object of study that is different from landscape as understood in other disciplines. Indeed, despite the semantic ambiguity linked to the visual and palpable connotation of the term *landscape*, a consensus has emerged on the object of study, and subsequently on the methods and objectives of this field of research. The first part of this chapter identifies the landscape as understood by the ecologist. In the second part, we show how the inputs of various disciplines are necessary to address problems of landscape ecology, particularly to understand human activities. In the last part, we address contributions of landscape ecology to the development of environment management. Thus, while having a conceptual and theoretical autonomy, landscape ecology has the objective of creating links with other disciplines, links that are essential to an understanding of the phenomena that are the source of environmental problems and their solution.

1. LANDSCAPE AS UNDERSTOOD BY THE ECOLOGIST

We have seen in the preceding chapter that landscape ecology developed continuously with research leading to ecology since the beginning of the century, and that it has accompanied the expansion of recent theories, especially in the related disciplines of physics and mathematics. Landscape ecology was initiated by certain social demands linked to environmental problems that have burgeoned in the last few decades. For the ecological researcher, the landscape is a level of organization of ecological systems, where some processes occur and are controlled. Nevertheless, the use of the term *landscape ecology* has long been disputed in this field as well as in related disciplines such as geography. In this section, we attempt to define the landscape as understood by the ecologist with regard to other landscape disciplines and within the field of ecology.

Why is the use of the term *landscape* in scientific literature so often disputed?

The generally accepted definitions of the term *landscape* are based on two concepts: space and perspective. Pitte (1983) gives one definition from Littré: *Extent of land that is seen in a single view. A landscape that has been seen in all its parts one after the other has not been really seen. It must be seen from a sufficiently high elevation, where all the objects dispersed around are brought within a single glance.*

Webster defines *landscape* as *A portion of land or expanse of natural scenery as seen by the eye in a single view.*

Nevertheless, the origin of the term *landscape* is the juxtaposition of two words: *land*, which is a delimited portion of territory, and *scape*, which signifies a collection of similar objects. Jackson (1986, in Green, 1996) concludes that the original meaning of *landscape* is a collection of lands, a rural system. The term had no sense of aesthetics or point of view, and it was very similar to the sense that is presently given to it by scientists.

1.1. Landscape, a central concept of many disciplines

The landscape occupies a major place among the concepts related to our environment. There is almost no discipline—from geology to literature to pictorial art—that has not used this concept. Many geographers, for their part, see the landscape as a central concept of their discipline. To study the landscape is to study a system of elements in interaction. Such a concept of system may pretend to universality, since it applies to any complex phenomenon (Phipps and Berdoulay, 1985). This reflection of Phipps and Berdoulay legitimizes the appropriation of landscape by various disciplines as a unifying term.

Before being the focus of ecology, the landscape was used in various disciplines, including painting, architecture, literature, and geography (Berdoulay and Phipps, 1985; Inrap, 1986).

It was in the first centuries AD that in China, for the first time in the world, there appeared a landscape aesthetics in the full sense of the term, that is, represented explicitly in words, literature, painting, and gardens (Berque, 1995). In Europe, the idea of landscape appeared only in the 16th century, the pictorial representation preceding the existence of words to define it. The 19th century saw a considerable development of the notion of landscape. The landscape painters (Turner, Cezanne, Pissaro, Van Gogh) described the harmony of landscapes, rural life, changes in land use, and, finally, their own sensations. The painted landscapes included vegetation and animals, but also buildings, roads, and lanes that testified to human activities (Fig. 1). The work of Corot, a landscape painter of the 19th century, was the transition point between a classic representation of landscape and the poesy of nature loved simply for itself, by imaginary recomposition (Pomarede and de Wallens, 1996).

Luginbuhl (1989), in his work *Paysages, textes et représentations du paysage du siècle de lumière à nos jours,* cautions the reader against the vision expressed by these artists of an idyllic countryside, a place of harmony in happy village societies that know how to preserve nature and conserve its aesthetic qualities. This vision is false, he says, because the rural landscape is a social product that has developed not with a quest for the aesthetic, but with economical and social conflicts.

Architects developed a landscape theory at first related to this romantic vision of the landscape. Landscape architects, collected around the Ecole Nationale Supérieure du Paysage de Versailles, were often confined to the domain of the visible and of plastic creation. Donadieu (personal communication) recognizes the multiplicity of land use and recommends

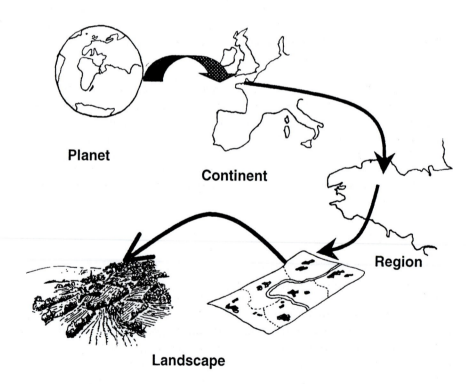

Planet

Continent

Region

Landscape

Fig. 1. The landscape as a level of organization of ecological systems, located above the ecosystem but below the region and continent (Forman, 1995)

the integration of various objectives in landscape projects. Nature, he says, must be managed as a faunistic and floristic inheritance as well as cultural territory, economic or ecological territories, and scenes of social practices.

Geographers, historians, ethnologists, and sociologists have recognized that the landscape results from relationships between nature and society (Chatelin and Riou, 1986; Weber, 1983). Pitte (1983) defines it as the expression of the combination between nature and human technology and culture perceptible to the senses on the earth's surface. Some authors developed methods of description (Wieber, 1985) and others analysed the establishment and functioning of landscapes (Meynier, 1976; Bertrand, 1978; Blanc-Pamart, 1986), a significant part of that analysis being ecological.

1.2. Definitions

The first ecological definitions of landscape are those of Bertrand (1975) and Forman and Godron (1986).

According to the geographer Bertrand, the landscape is a mediator between nature and society based on a portion of material space that exists as a structure as well as an ecological system, and thus independently of perception. He emphasizes that, to be able to define the landscape from a direct reading of a portion of land, we must fall back on one of the dogmas of traditional geography.

The definition of Forman and Godron, given in primary texts on landscape ecology, is close to that of Bertrand. These authors define landscape as a portion of heterogeneous territory composed of sets of interacting ecosystems that are repeated in a similar fashion in space.

The definition of landscape used in the rest of this work is a synthesis of the two definitions above: Landscape is a level of organization of ecological systems that is higher than the ecosystem level. It is characterized essentially by its heterogeneity and its dynamics, partly governed by human activities. It exists independently of perception.

1.3. The scale of the landscape

The size of landscapes varies and the definition of landscape units brings up problems of scale that are at the core of ecological principles (Meentemeyer and Box, 1987). The definition adopted above, which highlights the heterogeneity and dynamics of systems, can be applied over a very wide range of scales, from the continent to the micro-site. In the term *landscape* as used in our field, however, we must give weight to human activities or the scale of human perception, which excludes regional and continental scales on the one hand and very local scales, on the other hand, of a square metre, for example (Fig. 2). In this work, we define the landscape

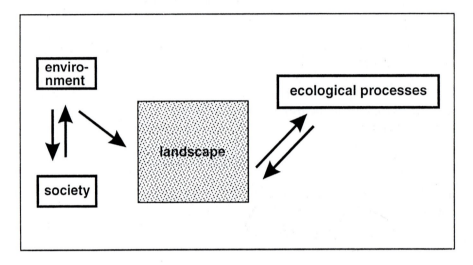

Fig. 2. Scientific approach adopted in landscape ecology: the landscape is the result of the dynamics of the environment and the society that develops in it. The structure, organization, and dynamics of a landscape constantly interact with the ecological processes that occur within it.

as a space involving human activities, which reduces the range of scales to one from a few hectares to a few hundred square kilometres.

Most tools used to analyse space and concepts linked to the ecological processes in heterogeneous environments can be applied over a much larger range of scales. For example, Wiens and Milne (1989) defined the landscape of a Coleoptera. This is a space perceived by a Coleoptera, a tenebrionid in this case, during the course of its movement. These insects live in deserts and in prairies. In New Mexico, they live in desert areas in which bare soil alternates with grassy tufts. Wiens and Milne retain as the chief raison d'être of landscape ecology its capacity to comprehend the heterogeneity of the environment and to reveal the way in which the spatial structure influences a large number of ecological distributions and processes. It thus seems entirely natural to them that the landscape of the Coleoptera is defined as a set of patches of vegetation distributed on rocky soil, and to study the relations between their spatial distribution and the movement of insects. They define these spaces of around one square metre as micro-landscapes, emphasizing their value as easily handled experimental landscapes. These micro-landscapes are useful tools for testing hypotheses on the relationships between spatial heterogeneity and ecological processes, using experimental protocols. Such experimental landscape models nevertheless have a limited

Plate 1. Fragmentation of wooded habitat by exploitation of forest, aerial photograph, Oregon (USA). The private forest is exploited intensively and the result is a mosaic of recently exploited parcels and mature parcels, with few intermediate parcels.

Plate 2. Two cartographic representations of a bocage landscape. (a) Photograph of bocage landscape in the Pays d'Auge, the identifiable elements of the landscape are hedgerows, grasslands, cultivated areas, and apple trees.

Plate 2(Contd....)

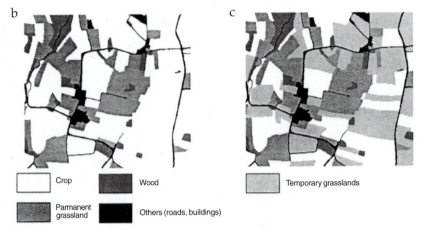

Crop
Wood
Parmanent grassland
Others (roads, buildings)
Temporary grasslands

Plate 2*(Contd....)*
b) A primary map of this type of landscape identifies cultivated areas, permanent grasslands, woods, and other elements (roads, buildings). (c) A secondary map identifies temporary grasslands among the cultivated parcels. The heterogeneity and diversity of the landscape vary as a function of the observer's perspective.

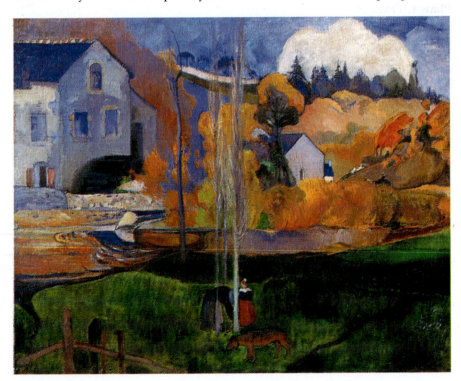

Plate 3. Paul Gaugin, *Le moulin David*. The heterogeneity of the landscape mosaic is remarkable here, as well as the size of hedgerows and the movement between landscape elements

value for two reasons. One, the transfer from micro-landscapes to landscapes on a "human scale" poses theoretical and practical problems that are still to be resolved. Two, the objects that make up the heterogeneity of the mosaic (grass tuft) are normally not directly manipulated by land users. These studies are nevertheless useful in landscape ecology research to the extent that they help to perfect tools and definitions of hypotheses that can be tested on other scales.

Although we recognize that heterogeneity is present in all scales, that it influences ecological processes, and that methods of measurement can be the same, we restrict landscape ecology to the scale of human activities. This limitation, originating from the same source as does the discipline, as presented in the preceding chapter, results in the interdisciplinary position of landscape ecology at the crossroads of ecology and human sciences.

2. LANDSCAPE ECOLOGY: AN INTERDISCIPLINARY APPROACH

The dynamics of landscapes depends on the relationships between societies and their environment. It creates structures that change in space and time. The spatio-temporal heterogeneity that results from this controls many movements and flows of organisms, matter, and energy. Therefore, to understand the mechanisms of maintenance of species and perenniality of flows of water or nutrients, it is essential to take into account the determinants of establishment of heterogeneity of environments. The approach adopted in landscape ecology therefore integrates the object of study, i.e., the landscape, with its determinants, i.e., the environment and society, and its effects on the ecological processes studied (Plate 3). The definition of spatial objects of study, components of the landscape structures, is common to several disciplines and is a crucial step in the research approach. We can thus study, on the one hand, the way in which plants or animals react to the quality of landscape elements and their arrangement, and on the other hand how the activities modify these elements, their qualities, and the way in which they are assembled.

In this second part, we give examples of the input of various disciplines to the problems of landscape ecology. Whenever possible our illustrations are drawn from a study begun in 1991 on the marshes and polders of the Bay of Mont Saint Michel (Burel et al., 1995). The objective of our work is to understand how agriculture, in shaping landscapes and the quality of habitats, affects the biodiversity and quality of surface waters. The human sciences—agronomy, geography, and history—define the structure of the landscape and its dynamics as a function of the ancient and recent history of societies. The landscape can thus be characterized by its organization, heterogeneity, diversity, and dynamics. Ecologists use this knowledge of the landscape to study ecological organization. Here it is a matter of linking flows of organisms or nutrients, spatial distributions of species or populations, to the spatio-temporal structure of the landscape. The traditional methods of field ecology are used—*in situ* measurements, capturing of animals, and floristic or faunistic surveys. One objective, which we will elaborate further, is to model the dynamics of processes studied in a changing environment. For this, we will use social and technical factors determining the dynamics of the landscape to implement the model. From the identification of control mechanisms of processes studied, we can define some principles of land management and development needed to reach a particular environmental objective (improvement of water quality, preservation of biodiversity). The social feasibility of these measures proposed by scientists must be addressed by questions from ethnologists and experts in law and economics.

2.1. Integration of the history of environment and societies: contribution of geomorphology, paleoecology and history

The history of societies, their technology, their culture, and their organization explains a good part of the present state of land distribution and occupation. From the evolutionary trends of landscapes under the influence of climatic variations and human activities, we can understand the establishment of animal and plant communities, as well as the management of flows of water and nutrients. Historians and archaeologists are increasingly focusing on reconstruction of the genesis of landscapes, and their approach may therefore contribute to the approach of landscape ecology.

For example, in the Mont Saint Michel bay, archaeologists, sedimentologists, and historians have reconstructed the landscape history over 8000 years, using data from palynological studies, historical excavations, and the study of ancient documents. The Mont Saint Michel bay is considered by sedimentologists to be one of the most beautiful models of sedimentation existing in the world (Larsonneur and L'Homer, 1982). It also has the advantage of being a model of coastal marsh systems established in the bays and along estuaries by the Flandrian transgression (Table 1) and reclaimed from the sea by the combined action of natural silting and human activities (polderization, draining).

The veritable archive of physical and biological data that is offered by the phenomenon of sedimentation allows us to study the conditions in which animal and plant communities were established in an environment in transition (from a marine to a terrestrial environment). In the littoral zone, whether it is a terrestrial or marine environment, the granulometric distribution of surface sediments that are flooded, permanently or occasionally, or exundated, their organic matter content, and the date at

Table 1. Principal stages in establishment of sedimentary deposits in the Mont Saint Michel bay

Subdivision of Holocene (Flandrian)	Start of climatic periods (years BP)	Dating (years BP)	Variation of sea level	Nature of dominant sedimentary deposit	Observations
Boreal	9500				
		8000	Very rapid transgression		Sea reaches bottom of bay in western part
Atlantic	7500				
		7500	Further transgression	Sea sand at Cardium	Maximum of Flandrian extension to west
		6500	Beginning of the slowing of transgression	Silt	Development of peat in valleys to northeast. In the western part, extension of schorre to Hydrobia towards north
		6000			The sea invades the bay again. Sea level lower than present level by 6 to 7 m
Subboreal	5700	5850 to 5400	Regressive phase	Peat	Marshes to Cyperaceae towards Mont Dol. Development of schorres to the west and east
		5000 to 4400	Important regressive phase	Silt	Sea level 3 m below present level
Subboreal sub-Atlantic	3600	3900 to 3450	Slowing or halt in transgression	Peat	Major development of peat bogs and schorres
		3300 to 3000	Transgressive phase	Silt	Formation of littoral cordon to the west, establishment of peat bogs
Sub-Atlantic	3000	3000	Regressive phase	Peat	
		2300			Formation of second cordon near Mont Dol with lacustrine marshes between the two cordons
		2000	Resumption of transgression	Silt	Sediments of schorre extend and reach the altitude of higher present seas
		1400	Pulsation, transgression		

BP = Before present.

which they were deposited partly determine the distribution of plant and animal species and their aggregation into communities, in the absence of anthropic pressure.

From identification of this pool of original species and analysis of flows of their aggregation as a function of environmental changes that have occurred over 10,000 years, we can evaluate the real importance of human activities in the organization and dynamics of plant and animal communities characteristic of present landscapes of the bay.

2.1.1. Major steps in the establishment of a study site

The recent history of our planet, particularly the end of the Quaternary era, is interesting in that it is characterized by a period of transgressions that have been intensively studied in Europe, especially on the coasts of Holland, England, and France, and in that it is related to the historical period without discontinuity (Morzadec-Kerfourn, 1985).

In less than 20,000 years, the sea level has risen considerably. It was approximately 140 m below the present level around 18,000 years ago, at the end of the last glacial period (Weichselian). The global warming of the atmosphere, leading to the melting of vast indlansis that covered northern Europe, as well as a thermal expansion of the oceanic mass, caused a rapid rise in the sea level. Some 10,000 years ago, the sea level was from –50 m to –10 m. This Flandrian transgression later considerably slowed down, the rise of waters being estimated on average at 1.5 mm/year.

The modalities of more recent and present sedimentation, especially as influenced by human activities, are also well known (Langouet, 1994).

2.1.1.1. End of the Quaternary

The different stages of sedimentary history of the Mont Saint Michel bay are summarized in Table 1. The sea reached the bottom of the bay from the boreal period onwards. The maximum extension of the Flandrian Sea could be delimited by means of the sand deposits at Cardium.

The sea rose again by a series of successive oscillations leading to a sedimentation sometimes in the intertidal domain and sometimes in the supratidal domain. Any transgression was represented by deposits of silt and any regression by formations of peat.

The thickness of the sediment deposited during these last few millennia since 7500 years before the present (BP) was on average about 15 m, which represents an accumulation of material of close to 10,000 million m^3 for the entire 500 km^2 of the bay, with a preferential carriage of fine particles towards the bottom of the bay, where they are found in the present marshes and polders.

2.1.1.2. Historical period

The modalities of present sedimentation are strongly influenced by human activities (Table 2). The first activities undertaken in the Mont Saint Michel

Table 2. Influence of human activities on recent evolution of marshes. Chronological list of major development projects in the Mont Saint Michel bay.

Major dates	Nature of development
11th c.	Consolidation of littoral cordon isolating the Dol marshes
13th c.	The Duchess Anne dyke
1769	Quinette de la Hogue Concession (1000 ha south of Mont), dykes destroyed from 1815 to 1857
1856	Concession to the Mosselmann company of land located between solid land, the Sainte Anne chapel, Mont Saint Michel, and the Roche Torrin
1858	Canalization of Couesnon over 5600 km
1858–1934	Construction of 50 km of dykes to the west isolating 2400 ha of polders (modification of the track of the outer dyke in 1914 to maintain the isolation of the mountain)
1859	Beginning of construction of the Torin dyke (4900 m constructed as against 6300 m planned)
1879–1884	Diversion of rivers located to the east (Guintre and Ardevon)
1878–1879	Construction of the insubmersible dyke linking the mountain to solid land
1966–1969	Barrage of the estuary on the Couesnon, place called Caserne, at 2 km from the mountain to protect 125 ha of farm land

bay date from the 11th century. Located in the western part of the bay, they enabled the consolidation of the littoral cordon isolating the Dol marshes. In the 13th century, these works were followed up and ended in the construction of a dyke of about 40 km, the Duchess Anne dyke, which was completed in the beginning of the 16th century.

It was only from the 18th century onward, and mostly in the 19th century, that other major works were undertaken in the bay. They involved essentially the immediate surroundings of the rock. They included a fixation of the flow canal of certain rivers such as the Couesnon, which had earlier flowed freely from one part of the rock to another. Another development was the recovery of 3500 ha of polders in all (3100 in the west). It was estimated that almost 5000 ha were either polderized or transformed into schorre between 1850 and 1950 (Fig. 3).

2.1.1.3. Present state of the site

As a function of natural processes of sedimentation and transformations of land under the influence of human activity, the bay is a heterogeneous landscape made up of a terrestrial part and a marine part.

In the terrestrial part, there are:

—the Dol marsh, itself made up of two distinct sectors, the white marsh (silt substrate) of around 10,000 ha and the black marsh (peat substrate) of around 1000 ha;

—polders (3500 ha);

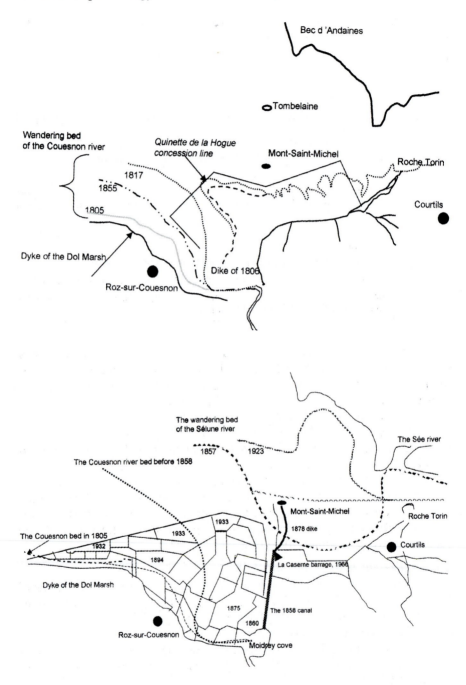

Fig. 3. Transformation of salt marshes of the Mont Saint Michel bay, from 1850 to 1950 (Larsonneur, 1989)

—peripheral marshes of the lower valleys of the Couesnon (600 ha) and of the Sée-Sélune together (250 ha).

The marine part, or the bay itself, opens wide on the English Channel between the points of Cancale and Champeaux, which are 20 km apart. It is characterized by a vast strand (20,000 ha) and the largest schorre area found in France (grassy or close to saline, 4000 ha). Along the median axis, the distance between the entry and interior parts of the bay is nearly 30 km.

2.1.1.4. Stages of human occupation of the marshes

Before being dyked, the Dol marshes were frequented for a specific activity from the protohistoric age: the ignigenous manufacture of salt bricks. In a landscape of schorre and tidal channels, inhabitants of the area (on a plateau located on the cliff) temporarily came into the flood-prone marshes to manufacture salt. They found the essential raw materials there (silt for making moulds and stoves, sea water). Eventually, they brought to these work sites rocks to skim the silt, fuel to feed fires, and some small domestic items (pots, pitchers, etc.). On certain sites, we find the trace of activities during the Gallo-Roman age but there are no indications of permanent establishments. Only in the Middle Ages did groups of people settle in the exundated land.

From a few charters of the first third of the 13th century, we know that the elders of Combourg favoured the drainage and exploitation of the marshes and the establishment of dams against the caprices of the tides. From the 11th century onward, there had been a continual struggle against the sea.

In the light of observations based on cartographic analysis of hydrographic, topographic, parcelling, and archaeological data, we can draw a diagram for the occupation of this eastern portion of the Dol marshes, a diagram that summarizes four phases (Fig. 4).

In phase 1, the littoral cordon that rose over Cherrueix and a zone lying between Mont Dol and Vivier sur Mer must have been the site of the first settlements of this eastern zone of the Dol marshes. The disorganized parcels and the presence of places whose names end in -ville are the indications of this pattern. In phase 2, once a network for draining water and circulation routes were established, agricultural settlement of the land became possible. Place names ending in -iere, -erie, and -ais are the best indications of these settlements. However, the pastures related to this phase are associated for the most part with roads and lanes. They therefore date from the 11th to the 12th centuries. Phase 3 corresponds to occupation of the central zone, which is characterized by relatively unorganized parcelling. During this phase, which extended from the 12th to the 15th centuries, there was gradual encroachment on the marine zone and occupation of land lying slightly lower than the "plateau".

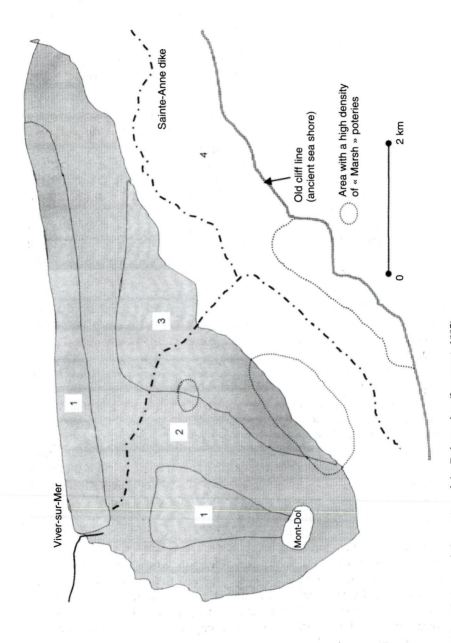

Fig. 4. Map of phases of occupation of the Dol marshes (Langouet, 1995)

Phase 4, which may have extended over a long time, resulted from the closure of the passage that allowed the Couesnon and the sea, beyond the Saint Anne chapel, to flow back into the valley of the old Banche during high tide. This corresponds to a series of major works: construction of the Duchess Anne dyke and the digging of canals. The inversion of natural water flows and the establishment of a particular type of parcelling characterize this phase. There must have been abundant manpower. People could live on the site in more or less flimsy dwellings. Later, within this parcelling, artisanal activity developed during the 15th and 16th centuries.

Agriculture seems to have been practised at the base of the cliff only in the 16th century, when the construction of the Duchess Anne dyke was completed, and beyond the dyke in the 19th century, when the polderization activity was begun. The land thus acquired from the marshes was first used for high value added production (vegetables), then occupied by permanent grasslands after World War II; a change towards vegetable production took place around the 1960s.

2.1.2. Conclusion

From this geomorphological and historical synthesis, we can draw a lesson on the continuous transformation of the landscape over various periods of time. Over the long term, there is a succession of processes that the developer cannot control. There no longer seems to be a physical determination of land use, although the soil characteristics fix the limits of uses, at least in a given technical and economic range. For the population biologist, this signifies that we must undoubtedly abandon the postulate of equilibrium between landscape structures and the distribution of species. We will have occasion to return to this point.

Studies of this type, taking into account large areas over a long period, of the order of a millennium, are increasing in number. Berglund (1991) traced the history of landscapes in southern Sweden over a period of 6000 years. The disciplines involved are paleoecology, plant ecology, prehistoric and medieval archaeology, history, and human geography. One of the objectives of this multidisciplinary research is explicitly to contribute to the management of the natural environment and the cultural landscape. Such an approach was also developed at the Institut Mediterranéen d'Ecologie et de Paléoécologie at Marseilles (Carcaillet, 1997; Berger et al., 1997).

2.2. Role of techniques implemented in land use: input of agronomy and anthropology

Techniques used by a society, whether for agriculture, forestry, industry, or conservation of nature, determine the land use and quality of habitats exploited. Knowledge of these uses is essential not only for understanding the establishment of landscapes, but also for evaluating their future quality. Anthropology, ethnology, and agronomy are the disciplines that address such questions.

2.2.1. *The agronomist's approach*

The agronomist, we will see later in this work, contributes a technical perspective to the reading of the landscape (Deffontaines, 1996). Depending on the levels of organization observed, he or she defines the agricultural practices for the field or field boundary, the technical itinerary, and the systems of exploitation for the entire farm. The agronomist's spatial reading of human activities relates the organization of the landscape mosaic and the "ecological" quality of the habitats with the choices of production, technical choices, and social actors.

For example, INRA researchers have shown how agricultural activity, in a given geomorphological and climatic context, created the landscapes of the Vosges (INRA et al., 1977). A very strong organization was established around farms (Fig. 5), with partition of the land according to various uses constituting landscape modules in which the farm buildings, located at mid-slope, were the pivot.

Below these buildings are mown grasslands, maintained with great care (the Vosgien mantle), irrigated by rivulets or canals carrying water and fertilizers (liquid manure) coming from the stables. Above the buildings are the farmed lands and, further above that, the pastures.

This segregation of land, here clearly visible in the landscape, is found in many other regions, especially mountainous regions, including the Dome mountains, in Auvergne (Bazin et al., 1983). What is remarkable, from the

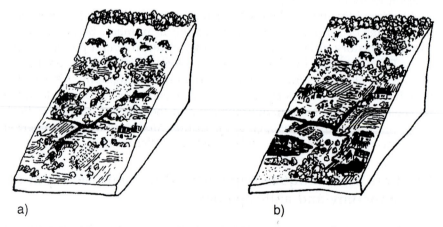

a)   b)

Fig. 5. Organization of Vosges model. (a) Until 1950, spatial organization of activities was strict around villages and along the slope. (b) A decline in agricultural activities led to a change in the landscape organization. Trees colonized the pasture land on the upper slopes and fields were abandoned on the lower slope.

ecological point of view, is the transfer of nutrients from some parts of the landscape (pastures) to others (mown and cultivated fields). Throughout history, human societies have profoundly altered the abiotic environment, on fine scales, which may explain the present distribution of vegetation.

This highly organized system is largely dependent on agricultural production, demographic pressure, and techniques in use. The evolution of these three factors led to changes in landscape structure and composition. This was demonstrated by Balent (1987) in his study in the valley of Oo, in the Atlantic Pyrenees. The land use in the 1950s was based on collective grazing by all the herds on valley land in the winter. In summer, the animals gradually gained the upper pastures, while the farmers individually reappropriated the valley lands for cultivation or fodder production. The inherent difficulties of mountain land use have led to a reduction in agricultural activity that has resulted in infringement of part of the valley parcels, the abandoning of collective winter practices, and spontaneous forestation of part of the upper pasture slopes.

Analysis of systems of polder production (Acx, 1991) shows that polders are relatively homogeneous from one farm to another. The planting comprises vegetables in the open fields, maize, and wheat. However, there are significant differences in the inputs of pesticides (insecticides, herbicides) for a single type of crop, depending on the farmer. That is, there is a high environmental heterogeneity due to farming practices that is not obvious but must be taken into account in the modes in which various species use the polders. In contrast, the fertilization is remarkably homogeneous. Responses (Charrier, 1994) from the farmers using marshes located between the polders and the ancient cliff reveal that they use marshlands as well as the much less fertile plateau. The dynamics of use of these two spaces is linked, which partly explains the relative withdrawal from the edge of the plateau.

2.2.2. The anthropologist's and the ethnologist's approach

The anthropologist studies a landscape only to the extent that it plays a role in social relationships. The ethnological analysis of land really begins only when the ethnologist attempts to gain access to the indigenous point of view and probe the determinations on the basis of which the people studied construct their space, delimit it, occupy it, transform it, and in sum imprint on it a revealing mark of their identity (Lenclud, 1995).

For example, for the Dol marshes, the establishment of which has been presented in this chapter, ethnological analysis reveals conflicts in land management and the attempts or aims of various actors (Boujot, 1995). According to Boujot, the present state of the marsh, totally exundated and cultivated, is much less the result of a relation of the local culture to its surrounding nature than the product of a series of social and political conflicts around the government of this enclave that has been reclaimed from the sea. Drainage and irrigation projects were undertaken from the 16th century for collective control of this land. Recently, concepts of "natural state" brought these developments into question. The regional administration (Diren) attempted to allow water back into the area in order to return it to its natural state. For local ecologists, the poplar plantations and the return of trees are a means of controlling environmental damage due to agriculture. Although the idea of nature seems commonly agreed upon, here as elsewhere, nevertheless the relationships to water, air, land, trees, animals, etc.—the relationships said to be "with nature"—cannot be learned as such, because they are not formulated as such. Analysis of these conflicts between the different agents of development of a territory, as well as all that is demanded from the idea of nature, will lead to the conclusion of C. Boujot, that the idea of nature refers to an ideological construction and a political thought rather than an object: the "nature" of the space on which a culture is deployed.

2.2.3. The concept of cultural landscape

The notion of cultural landscape has been defined in this conceptual framework. It involves linking "traditional" techniques to a state of landscape defined by a particular spatio-temporal structure and, often, a given biological richness and environmental quality. These cultural landscapes correspond often to an idyllic vision of landscape, a reflection of harmony between a society and its environment. The Union International pour la Conservation de la Nature has published a red list that names, at the global level, the "traditional", "heritage" landscapes threatened by technological and economic trends. Among these landscapes are the alpine grasslands and pollard trees of Norway (Austad, 1990), the Mediterranean routes of Crete and Sardinia (Pungetti, 1995), the agro-forestry plantations of Sweden (Berglund, 1991) and Portugal (Pinto Correia, 1993), and bocage landscapes of northern Europe (Bennett, 1996).

The study of these landscapes is rich in lessons about the relationships between a society and its environment and about their present organization. However, inclusion in a red list of threatened landscapes is more a matter of conservation of a cultural patrimony than an ecological approach to landscape.

Conservation of these landscapes requires the continuing maintenance of techniques of mowing, fertilization, management of field boundaries, and routing of animal herds. These aspects are found only rarely in agricultural policies and/or in the objectives of farmers. As in the case of the Dol marsh, it is a matter of ideological valuing of a reference state considered ideal from the social as well as environmental point of view, the recognition of a mythical state of equilibrium between nature and society, to which one attributes a character of stability in opposition to rapid changes of agriculture and landscapes in the past few decades. Such an attitude of "safeguarding" nature or landscape sometimes gives rise to the folklorization of peasant societies denounced by Lizet (1991). This approach also tends to dissociate itself from the dynamics of the environment. Indeed, if we can try to maintain techniques and practices, we can survive the evolution of climates, extinction, or invasion of a certain number of species. For example, the desire of American historians to reconstruct the battlefields of the Revolutionary War in the condition in which they were seen by General Washington and his armies is frustrated by the fact that the dominant species of the deciduous forest of that time (chestnut, elm) have disappeared from the American continent (Emily Russel, personal communication).

The set of approaches presented in this paragraph provides information on the relations between society, its culture, its techniques, and the establishment and management of landscapes. They must be taken into account in understanding and modelling the dynamics of landscapes and evaluating the social feasibility of environmental policies and thus the recommendations for land management that result from research in landscape ecology.

2.3. Recognition of past and recent landscape structures: contribution of geography

Even though landscape ecologists have integrated the study and analysis of land structures in their research approach since the 1980s (Turner and Gardner, 1991; Hulshoff, 1995; Plotnick et al., 1993; Burel and Baudry, 1990), this subject is essentially in the domain of geography.

Mapping of a given space with all the associated norms is the prerogative of geographers. The use of remote sensing data acquired by various means— spatial, aerial, terrestrial, or marine (Bonn and Rochon, 1992)—allows analysis of elements, units, and types of landscape. This vertical view of landscapes, however, remains incomplete because it obscures the third dimension, and it must be complemented by an association with other spatial data.

The documentation of remote sensing covers vast geographic areas that a single human perspective cannot perceive simultaneously. Moreover, the multi-temporal characteristics of records from afar facilitate diachronic studies, or even a study of modifications of the landscape (Bariou, 1978). The multi-scale dimension can also be considered with remote sensing data. Data from different satellites (e.g., Meteosat, Landsat TM, SPOT) and from aerial photographs complement each other to give at each degree of precision a type of landscape and a set of information that is associated with it (Table 3).

Legrand (1995) conducted a study of the dynamics of land cover in the polders of the Mont Saint Michel bay, from the time they were created to the present. The cultivation was systematic following polderization, one of the major products being seeds for the Vilmorin firm. Grasslands appeared

Table 3. Multi-scale remote sensing study of the landscapes of western France (Morant, 1995)

Type of data	NOAA/AVHRR satellite	Landsat 5 TM satellite	Vertical aerial photographs
Scale	1/250,000 to 1/500,000	1/50,000 to 1/100,000	1/10,000 to 1/25,000
Type of scale	Regional	Infra-regional	Local
Type of division	Landscape units	Landscape units and sub-units	Landscape elements
Type of information identified or interpreted	Major divisions of rural landscapes	One type of parcelling	One hedgerow, one parcel
Type of landscape observed	Rural	Rural and agricultural	Agricultural and agrarian

during World War I and occupied up to 70% of the polder land area from 1950 to 1970. The agricultural evolution took place at the same time as for the entire region of Brittany, but it was more intense on these lands, which had great agronomic potential. In 1995, there remained only 5% of the grasslands, and the major vocation of the polder became the production of vegetables such as carrot, onion, and turnip. These crops had a high economic value and required intensive use of pesticides. The study provoked an original response in that, by association with the dynamics of other humid zones of the French coast such as the western marshes (Anon., 1974, 1990) or the Camargue, the collective consciousness attributed to this zone a stability of permanent grasslands, from their creation to the 1970s. The variability of the land cover was certainly favourable to the colonization of this new space by many animal and plant species.

2.4. Ecological functions

The preceding chapter illustrates the continuity of approaches throughout the history of scientific ecology. From autecology up to landscape ecology, the objects of study have become more complex, from the individual to the landscape, but the questions have remained of the same order. Many of the concepts and tools can be transferred from one level to the other and allow the discipline to be identified. In the rest of this work, we often refer to ecological theories and derive knowledge of the functioning of ecosystems mostly from data relating to the biology of populations and communities. Readers who wish to study these questions further may refer to a number of works (Blondel, 1986, 1995; Barbault, 1981, 1992) that present a synthesis in these fields.

All the disciplines of ecology do not converge towards landscape ecology. The domains most often studied are the dynamics of populations and communities, as well as transfers of matter and nutrients. These points are amply illustrated in the subsequent chapters of this book.

In the framework of studies conducted in the Mont Saint Michel bay, we have essentially tackled questions of biodiversity in an intensive agricultural zone and on land recently reclaimed from the sea in comparison with the neighbouring hedged zones, which have a more ancient human occupation and less intensive agriculture. The question *Is there a variation of species richness from one landscape to the other?* (Burel et al., 1998) has been posed at the level of communities and of some species for which the researchers have defined the importance of the connectivity of the network of dykes and hedgerows (Clergeau and Burel, 1997), and of the permeability of the matrix (Paillat, 1994) in the functioning of populations.

The setting up of a research programme on landscape ecology must take into account all the disciplines mentioned above in order to comprehend the mechanisms that control the ecological processes at the landscape level, predict their evolution, and propose elements of management or development.

Figure 6 presents the organization of research on landscapes to the south of the Mont Saint Michel bay and the way in which issues of landscape ecology can be approached through various disciplines.

This interdisciplinary approach is rarely so detailed in research that has its roots in landscape ecology. However, although many studies limit

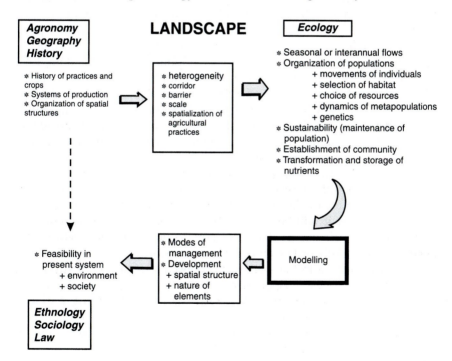

Fig. 6. Flow chart of research on landscapes of the Mont Saint Michel bay

themselves to the confrontation of two or three disciplines, because of material and other constraints, it is essential to maintain the spirit of the holistic quality of the process shown in Fig. 7, which describes the interdisciplinary dialogue that must be developed in landscape ecology.

3.   LANDSCAPE ECOLOGY: APPLICATION OF RESULTS OF FUNDAMENTAL RESEARCH TO CONSERVATION BIOLOGY AND LAND MANAGEMENT

The scale of ecological systems studied in landscape ecology is, by definition and in keeping with the origin of the discipline, that of involvement of human activities. From the start, researchers have faced issues of development and management of fauna and flora. Since it was founded in 1983, the International Association of Landscape Ecology (IALE) has brought together researchers from varied disciplines, as well as developers, rural and urban land managers, and professionals in the field of biology of conservation.

Efforts to use landscape ecology to improve conservation and land management were initiated in Western Europe by researchers and developers of the Netherlands, a country with a dense population in which public awareness of land deterioration and loss of biodiversity grew and coincided with intensified land use. According to Paul Opdam (Vos and Opdam, 1993), the Netherlands was an experimental area in which landscape ecology developed rapidly as an applied science to resolve problems related to conflicts of land use generated by opposing activities, such as agriculture and the creation of natural reserves. The objective was to provide developers with tools for evaluating the environmental effects of their plans and for comparing various scenarios. Most of the time, problems of development are represented spatially on plans that present the future condition of the landscape. In scientific terms, from the reading of these plans, hypotheses are elaborated on the evolution of ecological functions, defined in a spatially explicit manner, that will be induced by the proposed development. Even though the scientific expert does not have to take decisions, which often depend on policy, his or her objective is to measure the "sustainability" of ecological functioning in the scenarios proposed. The notion of sustainable development is in fact associated with any ecological reflection on development and management. It is not a matter of freezing ecological systems, which are, as we have seen, essentially dynamic, but of maintaining the ecological processes that characterize the system and ensuring the perenniality of renewable resources.

Landscape ecologists must test the different scenarios, develop tools to model their effects over the short and long term, and evaluate whether or not the expectations of a particular development are met. In other words, they must translate the spatial structure of a plan into spatial relationships and predict their effects on the functioning of the landscape.

For 20 years, technological and conceptual progress has allowed landscape ecology to respond ever more effectively to environmental questions. The field has emerged more prominently and it has an increasing voice in projects and studies conducted before development projects.

3.1. Landscape ecology and landscape management

In France, applications of landscape ecology have mostly developed in the framework of impact studies. Impact studies, which were made compulsory by legislation for environment protection in 1976, must evaluate the environmental impact of projects and therefore the modifications they are expected to cause in landscape structures. Very quickly, as we saw in Chapter 1, tools to spatialize the processes and link them to the heterogeneity of the landscape, after a development is proposed and before the project is begun, had to be redefined.

These methods found their first applications in the context of impact studies of land consolidation (Lefeuvre, 1979). The contribution of landscape ecology here was not only to consider the quality of the habitats conserved or created, but also to observe the exchanges between these elements and to be able to evaluate their perenniality, or even their restoration, according to various scenarios (Baudry and Burel, 1984). The points on which the researchers and developers focus are essentially the maintenance of biodiversity and management of air, water, and nutrient flows. There have been, and still are, frequent exchanges between the fundamentals and applications in this field. This will be developed further in the fourth part of the book.

At present, the applications are diversifying because of an increasingly wide diffusion of knowledge. The principles of landscape ecology are applied in impact studies of highways and other roads (Pain, 1996) and in urban plans (Fabos and Ahern, 1995).

3.2. Application to land management

The distinction we make between land development and management is based on the legislative tools involved, as well as the difference of modalities of application. The first involves planned modifications of the landscape structure (uprooting of hedgerows, road construction), which involve a relatively large area of thousands of hectares and a combination of agencies (users, inhabitants). The second involves individual local interventions, coordinated by rules or incentives, to derive a function or a structure at the landscape level.

In this context, landscape ecology can help to define the establishment and pursuit of agro-environmental measures, plans for sustainable development, hedgerow replanting, management of abandoned land (Acx and Baudry, 1993), or water management policies.

The contributions of landscape ecology here are also essentially linked to the study of exchanges between elements, continuities of flows of organisms or matter. For example, the drawing up of specifications for management of permanent grasslands in order to conserve certain animal species makes sense only if all the parcels involved in this measure offer a space with sufficient resources to sustain a population or are sufficiently close to another favourable zone.

Questions of scale are at the core of this issue. Is it preferable to have heterogeneous landscapes with a very fine grain, which increases the diversity over small areas, or to have consolidated parcels in order to allow more demanding species in a given habitat to survive?

3.3. Applications in nature conservation

Concepts linked to nature conservation have evolved considerably in the past few years. The first conservation measures, dating from the beginning of the 20th century, were demarcations of natural reserves that were more or less protected depending on their status. The first relevant law in France was passed in 1930. It concerned the protection of natural monuments and defined procedures similar to those adopted for historical monuments: inscription and classification. These procedures correspond to two objectives: to establish an inventory of sites and to protect threatened sites by determining some limitations on modification of the status and exploitation of those sites. A third method was proposed in the law, a "protection zone" covering a larger space around the protected site. Laws passed in 1957 and 1976, relating to protection of nature, fixed the general rules governing classification of a portion of land as a natural reserve.

The national parks in France were created in 1960 with this same end in view of protection of threatened land. They had a double objective: to conserve natural sanctuaries in which only scientists are authorized to enter, but also to open to the public part of the territory in which some activities are authorized and regulated with the aim of allowing everyone access to the scientific and aesthetic resources of the park. Other regulatory measures have since been defined, for example, to protect the coast (in 1975), demarcate zones of international interest for migratory birds, and demarcate natural zones of ecological interest for flora and fauna. In all these cases, the idea was to conserve an area supposed to be large enough to ensure the continuance of rare species, or of particular ecosystems. The extent of protection and management depends on the status of the protected land or inventory (Lefeuvre et al., 1979).

The evolution of ecological theories and especially recognition of the dynamics of ecological systems and processes of colonization and extinction of species have called into question this static vision of the conservation of nature. It is now recognized that exchanges between zones of ecological interest are fundamental for the survival of many populations and essential for the recolonization of disturbed habitats over the short or long term in the context of global climatic changes. The ideas that prevail now are that measures for the protection of restricted areas must be accompanied by a reflection on the relay and reserve roles of the entire territory (ordinary land) and that buffer zones and corridors must ensure the coherence of a set of measures.

We will see in the third part of this book how the concepts of landscape ecology have contributed to the definition of this new policy of protection of nature, and how they are used to implement it.

Part II

Landscape Structure and Dynamics

3

Analysis of Spatial Structures

As space, spatial relations, and heterogeneity were taken into account from the very emergence of landscape ecology, a terminology developed for the types of objects encountered in a landscape. Several studies were undertaken to propose various methods for measurements relating to land. Some studies are presented in this chapter; for additional information, the reader is referred to the exhaustive bibliography (Turner et al., 1991). Geostatistics (Haining, 1990) and geography (Ciceri et al., 1977) are also important sources. The chief applications of these fields are presented in this chapter.

Beyond, or before, presenting the techniques, it is essential to understand what space is from an ecological point of view and the consequences of that understanding with regard to spatial analysis. Two essential points will be seen: (1) the breaking up of a landscape into its elementary units and (2) the variation of a single measurement within a landscape depending on the position from which it is taken. Although it is important to develop methods of spatial analysis that have only the objective of providing a measurement on a map, it is still more essential to provide an ecological meaning to these measurements, i.e., to look at how the structural descriptions are linked to processes such as the moving about of species, the dynamics of populations, or physicochemical flows.

From looking even at a few landscapes, we can understand the importance of developing concepts and techniques to study landscape structure. Plate 4 presents six landscapes.

The first is a hedged landscape in Normandy. It is immediately characterized by the presence of hedgerows that form a network and permanent grasslands that constitute the major land cover.

The second is also a hedged landscape, in Brittany. Here annual crops (tilled fields) are dominant, and the hedgerows are relics. There are recently constructed farm buildings.

The third is a landscape in Lorraine, in which open fields are dominant on the hillsides, while the edge of the plateau is wooded and the valley is occupied by houses and farmed land with trees. This landscape is located opposite the Pays d'Auge, on the eastern border of the Parisian basin. The geomorphology is similar, but the history and culture are very different. Lorraine is characterized by grouped dwellings and triennial rotation regulated by the village community. Normandy is characterized by scattered dwellings and individual parcels of land.

The fourth is a landscape in the Dordogne valley; hedgerows, groves, fields, orchards, and houses are seen, apparently not following a perceptible pattern. The elements follow one another without regularity. Only the distinction between the farmed valley and the wooded slope marks a strict organization linked to the relief.

The fifth is a Mediterranean landscape in the Duyes valley, close to Digne. The land uses are diverse, with perennial species (lavender) mingling with annual crops. Wooded areas occupy the relief. The straight borders of these woods indicate that they were planted. Lower in the valley, the limits are more fuzzy and may result from colonization by woody species in the context of abandonment.

The sixth is a homogeneous landscape in a prairie zone in Washington State (USA). Cereal crops have replaced the prairie vegetation, which remains in a fragment at the centre of the photo. The plant diversity is considerably reduced, but the physiognomy of the landscape has changed little.

To compare these landscapes and to test relationships between ecological characteristics (types of flora and fauna) and landscape structures, the use of metric variables is essential. That is the focus of the present chapter. Before taking measurements, we must locate the elements, name them, and distinguish one from the other or put them in a single group. For example, we must decide whether the diversity of crops (wheat, maize, cabbage) must be taken into account or whether an aggregate variable such as "crops" will suffice.

The objective of this chapter is to present a set of methods to describe landscape structures. The beginnings of landscape ecology were marked by a significant production of indexes designed to measure structures in order to quantify them. Examples of many such indexes are found in the journal *Landscape Ecology*. Haines-Young and Chopping (1996) analyse these indexes and conclude that there is still an urgent need for research on the links between them and ecological processes. The use of indexes to evaluate development alternatives may even lead to errors. The presentation of these indexes is accompanied by tests showing the limits of their validity and the precautions to be taken while using them.

The major problem with such measurements is their attempt to reduce a spatial structure to a number. Very often, a large quantity of spatial information is lost, especially in the case of measurements relying on calculation of averages. The average area of woods in a landscape may be identical for class distributions of very different area. The same is true of average distance between the woods. The availability and easy use of geographic information systems now allows us to manage spatial information and to propose new methods for analysing landscape structures. Using these new approaches, it is possible to incorporate hypotheses on the "perception" of landscapes by different animal or plant organisms. Recent research involves, in large part, displacements of animals, which frequently are used as examples. Nevertheless, the spatial distribution of plants and the flows of matter (nutrients, pollutants) are also controlled by landscape structures and are the focus of many studies.

The chapter is organized in the following manner: a presentation of concepts relating to landscape elements and then to landscape structures, followed by a detailed description of measurement of heterogeneity, fragmentation, and connectivity. Subsequently, there are two parts with succinct presentations of fractals and geostatistics. An example of typology of landscape structures closes these structural analyses.

Throughout this chapter, simulated landscape structures are relied on. This facilitates the diversity of analytic approaches and the manipulation of these structures, especially to make them evolve or in changing their elements.

1. CATEGORIES OF LANDSCAPE ELEMENTS

In one of the founding texts of landscape ecology, Forman and Godron (1981) proposed a distinction between the different elements that can be distinguished in a landscape (Fig. 1). The *matrix* is the dominant, all-encompassing element. Within it, there are *patches* (groves and habitations) and *corridors*, linear elements. A set of patches constitutes a *mosaic* and a set of corridors constitutes a *network*. Within patches and corridors, we can distinguish an *edge* that has strong interactions with the neighbouring matrix or patches, and an *interior* in which the interactions are weak or nil. The more elongated these patches are, the higher the edge-interior ratio is.

The spatial arrangement of the mosaic and networks is the landscape *pattern*. It can be useful in studying the differences or similarities between two landscapes from the structural point of view.

This terminology provides the framework needed to describe landscape structures and establish sampling procedures of fauna and flora in research testing the existence of a "landscape effect". Two points are worthy of discussion.

One is the pre-eminence of the visual. The elements seen within the landscape—patches and corridors—are primarily visual entities, stable at least over the medium term, and not functional entities. We will see ultimately that there are a multitude of differences that are not directly visible, due to human activities (e.g., fertilization of parcels), and passing diversities due to physiological processes (e.g., flowering).

The second is the notion of matrix as an undifferentiated space, neutral or hostile. This is a notion directly transposed from the island biogeography theory, in which the ocean is a radically different environment from islands

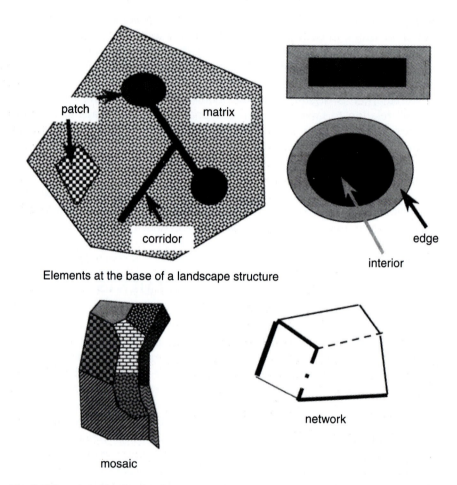

Elements at the base of a landscape structure

mosaic

network

Fig. 1. Categories of landscape elements

or continents, which is not the case in a terrestrial environment, where there is always a gradient of situations between the forest and the tilled field. Researchers nowadays prefer to speak of a landscape mosaic, i.e., a contiguous set of patches of different kinds. This complication of the representation of landscapes has been made necessary by biological investigations and has been made possible by technological developments. However, it must be emphasized that initial representations that may appear schematic have allowed us to locate simple situations, necessary to the emergence of a discipline and to the first tests of hypotheses.

This location of elements, like analysis of structures, is done within a perspective of links with ecological phenomena. The first question to be asked is, *What parameters can affect the presence, survival, displacement, and reproduction of an organism or of an animal or plant population in a landscape?* What, in effect, must be done to depict it on a map? What must be looked for during an analysis? There is a constant iteration between analysis of landscape structures and biological knowledge. Similar questions are posed in the study of relationships between human activities and the dynamics of landscapes. *What should be distinguished, and what should be aggregated?*

From the biological point of view, the first factor is the presence of a habitat, a favourable or at least acceptable environment. The essential theoretical and methodological problem is the need for a reference to a type of species, and to biological traits that characterize it (displacement, migration, feeding, reproduction). This perspective of analysis in relation to the functional value of landscapes is not the most common. The dominant approach, in landscape ecology, is to take off from a known ecological function and to attempt to link it to a measurement of structures. Since there is no question of analysing all the species, we must create typologies of species. This is the objective with which functional groups are defined (Lavorel et al., 1993; Medail et al., 1998) in relation to particular traits of the life history of species, such as their mode of locomotion or their feeding or reproductive strategy. We will again encounter this problem in the chapter on modelling.

2. FROM SAMPLE PLOTS IN A WOOD TO WOODS IN A LANDSCAPE

The study of plant or animal communities in woods and forests holds an important place in ecology. They are often considered representative of original communities, before the great clearings, an idea that it would be advisable to correct in the light of studies on the dynamics of non-anthropized landscapes. These forest species are an interesting model because, like all specialist species, they are sensitive to the fragmentation of their habitat (Farina, 1998).

Forest communities can be studied in various ways. The most conventional is to inventory plants, trap insects or small mammals, or plot bird calls and to relate these observations to the structure of the wooded cover or time since the last felling (Blondel, 1995).

Inspired by the island biogeography theory developed by MacArthur and Wilson (1967), several authors (Forman et al., 1976; Whitcomb et al., 1981) considered testing the effect of size of woods on the ornithological community. Forman and his colleagues (1976) indicated that large woods shelter more species than small woods (Fig. 2).

In addition to the internal structure of the vegetation of these woods and their area, we can also take into account their position in relation to one

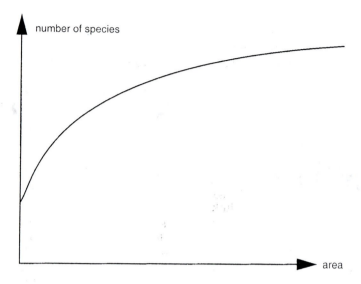

Fig. 2. Relationship between size of groves and number of bird species (simplified from Forman et al., 1976)

another to study the archipelago effects, which facilitate exchanges between groves (Fig. 3a).

Two complementary stages (Figs. 3b and 3c) can be considered in the analysis of integration of woods in the landscape. Additional information can be gained from taking into account wooded hedgerows that are linked to one another and serve as a corridor across the agricultural space. Finally, agricultural areas or buildings can themselves be differentiated and integrated in the analysis as a cause of proximity effect that can influence communities. This effect has been demonstrated by Matthiae and Stearns (1981).

The modes of sampling and data analysis depend on the choice of the observer. It must also be emphasized that the various modes of observation correspond to different questions. The first method (independent sampling of woods), covering the widest range of climatic conditions and vegetation structure, is based on a regional approach of communities. On the other hand, to test the archipelago effect, a small number of neighbouring groves must be analysed, the area to be studied being fixed almost as soon as the first wood to be studied is decided upon. Rather than the effects of internal variability, even though those always exist, it is the effects of spatial structure resulting from the relative position of groves that will be studied.

(a)

(b)

(c)

Plate 4. Six landscapes: six histories, six structures. (a) Auge, Normandy; (b) Rennes basin, Brittany; (c) Chatenois, Lorraine;

Plate 4 (Contd....)

(d)

(e)

(f)

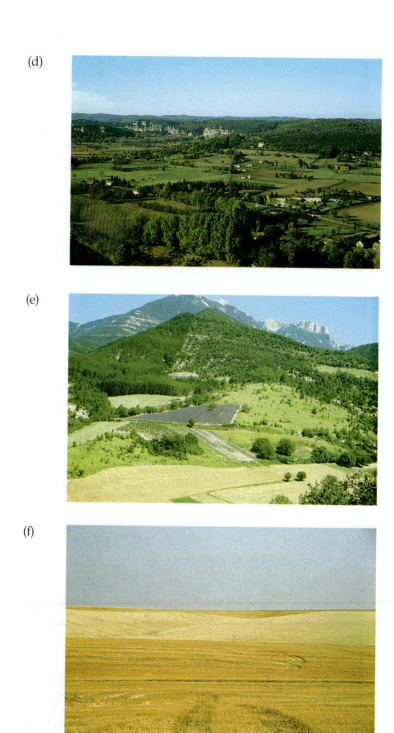

Plate 4. (Contd....)
(d) Dordogne valley, Perigord; (e) Diois, Provence; (f) eastern part of Washington State.

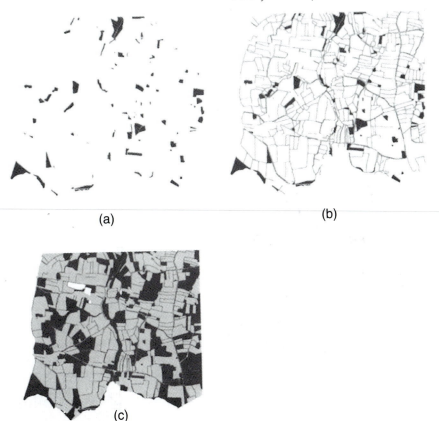

(a)

(b)

(c)

Fig. 3. Various representations of a set of groves corresponding to different questions and hypotheses. (a) Archipelago of wooded islands. (b) Wooded islands and network of hedgerows. (c) Wooded islands in the agricultural area.

3. TYPOLOGY OF PATCHES AND CORRIDORS

In the section above, a single category is considered: "woods". Often, a typology of corridors and patches must be established during the mapping process. Figure 4 gives an example of different possibilities of mapping a space with more or less detailed typologies.

We start with a simple cartography: only wooded and built zones are depicted, which implies that the organism we are studying makes no distinctions, in terms of the type of vegetation, for example, between the wooded zones, and is insensitive to the diversity of open areas. In the second map, the permanent grasslands appear. In the third, the crops are detailed, and we can make a distinction between winter and summer crops, but this could be a distinction of dates of flowering.

Roads, habitats,
woods, and fallow land

+ permanent
grasslands

Woods and fallow land

Houses and gardens

Permanent grasslands

Temporary grasslands

Maize

Other crops

+ permanent
grasslands,
temporary
grasslands,
maize and
other crops

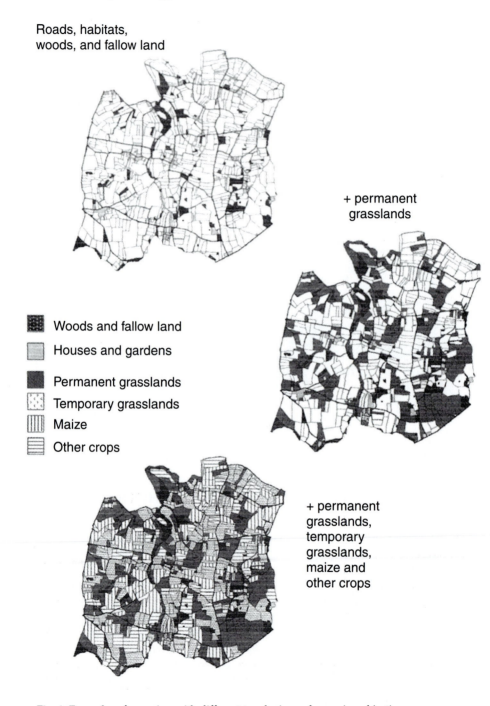

Fig. 4. Examples of mapping with different typologies and mapping objectives

Using geographic information systems (GIS), we can store various types of spatial information and thereby produce a set of maps for a single landscape. Table 1 gives a brief presentation of GIS, explaining the basic concepts needed to follow the analyses presented below. For more complete information, the reader is referred to GIS manuals and specialized texts, especially on applications to landscape ecology (Haines-Young et al., 1993; Johnston, 1998).

The coherence of typologies must be ensured. For example, a comparison of landscapes, in space or in time, must consider the same categories for

Table 1. Rudiments of geographic information systems (GIS)

What is a geographic information system?

A GIS allows us to store, organize, represent, and manipulate spatial information. The principle is to define objects located in space and to characterize them using a set of variables that arise from observation, investigation, or calculation. There are two types of GIS, defined as a function of the type of objects they represent: GIS in vector mode and GIS in raster mode.

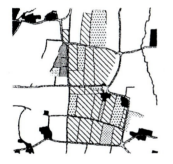

The vector GIS uses three types of objects:
1. polygons, which are two-dimensional objects, such as the elements of a mosaic, parcels;
2. arcs, which are lines, often borders of polygons; and
3. points.

Note: a hedgerow may be a line, i.e., the limit between two parcels, or it may be a polygon if we want to represent it so that its two borders are distinct. It is all a question of scale of representation.

This type of GIS allows representation of polygons and distinguishes them by means of a framework, as in the map above.

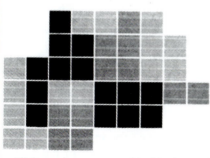

The raster GIS represents only pixels, square elementary units. Each pixel is represented by a colour. In the second figure, the pixels have been detached from each other so that they can be clearly seen. This format is that of satellite images or television or computer screens.

Rasterization is done from a vector map using a grid in which each square is attributed to one of the objects or the other. The size of the squares (rasterization grain) determines the precision of the map.

All the analyses presented in this chapter are drawn from rasterized maps. In a GIS, all the elements must be identified. The pixels are located by their X and Y coordinates. Generally, their primary characteristic is their identification as a parcel or a hedgerow, objects that are themselves identified. This is how the database containing information on the objects observed is linked to the rasterized map.

each case. Suarez Seoane (1998) has demonstrated the dependence of results from the analysis of landscape structures on the typologies followed. A preliminary decision to be made is the definition of the basic unit of the map: either an elementary unit located in the field or a synthetic unit. These decisions involve considerations that are ecological (species or process studied) as well as technical (number of units, resolution of map).

At this stage, once the mapping is done, what are the parameters that we wish to measure or to take into account in measurements?

Till now, we have used only typologies of elements having clearly defined limits, at least at the mapping level. We will later come to use typological gradients, i.e., spatial representations of objects varying continuously. For example, the distribution of trees in a landscape may be such that the limit between wood and non-wood is fuzzy. The perception that an organism has of a landscape may also lead to a representation in gradients rather than in clearly delimited entities. This leads us to insist on the fact that the various maps are only representations of landscapes corresponding to particular questions or points of view. Kotliar and Wiens (1990) emphasize the importance, but also the difficulty, of clearly defining the patches relevant to our understanding of an ecological phenomenon.

4. BASIC CONCEPTS FOR QUANTITATIVE APPROACHES

Quantitative approaches of fragmentation, connectivity, and heterogeneity may give way to important developments, especially since various computer programs are now available. These concepts are all interlinked. They must also be fundamentally defined and measurements of them must ultimately be developed. The close relationships between concepts appear in the fact that a single graph can serve to illustrate several concepts. Each time, it reveals different information.

4.1. Size of patches and fragmentation

The first of these parameters is the quantity of habitat available. The habitat is the set of patches that an organism can use. The availability is the overall quantity (total area) as well as the quantity of a single block, which poses the problem of spatial distribution of that area, i.e., the evolution of large patches towards patches that are smaller and further apart. In Fig. 5, the two types of patches are not fragmented in part (a), and they are extremely fragmented in part (c).

The sensitivity of individuals of a species to fragmentation depends on their radius of daily movement and their scale of activity. The habitats

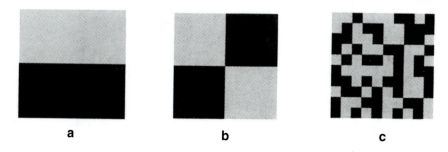

Fig. 5. Fragmentation. The three maps represent an increasing degree of fragmentation.

represented in Fig. 4c may not be perceived as fragmented by species that move over short distances. Fragmentation may play an essential role in the survival of populations. There may be enough space in the fragments for one or a few individuals, but not for a population.

4.2. Spatial relationships between patches: connectivity and connectedness

Movement between distinct patches is an essential process of landscape ecology. Organisms may move between patches of a single type or of different types. Movement between different types of patches corresponds either to different activities (feeding, reproduction, hibernation) or to the organism's ability to exploit several types of patches. The capacity of individuals of a population to leave one patch to colonize another similar patch is the fundamental process of survival of metapopulations. It is also an essential process of landscape dynamics after disturbance or abandonment of farm land.

We use the term *connectedness* to refer to the fact that two patches of the same type are adjacent and joined in space. We use *connectivity* to refer to the fact that an individual or propagules of a species can move from one patch to another, even if they are far away (Baudry and Merriam, 1988). Here again, the displacement capacity of individuals is an essential factor.

Figure 6 shows how different spatial configurations can cause variation in connectedness and connectivity. In (a), the black element forms a continuous patch. It is closely connected, and all its parts can be easily linked by the movement of an animal. In (b), the black element is fragmented. However, there are trajectories through which an animal can move continuously in this element. The connectedness is much weaker than in (a) and restricts the connectivity for an organism that cannot move within another element. In (c), the fragmentation is still worse, the connectedness is very poor, and only organisms having a limited need for space can use the fragments, especially if their behaviour does not allow them to move

(a) High connectedness, high connectivity, displacement with a patch

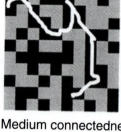

(b) Medium connectedness, high connectivity

(c) Low connectedness. Very poor connectivity, or connectivity ensured by movements across the matrix

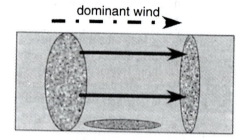

(d) Nil connectedness, connectivity ensured by physical mechanisms

Fig. 6. Variations of connectedness and connectivity. (a) High connectedness, high connectivity, displacement within a patch. (b) Medium connectedness, high connectivity. (c) Low connectedness. Very poor connectivity, or connectivity ensured by movements across the matrix. (d) Nil connectedness, connectivity ensured by physical mechanisms.

from one patch to another without spatial continuity. In such cases, connectivity can be ensured by flight, as with birds using an archipelago of groves. In (d), the connectivity between two elements of the same nature is ensured by a physical force, wind, that allows exchanges (unidirectional) between parallel elements, rather than between similar elements. There is no relationship between connectedness and connectivity.

One way of quantifying connectedness is to count pixel pairs of the same kind on the rows and columns of a grid.

4.3. The entire mosaic: heterogeneity

A landscape is presented first of all as a set of elements that are more or less fragmented or connected: this is the landscape mosaic that we recognize as a heterogeneously spatial set.

Heterogeneity has two components: the diversity of elements (patches) of the landscape and the complexity of their spatial relationships. In Fig. 7, the heterogeneity increases from (a) to (b), because the proportion becomes equal. In (b), there is an equal probability of coming across one or the other element, while in (a), there is greater probability of coming across the black element. There is better predictability of events. From (b) to (c), there is an increase in the number of elements, diversity, and thus the spatial heterogeneity. From (d) to (f), the parts of Fig. 5, illustrating fragmentation, are repeated. The diversity (two types of elements) and the proportions (50%) remain the same, but heterogeneity increases because of the spatial arrangements. More and more information is required to describe the relationships between the two types of elements. In (d), black and grey each occupy half the space and form a single block. In (e), black and grey each occupy half the space and each is in two equal blocks, distributed as four squares in the entire space. In (f), black and grey each occupy half the space, but they are highly fragmented: in the first line there is a succession of two small grey squares, then a black square, and so on. Here we see the effect of spatialization of the description of a landscape. The variety expresses how many elements are presented, the diversity gives their respective proportion of area, and the heterogeneity reveals their spatial relationships.

In the following sections, a set of measurements for heterogeneity, fragmentation, and connectivity is presented in detail. This sequence was

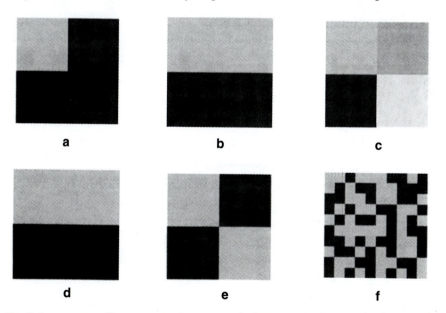

Fig. 7. Components of heterogeneity. From a to c, the heterogeneity increases by change in the proportion and number of elements. From d to f, it increases by variation of the spatial distribution.

adopted because it proceeds from essentially structural elements to concepts of spatialized functionality. However, we must first look at the concept of scale, which cannot be ignored in a study of space and time.

4.4. Concepts of scale and hierarchy

The scale of a map must be known for that map to be useful. A motorist's highway map does not contain the same information as a topographic map used by a hiker. These maps differ in two respects: (1) The space represented, the extent, is much larger in the highway map. (2) The resolution, i.e., the level of detail or grain of the information, and the size of the smallest object represented are different. The resolution is much larger in the topographic map, up to 1/25,000. Buildings are shown, as well as hedgerows and walls, all of them details that would be useless on a highway map. The first is a large or refined scale, and the second is a small or coarse scale without details (1/1,000,000 is much smaller than 1/25,000).

These two parameters, extent and resolution (or grain), need to be specified as soon as the scale is mentioned. In Fig. 8, maps (a), (b) and (c) represent a single type of landscape element for which a minimum rate of presence must be considered, seen across spatial cuts that are increasingly

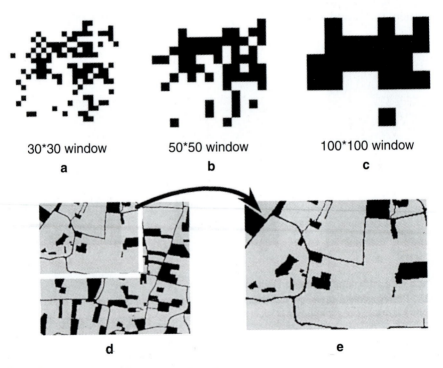

| 30*30 window | 50*50 window | 100*100 window |
| a | b | c |

d   e

Fig. 8. Scale: variation of the extent and resolution

coarse: these are windows in which the pixels are aggregated as 10 × 10, 50 × 50, and 100 × 100. With a fine grain, all the small patches are revealed. With a coarse grain, they are no longer perceptible. On the other hand, the coarse grain will make apparent the spatial continuities in a highly fragmented space. Maps (d) and (e) have differing extents: (e) is a part of (d). The same parameters apply to time.

It is advisable to note that the two scales taken as examples with regard to maps refer to different activities in space. One is movement on a highway, rapid and over some hundreds of kilometres, and the other is slow movement that covers much shorter distances. A link can first be established between scales of spatial (or temporal) analysis and the way in which space is used. That is, generally restricted spaces with many details are mapped for a species that moves slowly, and larger spaces with fewer details are mapped for species that move quickly. When space and time are taken into account simultaneously, there is often a correlation between spatial and temporal scales (Fig. 9). Here this correlation is used to define levels of organization in a landscape. An example is presented in Fig. 9 of Chapter 1, showing emergence of landscape ecology in the history of ecology by way of control of hedgerow flora. The microclimate varies rapidly according to the mode of hedgerow maintenance, whereas modifications of farm structure (presence or absence of hedgerows) constitute a phenomenon of much greater amplitude that takes several years.

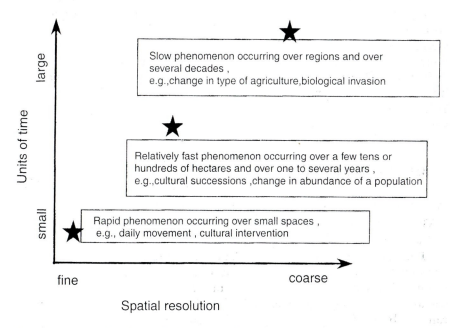

Fig. 9. Diagram of the relationship between scales of space and time

In varying the scales, or at least the extent or resolution, we often cause variations in the phenomenon observed. Roads, water courses, and coastlines, for example, are more sinuous on a large-scale map than on a small-scale map. The movement of an animal will be over a shorter distance (and slower) if we observe its movement all the time rather than every ten minutes. In effect, all the detours made during the hour will not be taken into account. Some will be taken into account with more frequent observations. Here we have the phenomenon of **scale dependence**. We will come across many examples of this phenomenon later in the book.

Each of the phenomena taken as an example has its own scales of space and time of functioning. It has autonomy that gives it a level of organization. However, this autonomy is only relative in the hierarchy of levels of organization. The hierarchy theory (Allen and Starr, 1982; O'Neill et al., 1986; Baudry, 1992) predicts that more inclusive levels have a lower rate of functioning than lower levels.

This theoretical framework has serious practical implications. It allows an actual division of levels of observation, which we often associate with a higher level of control. The case of control of colonization of hedgerows by new species, presented in Chapter 1, section 4.1, is an example. We will use this theoretical framework to study the landscape dynamics and factors of organization of mosaics.

The notion of *grain* and *extent* can be applied to species. According to Kotliar and Wiens (1990), the *grain of a species* is the smallest scale at which the organism operates differentiation in space. At finer scales, it perceives space homogeneously and does not react to any structure. The *extent of a species* is the largest space (vital space) or the longest duration (life span) to which the organism responds. Between these two extremes, characteristic of each species, the organism can react to any set of patch structures, organized hierarchically. A major problem raised by ethologists (Lima and Zollner, 1996) is lack of knowledge on the way in which individuals of different species effectively perceive landscapes.

5. MEASUREMENT OF HETEROGENEITY

The landscape is often defined as a heterogeneous mosaic (Risser et al., 1983; Forman, 1995). Hence, we need to define the concept of heterogeneity and a method of measuring it.

From Fig. 7, we have proposed (Baudry and Baudry-Burel, 1982; Baudry and Burel, 1985) a measurement of heterogeneity derived from the Shannon formula. It is calculated from maps in raster format, i.e., maps made up of squares or pixels, homogeneous units, like those in Fig. 7. The measurement can also be made from a series of points on a transect.

5.1. Formula

Heterogeneity (H) is measured as follows:

$$H = -\Sigma p(i, j)\log p(i, j)$$

where i and j are the two types of two horizontally and vertically adjacent pixels, and i and j belong to the set U of types of units in a landscape. The taking into account of pixel pairs (i, j) is equivalent to a measurement of their spatial connectivity, within and between patches.

For a landscape comprising U types of units {1, 2, 3, ... U}, there are U × U classes of pairs of points (i, j) ordered.

Demonstration of H

(a) When there is an equal distributions of points sampled between the units, H is measured such that:

—H increases when the number of units increases, and the function $H = f(U^2)$ must be non-decreasing;

—if there is only one unit, $H = f(1) = 0$;

—if, on a single transect, there are n independent selections of N equidistant points, at each selection we obtain U^2 classes and there are $(U^2)n$ ways of distributing the n samples in the U^2 classes, consequently, $H = f(U^{2n})$. As the selections are independent, f must be such that $nf(U^2) = f(U^{2n})$.

The only function that satisfies the conditions above is the logarithm function.

—It is an increasing function on $R^* +$,

—$\log_a(1) = 0$ (a = base of logarithm), and

—$\log_a(x, y) = \log_a(x) + \log_a(y)$,

from which we get $H = \log_a(U^2)$.

All the classes have an equivalent role and each introduces a heterogeneity equal to

$$1/U^2\log_a(U^2) = 1/U^2\log_a(1/U^2)$$

$$H = \sum_1^{U^2} 1/U^2\log_a(1/U^2) = \log_a(U^2)$$

(b) Cases of equal distribution are rare. In general, the class (i, j) has a probability p(i, j) of being present, whence:

$$H = \Sigma_{i,j} \, p(i, j)\log_a[p(i, j)]$$

with a = 2 (H will be in bits in the information theory) or a = e (Naperian logarithms).

Certain concepts of the information theory can be used.

—Self-information of the class $(i, j) = -p(i, j)\log_a[p(i, j)]$ is the contribution of class (i, j) with global heterogeneity.

—Maximal heterogeneity is the case of equal distribution, $H_{max} = \log_a(U^2)$. In fact, in practice, there is no reason to distinguish the succession of points (i, j) from (j, i). There is thus a maximal heterogeneity less than the maximal theoretical heterogeneity.

—Redundancy, $R = 1 - H/H_{max}$. It varies between 0 and 1, and it is as strong as there is repetition of a motif in the landscape.

H can be broken down into two parts, a heterogeneity due to the continuity of patches $H(i, i)$ and heterogeneity due to fragmentation $H(i, j)$ with $i \neq j$. These are two essential components to distinguish landscapes. Let us note, finally, that the formula $H = \Sigma_i p(i)\log_a p(i)$ is the Shannon formula commonly used in ecology to measure diversity.

5.2. Properties

To make good use of a measurement, we must know how its value varies with variation in other factors that have not been taken into account in the measurement. The boiling temperature of water is known to change with the altitude; that does not mean that measurement of the temperature is false or poor, only that it depends on a larger context than the state of the water.

We will first study how the value of H varies when the size of the landscape studied varies and when the raster unit (size of elementary pixel) varies. In each case, we can analyse the overall variation of the heterogeneity and the variation of the components—diversity and connectivity. Chloe software has been used for the measurements (Baudry and Denis, 1995).

For this analysis, a basic map was constructed (Fig. 10) with three types of elements, distinguishing a northern part and southern part in which the proportion of these elements differs. Therefore, there can be said to be two mosaics, two different landscapes.[1] Can the differences be found from the measurements? How do the measurements vary when the two mosaics are studied simultaneously? This is the type of question we asked ourselves.

To analyse the variation of heterogeneity as a function of the size of the space studied, we started from the north, then from the south, increasing the space by 50 pixels at each step, following parallel bands as indicated in Fig. 10. We observed not only a variation in the heterogeneity as a function of the area studied, but also a very different behaviour of the curve depending on whether we start from the north or the south (Fig. 11).

---

[1]Technically, a mapping base of polygons was used, having identifiers that are nearly "arranged" in space. Two groups of polygons were constituted from these identifiers. To each group we attributed a different probability of presence of three types of elements. Using a uniform series of random numbers, we attributed one type of element to each polygon.

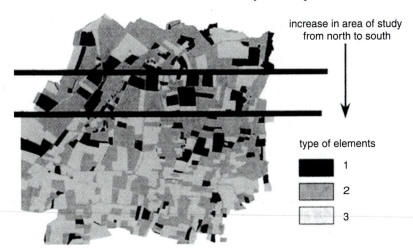

Fig. 10. Base map

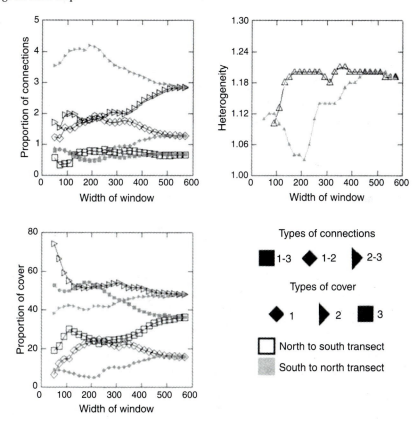

Fig. 11. Variation of heterogeneity, proportion of cover, and connectivity as a function of the area of analysis

When starting from the north, one immediately finds a high heterogeneity; starting from the south, one finds a zone made up of repetition of two elements. There is subsequently a sudden increase in the heterogeneity in the middle of the map (towards 300 pixels) as we enter the northern part. The two curves do not converge; they remain parallel beyond a zone of 450 pixels in width.

The value of heterogeneity thus depends on the extent of the space studied, and the existence of two differentiated zones is revealed only when we go from the more homogeneous to the more heterogeneous zone. This is not true in all cases and, in the present case, it is due to the effect of the inertia of the high heterogeneity of the northern zone.

These differences are found again in the study of proportions of various elements (Fig. 11), particularly for the element 1 very sparsely represented in the south. The north-south differences are more clearly indicated by the differences in behaviour of the two curves than by their individual behaviour. This phenomenon is found for connectivity between types (Fig. 11). Here it is the proportion of number of pairs of adjacent pixels belonging to different categories, a number that is involved in the calculation of heterogeneity.

Figure 12 represents the variation in the value of heterogeneity when we vary the resolution of the map from a fine grain to a coarse grain. For this operation, we make successive maps (Fig. 12) taking one pixel out of

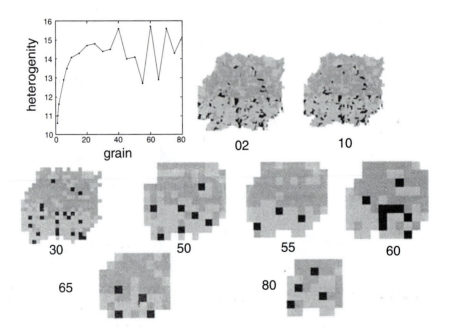

Fig. 12. Variation of heterogeneity as a function of size of base pixels

2, 10, ... 65, 80 in the initial map. This results in a significant loss of information. We observe first of all an increase in the heterogeneity, then a random fluctuation, and no stabilization. These oscillations are mainly due to the small number of pixels in the map. If we repeat the map at the same resolution, we obtain, on average, a stable curve. That is, the value of the heterogeneity is also a function of the resolution of the map.

The initial increase is due to the fact that the operation of "thinning" the pixels leads to a reduction in the size of the patches and thus the intra-element connectivity and, consequently, to an increase in the frequency of inter-element relationships. Then, the two types of connection are alternately reduced.

Here we see the two components of change in scale—size and resolution—and the *scale-dependence* relation a measurement may have. That is one reason why this measurement must be used with caution. However, symmetrically, the study of variation of the heterogeneity can be used as a function of the size of the territory studied; when a plateau is reached, there is a sufficiently large landscape that corresponds to the definition of Forman and Godron (1986) of a motif repeated in the space.

There are other reasons to use the measurement of heterogeneity with caution. The first is that identical values for it are obtained for inverse proportions between the types of landscape elements. If type 1 and type 3 on the map in Fig. 10 are inverted, we obtain the same final values. The second reason is that a single value can be obtained because there are more types of elements or because certain types are more fragmented.

In practice, the measurement is used relatively, to compare landscapes having the same types of elements or to study the evolution of a landscape.

The study of heterogeneity has a great conceptual advantage because it enables analysis of spatial structurations, their components, and the effects of scale. It has less practical interest, especially when we wish to relate the measurements to an ecological phenomenon.

6. FRAGMENTATION

Fragmentation is one of the most widely used concepts in landscape ecology. It is a basic concept in the development of the discipline, which is essentially spatial. It is applied to habitats as well as populations. Forman et al. (1976), applying the island biogeography theory to wooded islands, undeniably contributed a great deal to the consideration of spatial characteristics of habitats. They demonstrated a relationship between the size of woods and the diversity of avifauna. Their work is crucial not only because it demonstrates size-richness relationships, which constitute a fundamental ecological concept (notion of minimum sampling area), but also because it demonstrates the area thresholds allowing the presence of species. For large woods, fragmentation also appears to lead to a loss of species, because a single large wood shelters more species than the same area of small woods. The relationship is not clear for woods of medium size. Finally, the authors also emphasize that species having peculiar traits of life history (insectivores) are sensitive to fragmentation. They point out a group of ecotone or edge species that benefit from fragmentation up to a point. The distance between fragments, indeed, should not be too large. Beyond that, it seems that it is not only the quantity of habitat that controls the presence of species, but also the fragmentation and even the distance between fragments. Moreover, the different species, and even more so the different biological groups, will react in various ways to fragmentation.

The work of Forman et al. (1976) was followed by the studies of Burgess and Sharpe (1981) on fragmented landscapes of North America. The history of those landscapes since the arrival of the Europeans is known. The species in them were inventoried very early on and the territories mapped. Thus, we have a reference from which we can study the process of fragmentation as well as its ecological consequences. A famous example is the Cadiz region (Wisconsin) studied by Curtis, in which the forest area changed from 8724 ha of a single holding in 1831 to 841 ha distributed in 61 woods in 1902. Burgess and Sharpe (1981) give successive maps of this area and measurements of various parameters.

In Europe, studies on forest fragmentation were taken up by Helliwell (1976). They became a central axis of research (Opdam et al., 1993; Lauga and Joachim, 1992). Although works on forest fragmentation were more common, fragmentation of grassland habitats such as the North American prairie were also studied (Herkert, 1994). Beyond the effects of size and edges, fragmentation causes changes in the internal heterogeneity of habitats (Freemark and Merriam, 1986). The larger the forest, the greater the chances it will be diversified. Finally, fragmentation affects exchanges between islands and the probability of being colonized by propagules coming from other islands (Johnson et al., 1981; Johnson and Adkisson, 1985).

What is fragmentation, and how can it be characterized? The following discussion aims to examine this question from a structural point of view. As with processes, fragmentation is characterized by a reduction in the total area of a habitat and its breaking up into isolated pieces.

6.1. Global structural approach

The maps in Fig. 13 present a fragmentation process, which could be a deforestation. They represent a simulation during which the fragments of the forest are randomly destroyed. In stage 1 the forest is nearly the only element in the landscape and in stage 5 it is only a small percentage of the area. Note that a wooded corridor persists until stage 4.

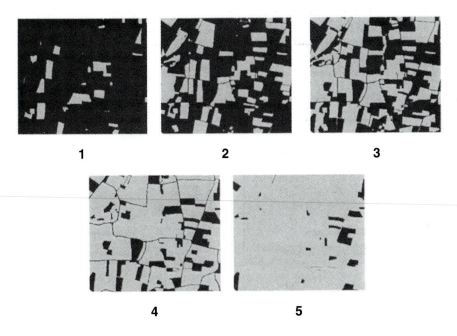

1   2   3

4   5

Fig. 13. Simulation of a process of fragmentation

These maps can be analysed using a set of spatial parameters (Fig. 14):

—The total area of the woods (in pixels) decreases constantly from one stage to the next.

—The number of patches (fragments) increases and then stabilizes, while their average area decreases abruptly, then more slowly.

—The perimeter of the patches (their contact with another environment) increases with increase in number of patches, then reduces with the reduction in their area.

—The surface/perimeter ratio, which is a measurement of the extent of edges, rapidly falls, i.e., the wooded area subjected to external influences occupies an increasingly large relative area.

—A constant reduction is observed in the connectedness between wooded pixels, along with an increase in the distance from non-wooded pixels to wooded pixels. Fragmentation leads to isolation.

—The overall heterogeneity of the landscape increases, then diminishes when the non-wooded space becomes predominant.

Thus, a linear reduction in woods over time (over different stages) corresponds to many non-linear phenomena, with acceleration, slowing, or inversion. Fragmentation is more than a loss of habitat, it is a modification in the quality of the habitat. There is reduction in contiguous patches of habitat, isolation of patches, and increase in the ecotone effect. One therefore

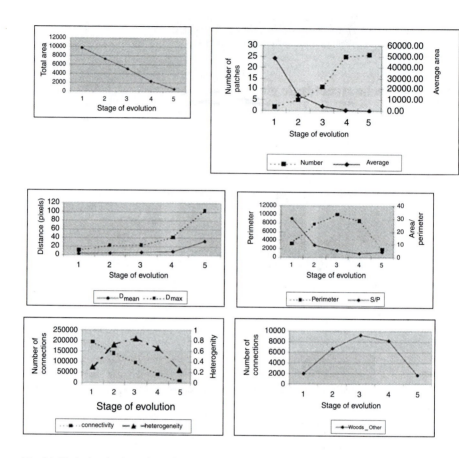

Fig. 14. Variation in the value of some spatial parameters during fragmentation

expects biological effects, presence of species, and abundance of populations to vary in a non-linear fashion in the course of fragmentation. Species having a large territory will be affected more quickly than those having a small territory, especially when the wooded area in this territory is large.

A primary approach of effects differentiated according to the resolution of species is proposed in Fig. 15. For this, we change the resolution of the spatial analysis. The maps at different stages have been divided into windows of various widths (10, 20, 50, and 75 pixels) with different proportions of woods (10%, 75%). While species that need only 10% of woods in their territory have habitats available at five simulated stages, those needing 75% of woods are rapidly deprived of a habitat, and more so if their territory is large. Thus, there is inherently a different perception of fragmentation in various species. This agrees with an observation of decrease in the species

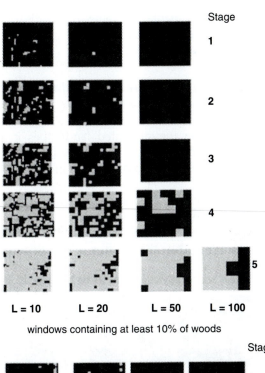

windows containing at least 10% of woods

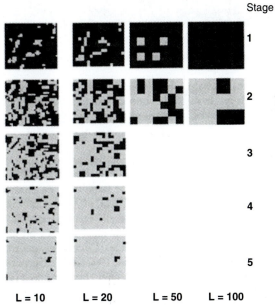

windows containing at least 75% of woods

Fig. 15. Maps of fragmentation by windows of different sizes with different proportions of woods

richness with increasing fragmentation: some species disappear, while others survive. In the following section, we present a more thorough analysis of spatial structures in this perspective.

6.2. Fragmentation and available habitat: analytical approach

The analyses are based on the simulated landscapes in Fig. 14, making another change of resolution: one pixel out of five is selected to make a map of 4200 pixels. When the map has a coarser resolution, the wooded corridors disappear (Fig. 16).

The first set of analyses provides overall characteristics of maps, as well as a reconnaissance of groups of pixels (clusters) of woods. Figure 17 presents the largest group at stages 2 to 5. At the first stage, there is only a single group, and then this number increases, till it falls suddenly at the last stage (Fig. 18).

Fig. 16. Configuration of maps with coarse resolution at five stages of evolution

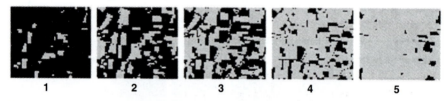

Fig. 17. Groups of pixels of woods at different stages. The largest group is in black, the others in grey.

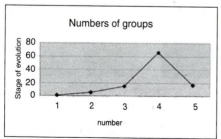
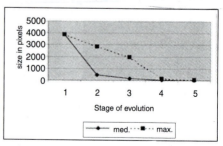

Fig. 18. Evolution of number of pixels and size of the largest group at different stages

The change in resolution in the maps does not cause loss of information with respect to the quantity of various types of pixels but leads to a significant change in spatial structures, and of connectedness in particular (Table 2). The smaller groups of pixels that ensure connectedness at finer scales disappear. The relative loss of connectivity is greater at advanced stages of fragmentation. Once again, we see that the resolution of maps analysed is an important factor to be taken into account.

Table 2. Abundance of woods and connectivity after reduction in map resolution; comparison with estimation according to initial maps and percentage of difference

Stage	Abundance	Estimated	% diff.	Connectivity	Estimated	% diff.
1	3883	3910	0.68	7457	7754.24	3.83
2	2892	2919	0.94	5137	5685.84	9.65
3	2033	2045	0.60	3272	3890.68	15.90
4	939	943	0.45	1268	1714.48	26.04
5	246	242	−1.62	334	446.8	25.25

During the following analyses, we have sought to find out to what type of window the different pixels of maps belong. Around each of the 4200 pixels, we have constructed windows 5, 10, and 15 pixels wide to analyse their contents. Three threshold rates of forestation have been retained: 10, 50, and 75%. These rates correspond to needs of various organisms for wooded spaces and to their differing displacement capacity.

The notion of grain of a species, defined above, leads to the idea that different species do not have the same perception of landscapes. Highly mobile organisms have a coarser grain than sessile organisms (Wiens, 1997). The needs of different species in terms of habitat are also an essential factor. Andren (1992) demonstrated that various species of corvids have a density peak corresponding to different stages of fragmentation. The jay (*Garrulus glandarius*) and the raven (*Corvus corax*) are specialists of the forest. The jackdaw (*Corvus monedula*) and magpie (*Pica pica*) prefer farm lands, while the hooded crow (*Corvus corone cornix*) reaches its optimum in mixed landscapes of forest and farm land.

Table 3 gives some examples of biological groups corresponding to large and small values of radius of activity and need for afforestation.

The number of wooded pixels at each stage is an estimation of the potentially available habitat. What we analyse is quantities of habitat

Table 3. Biological groups corresponding to various perceptions of fragmentation of woods

Width of window	Rate 10% (low need for woods)	Rate 75% (great need for woods)
5 (small radius of activity)		Marching forest insects
15 (large radius of activity)	Birds of ecotone areas	Mammals, forest raptors

"effectively" available depending on the modalities of perception. Well understood, it is here a perception by model species, just as the landscapes studied are modelled.

6.3. Characterization of pixels and their context

Each pixel belongs or does not belong to the category "woods" and for each type of window it belongs or does not belong to the category window of width W, having at least R% of woods. Thus, a pixel can be "not wooded" and yet belong to a type of window having a minimum wooded area. This will often be the case with pixels located in large windows with a low rate of wooded area. In that case, the quantity of "usable" habitat is greater than the quantity of habitat mapped. This is what happens for ecotone species that, in small woods, have a very high density because they use the surrounding space to feed in. It also leads to the existence of large, continuous groups of habitats, as illustrated in Fig. 19. Inversely, pixels can be wooded and not included in the windows.

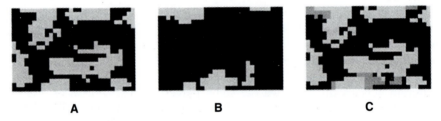

Fig. 19. Representation of pixels according to various modes of understanding: the pixels can be represented according to their own characteristics or according to the characteristics of windows in which they are included. In A, the wooded pixels are represented in black. In B, the pixels included in a window of 10 pixels width comprising at least 50% of woods are black. In C, the black pixels are those that are wooded and included in the type of window defined in B, and the wooded pixels not included in this type of window are grey.

With a representation of type B in Fig. 19, at stages 2, 3, and 4, for low rates of forestation, the map seems not to be fragmented; it seems to be occupied by a large area of woods (Fig. 20).

At stage 4, where there remain only 1268 pixels of woods, there is a single habitat group of 3717 pixels for windows of width 15 and rate 10% (Fig. 20). For this rate, we see (Fig. 21) that the process of fragmentation is slow to the extent that the window is large, i.e., that there is a possibility of putting distant pixels in the same window. Inversely, the reduction in habitat and the fragmentation are very rapid for high rates. At stage 3, only 58 pixels remain in windows of width 15 and rate 75%. The phenomenon is slower for narrow windows that can be filled by small patches.

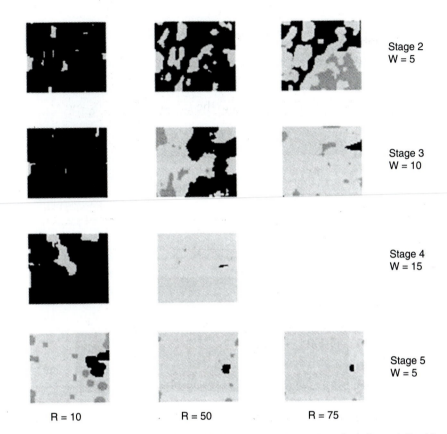

Fig. 20. Some examples of groups of pixels belonging to various types of windows defined by the width (W) and the rate of forestation (R) at different stages of fragmentation

6.4. Conclusion

Fragmentation is primarily a spatial phenomenon that leads to a non-correlated modification of various parameters describing structures. Subsequently, it is a phenomenon that inherently is perceived differently by various species. Figure 21 takes into account the fact that "internal" species (large forest mammals) see their habitat diminish very rapidly (21c) and are the first to disappear, while ecotone species (21a) perceive the changes only much later. This conforms to historical observations made during progressive clearing activities in North America (Mankin et al., 1997). Between 1800 and 1880, the hart, the black bear, the puma, and the beaver disappeared from the American Midwest. Middleton and Merriam (1985), when taking a census of species of forest fragments in the Ottawa region two centuries after the beginning of colonization, made the same observation.

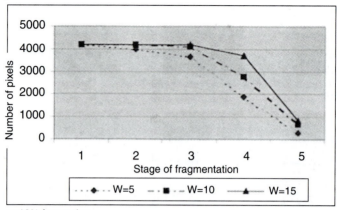

a: 10% forestation

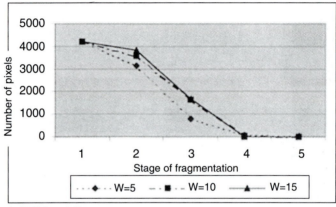

b: 50% forestation

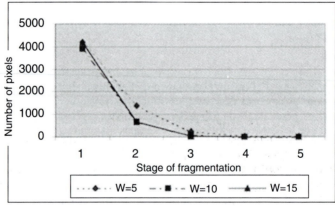

c: 75% forestation

Fig. 21. Maximum size of groups of pixels belonging to different types of windows. In a, species requiring only 10% of forestation are in the windows; in b, 50%; and in c, 75%.

7.   CONNECTEDNESS

Connectedness is also a central term in landscape ecology. One of the first hypotheses tested was the following: if spatial structures are important in the regulation of ecological characteristics of a landscape, a strong correlation must exist between elements of the same nature that are linked together (Merriam, 1984; Baudry, 1988). Simulation approaches (Fahrig and Merriam, 1985) are consistent with the results of empirical tests (Burel, 1989).

The creation of corridors to increase connectedness within a landscape at first appeared to be a panacea for problems of fragmentation and developers promptly snatched at it (Bennett, 1990; Hudson, 1991). However, many questions remained as to the efficiency of corridors, and even their harmfulness (Henein and Merriam, 1990; Noss, 1991; Hobbs, 1992), as we shall see in the subsequent chapters.

Inversely, there must be structures that hamper, even prevent, displacements and flows in a landscape. Such barriers may have negative effects, for example in preventing groups of individuals of a single species from meeting each other. This would be the case with highway constructions. They may also have beneficial effects in arresting the circulation of pollutants, protecting against wind (windbreaks), and preventing the spread of diseases. The barriers are often corridors such as hedgerows that protect against wind and favour the circulation of fauna or riparian zones at the bottom of a valley that can be biological corridors as well as buffer zones protecting water courses from the input of nitrates. Corridors and barriers reflect the percolation theory.

In this chapter on spatial structures, the measurements of connectedness given in the preceding sections are supplemented with (1) an analysis of corridor networks and (2) an approach starting from the structural permeability of a landscape, i.e., of its greater or lesser aptitude to allow a species or flow to circulate. This involves analyses using GIS. The existence of structures that may have a buffer effect will be analysed in a similar manner.

7.1. The corridor network

Corridors are often organized in networks of hedgerows, water courses, and roads. The description of these networks is based on some particular parameters, the number of connections and intersections.

The number of intersections is the number of nodes in the network, the number of places at which the corridors cross each other. At the intersections, we often have more complex vegetation, a greater quantity of available habitat that can lead to a particular biological richness (Lack, 1988).

The number of connections is related to the number of links between corridors at an intersection. There may be a different number of connections depending on the configuration of intersections (Fig. 22), and those configurations can also be qualified. A connection with a wood is an essential node that will link the network to the potential source of forest species. A T connection ensures links between three hedgerows, an L connection between two hedgerows, and an X connection, which is rare, between four hedgerows. An O connection represents a cul de sac.

When the corridors have different qualities, portions of the networks of similar quality can be represented as a cluster, which has a particular function. In Fig. 23, we go from a structure to what can be called a functional set of hedgerows having a particular quality. The route from node A to node B cannot be the hedgerow directly linking the two points; the animal must make a detour by way of "efficient" hedgerows. For carabids (Coleoptera), Charrier et al. (1997) have demonstrated that hedgerows with heavy plant cover serve as much better corridors than other hedgerows. Thus, we can differentiate and locate "efficient" hedgerows.

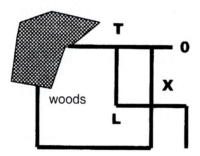

Fig. 22. Types and number of connections

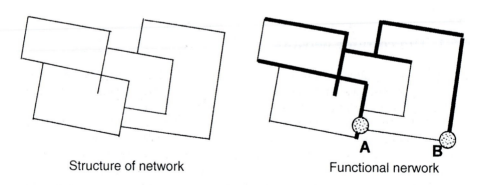

Structure of network   Functional nerwork

Fig. 23. From the structure of the network to functional groups

7.2. Effect of presence of wooded corridors on connectivity between groves

Taking stages 4 and 5 of fragmentation, we add a hedgerow network. For each stage, the maps 1 and 2 have the same number of hedgerows, with a different configuration, and map 3 has an additional third of hedgerows (Baudry and Burel, 1998). We calculate the distance between the woods with and without hedgerows, considering that the species can cross gaps three times as wide as the corridors (hedgerows). Then we construct maps from clusters of pixels that are "functionally" linked (Fig. 24).

At the most advanced stage of fragmentation, the presence of a corridor has a greater effect (Fig. 25). While the number of wooded pixels is multiplied by only 1.8, the number of "functional" pixels is multiplied by more than

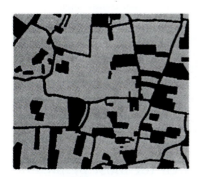

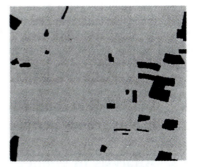

without a corridor

without a corridor

with a corridor

with a corridor

Stage 4

Stage 5

Fig. 24. Addition of a hedgerow network in landscapes with fragmented woods

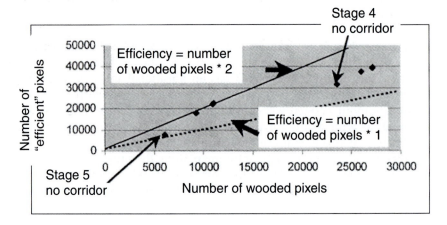

Fig. 25. Change in connectivity of wooded elements with addition of wooded corridors: representation of relationship between number of wooded pixels and connectivity.

2, even 3. At stage 4 of the fragmentation, the number of wooded pixels is multiplied by 1.1 and the number of functional pixels by 1.2 to 1.3.

The potential functional effect of wooded corridors depends on the general landscape context. The differences in size between the largest clusters is the least: they are 1.4 to 1.8 times as large at stage 5 and 1.2 to 1.4 times as large with the hedgerow network at stage 4. Once again, the effects of changes in structure can have non-linear consequences.

7.3. Analysis of connectivity by search for most permeable zones

The matrix was at first considered a hostile environment, but our conceptions have since evolved. We must consider that the space between patches of the type of landscape element studied is of various kinds. Cultivated areas are not always unusable for forest species. For example, a full-grown crop of maize offers shade and humidity to small forest animals (Merriam, 1989). Even in the absence of a material corridor, there may be spaces that serve as a corridor and therefore increase the connectivity.

Figure 26 presents a structural analysis of the permeability of a landscape for insect species associated with a herbaceous environment. We start from a map of grasslands (26a) and can represent the Euclidean distance between these grasslands (26b). Then, we consider that the intermediate spaces are more or less permeable or, inversely, introduce a low or high viscosity that works against movements of insects from one patch of the grasslands to another (26c). When we take this rugosity into account in the calculation of distances, we have a "cost" of displacement that is added to the Euclidean

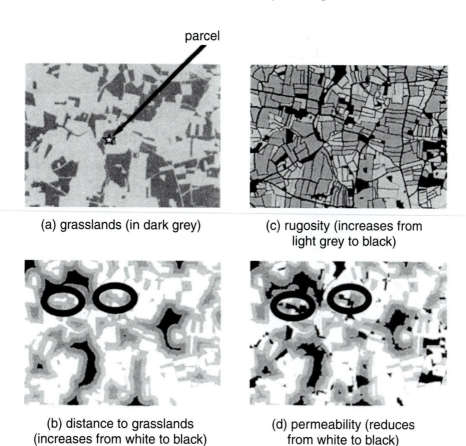

parcel

(a) grasslands (in dark grey)

(c) rugosity (increases from light grey to black)

(b) distance to grasslands (increases from white to black)

(d) permeability (reduces from white to black)

Fig. 26. Study of structural permeability of a grassland landscape. The grassland located by an arrow is that studied in Fig. 27.

distance to give a "functional distance" (26d). In the example, rugosity is high for all wooded structures (groves, hedgerows); for crops it varies according to their potential height, which decreases progressively from maize to straw cereals (wheat, barley).

Thus, we observe (in surrounded zones) that nearby spaces can be separated by elements that are difficult to permeate. The distance is small, but the rugosity is high, and the permeability is, in effect, low. An animal or propagule rarely has a null probability of crossing. While the connectedness is relatively high, the functional connectivity between two elements of the same type may become very low.

Taking into account the nature of elements between patches of a particular type leads us to a more functional approach to a landscape. We see the appearance of zones with less or more resistance to circulation in a landscape. We can thus pose hypotheses on relationships favoured between certain grasslands among which exchanges of insects are more frequent than among others. We can also find the favoured directions of departure from a given patch of grassland. For example, there may be some directions that facilitate movement, such as the southern direction in Fig. 27. The apparent reason is the presence of grasslands in this direction.

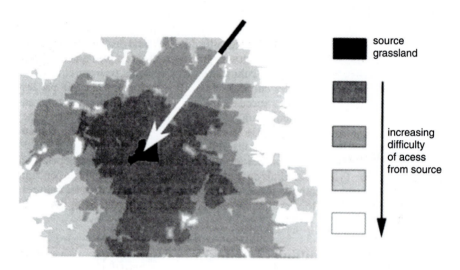

source
grassland

increasing
difficulty
of acess
from source

Fig. 27. Structural analysis of directions favoured at the outset from a patch

This type of structural analysis is only aimed to pose hypotheses, as for example whether floricolous insects present in a grassland disperse at random or in a favoured direction.

The difference from the hypothesis resulting from Fig. 26 is that here we are interested in the frequency of departures in certain directions and not the frequency of exchanges, and thus of arrivals, as a function of certain directions. The fact that insects favour some orientations in their departures signifies that they "perceive" the permeability. Origins differentiated according to the departures signify that the insects have adapted to different permeability values.

7.4. Variation of connectivity over time in an agricultural zone

Agricultural landscapes are often perceived as having a matrix of crop lands that varies little. In fact, the cover and use of these lands varies greatly as a function of crop successions, changes in farming systems, or, simply, the growth of plants. In the case studies that follow, we use the matrix of stages 4 and 5 of fragmentation and attribute three types of crops to different parcels. The simulation refers to changes that take place over two successive years as a function of crop successions. At stages 4 and 5, only the quantities of woods and crop 2 change; the quantities of crops 3 and 4 remain the same. They succeed each other between year 1 and year 2. We analyse the connectivity between the patches of crop 4, again taking into account the cost of crossing the other types of land cover. We suggest that crop 3 is five times less permeable than crop 4 (target crop), crop 2 is 20 times less permeable, and the woods are 50 times less permeable. The Idrisi® cost-push module is used to generate the costs of displacement. Since the different species have varying perceptions of the costs of crossing a landscape between two favourable patches, two threshold values (A and B) are used to compare the connectivity of different landscapes (Fig. 28). For each landscape and each threshold, we create the map of connected pixels and discover the largest. The size of the largest cluster is used as a measurement of the connectivity.

We can pose some hypotheses about the results:

—An increase in the agricultural area (from stage 4 to 5) in a landscape decreases the cost of displacement and increases the connectivity.

—Connectivity is greater for species (from A to B) having a better ability to cross crop areas (higher threshold).

—The crop succession from year 1 to year 2 is neutral with respect to connectivity.

After simulation, we obtain the following results (Fig. 28):

—Neither the proportion of pixels connected nor the size of the largest cluster increases significantly between stages 4 and 5.

—The number of pixels connected increases at the level of perception A to B, though in various ways. The number of clusters and the size of the largest changes between year 1 and year 2 of the rotation.

—The crop succession has a significant impact on connectivity. For example, in the comparison between 4B1 and 5B2, the latter is in the inherently more permeable landscape, but the largest cluster there is smaller (75% the size of the largest of 4B1). 4B1 and 4B2 differ in the number of clusters rather than in the size of the largest.

Thus, the movement of species living in farm lands may be much more restricted by crop rotation and the spatial distribution of crops than by the area of cultivation. Since, because of rotations, a crop may grow on many places in a landscape, over time, all the parcels on which crops are grown may end up being connected. In a way, a species may, after some time, be present in many places that are not apparently correlated with the landscape structure. In fact, the distribution of species associated with crops depends on at least two factors: (1) the possibility of finding a refuge outside the crop during a period that may be longer than the time the crop takes to

Stage 4

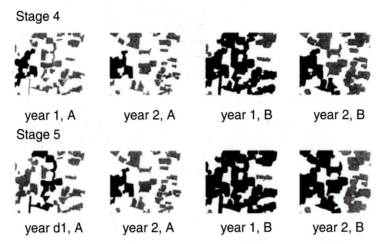

year 1, A   year 2, A   year 1, B   year 2, B

Stage 5

year d1, A   year 2, A   year 1, B   year 2, B

(a) Group of connected pixels, given the permeability of the landscape: the largest group is in black

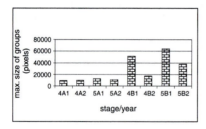

(b) Size of largest cluster in various situations

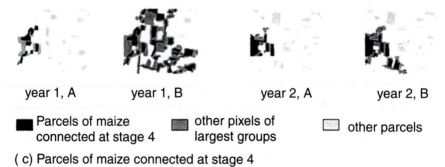

year 1, A   year 1, B   year 2, A   year 2, B

■ Parcels of maize   ▨ other pixels of   ▢ other parcels
  connected at stage 4   largest groups

(c) Parcels of maize connected at stage 4

Fig. 28. Evolution of connectivity in cultivated patches as a function of the crop succession and differences in perception of species. The wooded spaces decrease from stages 4 to 5. The differences between years 1 and 2 are that crop 4 (target habitat) follows crop 3 and vice versa.

return and (2) the way in which the crops are spatially organized across the collective, but not concerted, action of farmers of a region.

7.5. Conclusion: the many facets of connectivity and connectedness

The examples presented show that various approaches can be taken to studying connectivity within a landscape. Depending on the type of species and their mode of displacement, there are different levels of connectivity. The simulation of structures generated by displacements is an important tool for testing hypotheses and, especially, for elaborating an approach to construct a plan for sampling biological material in a landscape to test effective displacements.

8. RETURN TO SCALE DEPENDENCE: CONTRIBUTION OF FRACTAL GEOMETRY

Since the publication of B. Mandelbrot's work on fractal geometry (Mandelbrot, 1982), many related publications have been produced in various fields, especially in ecology and landscape ecology (Rex and Malanson, 1990; Milne, 1991; Baudry, 1993). We present here the basic elements of this approach, in order to demonstrate its advantages.

8.1. What is a fractal object?

The archetype of the fractal object is the rocky coastline of Great Britain or Brittany. Maps of these coastlines at various scales from 1/1,000,000 to 1/25,000 have important similarities: a highly convoluted edge and a multitude of curves. A person who walks on a footpath along the coast perceives the curves of the footpath as well as the more numerous curves of the land's edge, cut into peninsulas, bays, peaks, rocks, coves, and sand formations. Thus, an enlargement of any portion of the coast taken at a given scale will cause the appearance of new convolutions. These convolutions are all similar.

Such figures present two important properties: (1) a ratio of homothetie between the lengths at different scales and (2) dependence between the unit of length measurement and the length. The following sections illustrate these properties.

Whatever the scale of analysis (magnification of object or map), we perceive similar forms linked by a *relation of homothetie*. That is, we can establish a ratio between the areas or perimeters perceived at these different scales. The von Kock curve, or snow flake, is an example. This curve is constructed by starting from an equilateral triangle. On the middle third of each side of the triangle (of unit length 1), we place a new equilateral triangle, of side equal to 1/3 of the unit length. We thus obtain a Star of David. The ratio between the length of a segment at one stage and the length of a segment at the preceding stage is 1/3. Continuing in this fashion, we obtain a figure with a complicated outline (Fig. 29). The first triangle has a perimeter equal to 3. The second figure (first stage carving) has a perimeter equal to 4. Each side of the first triangle has been replaced by four segments of length 1/3 ($4 \times 1/3 \times 3 = 4/3 \times 3 = 4$). In the following stage, each of these four segments will itself be replaced by four segments of length equal to $1/3 \times 1/3 = 1/3^2$ and so on. The length of the perimeter is generally equal to $3 \times 4k/3k$ for the kth stage. This length thus increases by a factor 4/3 at each stage.

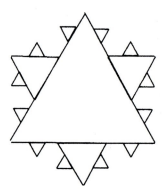

Fig. 29. Construction of the von Kock curve

—Scale dependence in measurement of lengths

A significant consequence of the sinuosity of coasts is that their length varies according to the unit of measure, as shown in Fig. 30. This is characteristic of fractal objects (Mandelbrot, 1982). A Euclidean object such as a straight line, on the other hand, has the same dimension (1) no matter what the unit of measure. In Fig. 30, the length of a curved object is measured using different units of measure. The integer 1 is the base unit, 2 is twice the base unit, 4 is four times, and 5 is five times. Each measure is transformed into base units. We see that the straight segment always has the same length (a kilometre in a straight line is always a kilometre, whether it is measured in metres, decametres, or kilometres), while the apparent length of a curvilinear object decreases. There is a relation of dependence between the length measured and the unit of measure.

Fractals frequently occur in nature. They are the product of the complexity of phenomena that transform them. The crown of a tree is not a ball, a cone, or an oval. It has an irregular surface. The crest of a mountain is fractal, as well as the spatial distribution of many organisms and ecological resources. The heather in the photo in Fig. 31, for example, is distributed in patches of varying size. If the observer does not know the size of the plant,

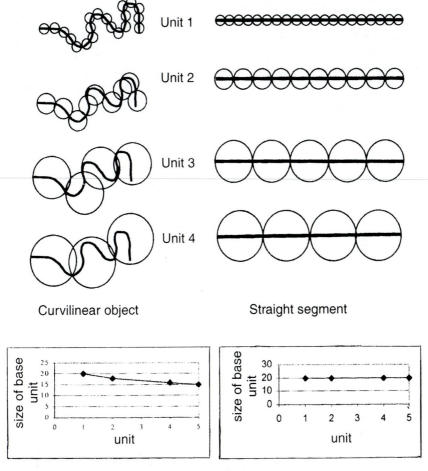

Fig. 30. Dependence of length on the unit of measurement when the length of a curvilinear object is measured

there is no means of discovering the dimension of the photograph. These images "without dimension" have inspired various nature photographers. The beauty of fractals is an important aspect of their appeal (Peitgen and Richter, 1986).

8.2. Methods of measurement

There is a wide variety of methods of measurement (Peitgen and Saupe, 1988; Sugihara and May, 1990). Two simple methods are presented here. The first involves essentially curvilinear objects and the second involves objects located on a flat surface, such as patches in a landscape.

Fig. 31. Heather

(1) Ratio of homothetie and fractal dimension

The division of the von Kock curve from one stage to another is only a change in scale. The factor of increase in length from one scale to another provides us a measure of the length at different scales. This is therefore a scale-dependent measure.

The equation linking the perimeter of a polygon to the unit of measure is, generally (Sugihara and May, 1990):

$$L(U) = CU^{1-D}$$ (equation 1)

where U is the unit of measure, C is a constant, and D is the fractal dimension.

In the case of the von Kock curve, we have $L(U + 1)/L(U) = 4/3$, from which, according to equation (1),

$$4/3 = 1/3^{1-D}$$

It follows that $4 = 3D$ or $D = \log(4)/\log(3) = 1.26$.

Many applications can be found in Hastings and Sugihara (1993) for the use of these laws in the study of fractals. One application is estimation of the fractal dimension of a polygon using the ratio between the perimeter and the area. The relation is:

$$P = CA^{D/2}$$ (equation 2)

where P is the perimeter, A is the area, C is a constant, and D is the fractal dimension.

(2) The grid method

A grid is laid on a plan containing objects that we wish to calculate the fractal dimension of, and we count the number of squares covering an object at least partly. By varying the dimension (S) of the squares of the grid, we obtain different units of measure. In order not to introduce a bias due to the positioning of the grid, we vary that position (Fig. 32). Consequently, for each value of a unit of measure, we obtain several values of the measure. The slope of the log*log relation of the inverse of units and values of measure gives the fractal dimension of the object.

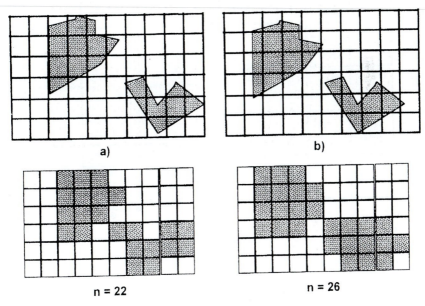

Fig. 32. Effect of displacement of a grid in the measurement of the number of squares in a grid filled by an object

A simple example of measurement of the fractal dimension, without multiplication of the grid, is given in Fig. 33. In this case, the fractal dimension is 1.67.

This method can also be applied to curvilinear objects such as coasts. Peitgen et al. (1992) thus calculated that the fractal dimension of the coastline of Great Britain is 1.31.

8.3. Examples of fractals of landscape elements

Fractal geometry has many applications in the study of landscape elements. Krummel et al. (1987), using the perimeter-area method, demonstrated that

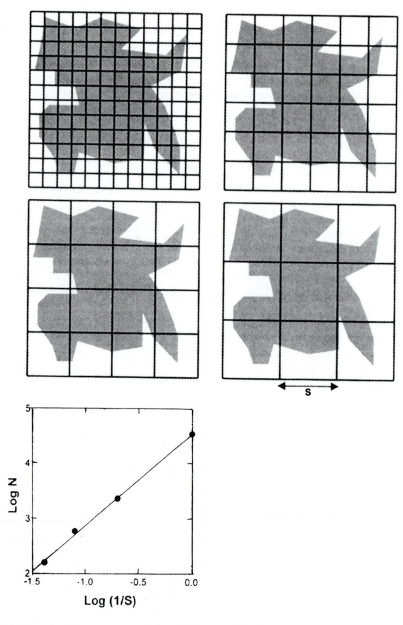

Fig. 33. Example of measurement of fractal dimension by the grid method

the wooded fragments of the alluvial plain of the Mississippi River, which is heavily farmed, have a fractal dimension less than the extensive forests of the heights. They concluded that the borders created by farmers in agricultural areas are much straighter than those determined by geomorphological structures. They also observed that the shapes resulting from the two different processes can be distinguished by their fractal dimension.

Baudry (1993) measured fractal dimension of a certain number of elements of an agricultural landscape in Normandy including grasslands, those between tilled lands and patches of bramble resulting from a decline in the pasture pressure. These last elements may appear to be distributed randomly in the landscapes. However, their distribution is remarkably self-similar for grains of analysis ranging from 0.25 to 16 ha (Fig. 34). The fractal distribution, measured by the box method, is 0.86. This corresponds to highly fragmented, inherently unstable elements. The same is true of bramble patches that may be mown from time to time and grow again (Asselin and Baudry, 1989), unlike groves, in which vegetation grows over a long time.

Van Hees (1994) used a fractal model to distinguish different regions of Alaska according to the complexity of forms of vegetation patches but could not establish a clear distinction.

Leduc et al. (1994), estimating the fractal dimension of the forest cover of a region in southern Quebec, determined that it changes according to the scales of analysis. These differences could be due to methodological problems, including the relation between the resolution of the analysis and the resolution of the phenomenon studied.

8.4. Fractal dimension of resources

Beyond the description of landscape elements, we must explore the advantage of fractal geometry in ecology. From the perspective of the study of populations, the major application is the analysis of resources. It is based on the preliminary approach of fractals (Fig. 31) that the quantity of small objects that can be placed on a fractal is proportionately higher than that of large objects. In other words, if we place 10 objects of dimension 10, we can place more than 100 objects of dimension 1, and so on, as high as the fractal dimension is high.

Morse et al. (1985) tested this idea in measuring the fractal dimension of various woody plants to deduce the density of arthropods potentially present, as a function of their size. Their calculations indicate that a decrease in the size of arthropods of a power of 10 must lead to an increase ranging from 560 to 1780 times the number of individuals. The empirical data available tend to confirm this prediction. Similar results were obtained by Shorrocks et al. (1991) in a work on the distribution of arthropods associated with lichens.

Milne (1991, 1992, 1997) developed this approach of available resources as a function of their fractal dimension and needs of species, expressed in an allometric relation (Peters, 1983) to their weight.

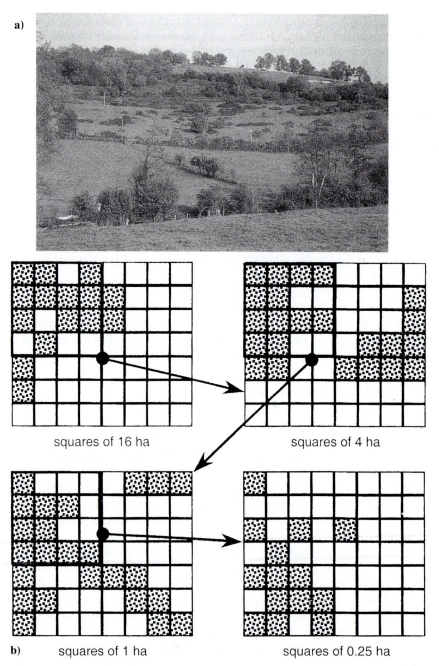

Fig. 34. Distribution of bramble at various scales in municipal land in Pays d'Auge (Lower Normandy). (a) The photograph shows that brambles are present only in certain places, and when they are present their distribution is irregular. (b) Mapping at various scales emphasizes this fragmented distribution.

8.5.   Fractal domains

The explanatory hypothesis of the self-similarity of fractals is that a single process or even a single set of processes organizes the form at all scales. Thus, the same physical forces carve out the coastal landscape into bays and peninsulas and shape the rocks. We can infer from this that different fractal dimensions correspond to different processes (Sugihara and May, 1990). Since it is evident that the structure of landscapes depends on a combination of climatic, geomorphological, political, technical, economic, cultural, and biological factors, it may be interesting to find out whether the fractal dimension of an object changes with the scale of investigation in order to detect possible changes in the factors of organization and subsequently discover these factors.

The distribution of perennially grassy areas in Lower Normandy offers an example of change in fractal dimension with change in the scale of analysis (Fig. 35). At the regional level, the climate and geomorphology control the distribution of grassy areas. These areas are larger in the cold and rainy zones and in places that have a strong relief. In the municipalities

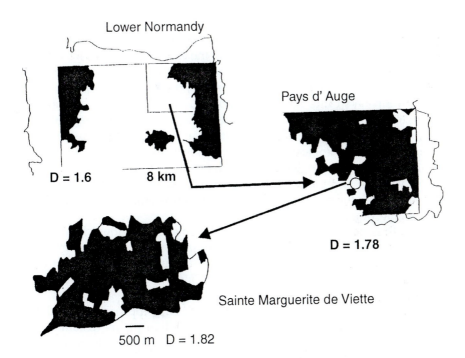

Fig. 35. Distribution of perennially grassy areas in Lower Normandy in 1979 (general agricultural census, Baudry, 1993, with permission).

of the Pays d'Auge or parcels of one of these municipalities, the physical context is relatively homogeneous, and it is thus the strategic choices of farmers that determine the proportion of grassland (Laurent et al., 1994; Deffontaines et al., 1995).

8.6. Conclusion

Fractal geometry seems full of promise in landscape ecology (Milne, 1997), but many methodological problems remain (van Hees, 1994; Leduc et al., 1994; Milne, 1997) and the interpretation of results is delicate.

The chief question concerns the functional significance of descriptions, as for most of the other analyses presented in this chapter.

The allometric relations between the size of organisms and their physiology may serve as a basis for identification of the density of resources in a landscape and, perhaps, for extrapolation of these results from one species to another.

9.   ELEMENTS OF GEOSTATISTICS

Geostatistics was developed by geologists to take into account variations of spatially correlated phenomena, such as the distribution of rocks or soils. Since landscape ecology relies largely on the existence of a mosaic made up of discrete entities with marked transitions, the use of these methods has not been well developed till now. Moreover, the spatial dimension seldom having been taken into account in ecology, the geostatistics has not been integrated. Still, there are notable exceptions. For example, Legendre and Fortin (1989) use geostatistics for spatial analysis within communities. They observe that the ecological data are often internally correlated, i.e., that the value of a variable in a given place is dependent on or correlated to values of this variable in the neighbourhood. This is linked to the existence of an environmental gradient, as well as to mechanisms of dissemination of plant and animal organisms.

It is thus useful to present some rudiments of techniques in a work on landscape ecology. These basics may be useful in studying the gradients, especially in non-anthropized landscapes (Burrough, 1987). We refer to this work, as well as that of Ciceri et al. (1977), Cliff and Ord (1981), and Haining (1990), for a description of all the techniques.

The construction of a variogramme is the first approach. A spatialized variable is considered, along with its value in two points distant from h: $Y(x)$ and $Y(x + h)$. If we have n_h points distant from h, we have a function called the intrinsic or semi-variance function.

$$g(h) = 1/2n_h \, S(Y(x) - Y(x + h))^2$$

The square of the difference indicates how Y varies between two points distant from h. We obtain a variogramme taking regularly increasing values of h in different directions. The differences between variogrammes constructed in different directions indicate effects of anisotropy. The result is a graph of the kind presented in Fig. 36.

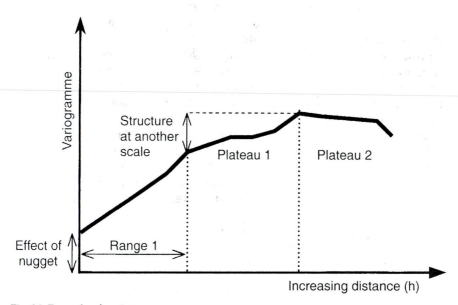

Fig. 36. Example of variogramme

The slope is very steep at the start. A horizontal variogramme indicates an absence of spatial structure. The range (Fig. 36) is the interval beyond which the spatial correlation becomes weak. A variogramme may present boxed structures corresponding to phenomena of spatial structuration at various scales. Another important concept is the "nugget effect", in which measurements taken at very short intervals are very different from zero (as in the distribution of nuggets in sand).

The variogramme can be used to operate a Krigeage, i.e., to construct a mathematical function adjusted to the variogramme. We can thus interpolate the value of points for which there is no measurement and make a continuous surface.

10. TYPOLOGIES OF LANDSCAPE STRUCTURES

The overall characterization of a landscape by measurement of heterogeneity, fragmentation, and connectivity is only a preliminary approach. Very often, it is necessary to analyse internal variations of the landscape. We can picture this landscape (represented by a raster mode map) in a grid of windows and characterize each of the windows. We thus have a grid characterizing the internal heterogeneity of the landscape. The size of windows as well as the position of the grid may cause large variations in the results, as illustrated in Fig. 32. Depending on the position of the grid, a wood may be entirely within a window or be divided among four windows; according to the position; therefore, we obtain different types of windows. Similarly, an increase in the size of the windows modifies the perception or the representation one has of a landscape. One solution would be to have windows of a size corresponding to the area of displacement of the species considered and to centre the windows on the points of observation. This type of solution is impossible in most situations. The observations are not equidistant, and therefore they cannot be used to fix the position of the grid. There is no reason to think that there is only one optimal area of perception. There may very well be several solutions depending on the activity of the organism. According to Kotliar and Wiens (1990), there is a hierarchic spatial organization of landscape elements proper to each species.

The typologies of windows can be made conventionally by characterizing them using a set of variables, conducting a factorial analysis on this table, and drawing up a hierarchic classification of the major factors. The characteristic variables may be the composition of pixels in nature, with a more or less detailed typology of patches and corridors and/or of structural variables such as the number of connections between the various types of pixels. Global variables such as internal heterogeneity at the window can also be used. We can thus test (1) the relationships between the typologies obtained and the presence of a particular species or group of species and (2) the differences between landscapes or evolution over time.

The typologies of a landscape in northern Spain at three different periods, drawn up by Suarez Seoane (1998), are presented in Table 4. This example allows comparison of results obtained by varying (1) the typology of basic elements and (2) the size of windows of analysis.

10.1. Basic data

The study involved the municipality of Chozas de Abajo, in the province of Leon. It has an area of around 10,000 ha (10,009 ha is the total area of the municipality, 8890 ha is the area of the zone studied). During the second half of the century, transformations of the agricultural landscape were rapid; the intensification of agriculture was represented by a modernization of farm structures (irrigation, consolidation of parcels) and mechanization. When Spain was integrated into the European Community in 1987, large areas of farmland were abandoned. A map of the land cover was drawn at three periods—1956, 1983, and 1995—from aerial photographs and field

Table 4. Typology of landscape units in maps of Chozas de Abajo (Suarez Seoane, 1998)

UP	1956	1983	1995
1	2.32	2.33	2.33
2	0.37	0.37	0.37
3	54.69	29.78	2.92
4	2.57	4.04	1.45
5	0	22.85	5.47
6	0	0	10.9
7	0	0	2.11
8	0	7.59	12.16
9	19.9	17.16	5.11
10	0	0	17.16
11	0	2.04	1.04
12	2.06	7.61	2.05
13	8.08	0.22	7.5
14	0.18	0	1.59
15	0	0	21.58
16	2.5	1.53	1.55
17	7.32	4.47	4.35
18	0	0	0.37

verifications for the last period. The maps are drawn at a scale of 1/25,000. The parcels are highly fragmented, and the average parcel size is less than half a hectare. Farms have an average area of 4 to 6 ha. In light of the difficulty of mapping all the parcels, Suarez Seoane (1998) constituted landscape units. These units were characterized by two typologies of land cover, more or less detailed. The evolution of the area of different units is given in Table 5.

Table 5. Evolution (%) of area of landscapes between 1956 and 1995

	1956	1983	1995
Villages	2.32	2.33	2.33
Lagoons	0.37	0.37	0.37
Crops	77.06	81.42	32.11
Grasslands	10.32	9.88	11.13
Forests	9.82	6	6.26
Crop-abandoned mosaic	0	0	26.22
Total abandoned	0	0	21.58

10.2. Methods

The digitized maps were put into the Idrisi® format, so that an analysis could be done with windows of different sizes. Four window sizes were used: 15, 25, 35, and 50 pixels a side, or respectively 495, 825, 1135, and 1650 m. For each window size, several successive grids were used: two grids for windows of 15, three for 25, four for 35, and five for 50. The windows containing at least 80% of pixels with information (windows around the map) were eliminated. The analyses were conducted on a total of 4134 windows.

These windows were characterized by a factorial analysis of correspondences (FAC) followed by a hierarchic ascending classification (Lebeaux, 1985) that resulted in the constitution of classes of windows. For each typology (A, B), all the windows, from different years and from different sizes, were analysed simultaneously. First, the data were re-coded. The extent to which each window belongs to a particular frequency class of landscape units (0–10%, 10–25%, 25–50%, 50–75%, and 75–100%) was calculated using fuzzy logic (Klir and Folger, 1988). That is, a window belongs more or less to a class of numbers. This approach offers the advantages of re-coding into classes, notably the taking into account of non-linear relations between variables (Lebart et al., 1977; Benzecri and Benzecri, 1984), while avoiding an arbitrary division into discontinuous classes (Gallego, 1982). In concrete terms, if a type of unit represents 23% of the number of pixels of a window, this window will belong to the class 10–25% and to the class 25–50%. Burel et al. (1998) have used a similar method, but they characterize the windows by the number of connections between pixels of each type of element. However, that is technically possible only when the number of types of elements is small; otherwise the number of types of connections becomes very high in relation to the possibilities of the database (limited number of fields). Here, the measure of heterogeneity of the window has been used.

10.3. Results

The classes of windows: With the typology A of units, the five primary factors of the FAC explain only 17.7% of the variance, at the following values: factor 1, 4.5%; factor 2, 4%; factor 3, 3.5%; factor 4, 2.9%; and factor 5, 2.8%. For the classification, only the first three factors were used. This classification produced 10 classes of spatial structure. Figure 37 gives the classification tree, and Table 6 gives the composition of the various classes.

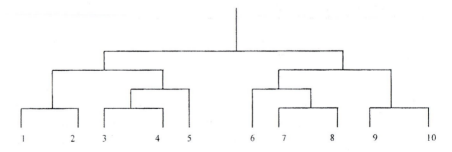

Fig. 37. Classification tree of windows of maps characterized by typology A

Table 6. Characteristics of classes resulting from classification of windows from maps of typology A

UP types or classes	1	2	3	4	5	6	7	8	9	10
UP1	0.49	0	7.54	7.15	**11.03**	0.24	1.57	0.26	0.08	0.16
UP2	0.20	0.47	0.31	0.29	0.24	0.28	0.35	0.50	0.39	0.73
UP3	1.46	0.07	3.30	13.57	43.24	32.54	53.69	73.83	0.01	1.72
UP4	1.31	5.53	0.36	0.23	0.08	9.46	0.69	0.19	0	1.16
UP5	6.51	**11.63**	1.17	2.48	4.72	23.77	12.18	0.64	0	16.63
UP6	**16.86**	1.34	**10.82**	14.75	0.01	0	0	0	0	2.56
UP7	1.43	**15.66**	0.35	0	0	0	0	0	0	0.12
UP8	19.70	27.95	7.19	4.20	0	0.02	0	0	0	5.67
UP9	0.57	0.11	9.50	7.20	16.20	2.84	14.73	14.52	0.77	40.53
UP10	0.09	0	10.34	2.73	4.53	0.01	0.07	0.01	97.67	10.90
UP11	0.35	0	4.32	0.02	0	0	0	0	0.01	0.37
UP12	0.63	0	6.60	5.28	7.59	0.14	1.86	1.33	0.21	0.65
UP13	9.15	4.75	8.64	14.87	11.73	7.07	11.96	8.12	0.85	4.15
UP14	2.64	2.81	0.31	0.81	0.15	0.34	0.51	0.21	0	0.03
UP15	**28.80**	7.94	**28.06**	24.74	0.03	0.01	0.11	0	0.01	12.53
UP16	1.66	1.53	0.24	0.37	0.39	2.38	1.32	0.09	0	1.47
UP17	7.73	**20.22**	0.19	0.32	0.06	20.90	0.97	0.31	0	0.61
UP18	0.43	0	0.75	0.98	0	0	0	0	0	0
Heterogeneity	1.43	1.31	1.76	1.36	1.57	1.12	1.11	0.47	0.08	0.39

(a) Composition in types of landscape units of window classes:

Class 1: heterogeneous class dominated by thyme plants and mosaic of rainfed crops

Class 2: class dominated by rainfed intensive crops with abandoned areas, rainfed crops with isolated trees

Class 3: mixed· class with thyme grasslands, rainfed crops in small parcels, and some irrigated crops

Class 4: thyme grasslands and, in smaller proportion, the mosaic of traditional rainfed crops and pastures at the bottom of the valley

Class 5: the dominant element is traditional rainfed crops, with crops irrigated by wells, villages, and grasslands at the bottom of the valley

Class 6: rainfed crops in small parcels are dominant, some intensive rainfed crops and oak groves

Class 7: traditional rainfed crops, some crops irrigated by canals, and pastures

Class 8: traditional rainfed crops in small parcels are dominant, some irrigated crops

Class 9: crops irrigated by canals

Class 10: crops irrigated by wells are dominant, intensive rainfed crops are well represented, some thyme grasslands and irrigated crops

(b) Description:

With the typology B of units, the five first factors of the FAC explain 41.09% of the variance, with, respectively: factor 1, 12.77%; factor 2, 8.99%;

factor 3, 7.18%; factor 4, 6.35%, and factor 5, 5.80%. The slightest stretching of the basic information allows better characterization of the windows. For the classification, the first three factors were retained (28.94% of the variance).

This classification produced eight classes (Fig. 38), described in Table 7.

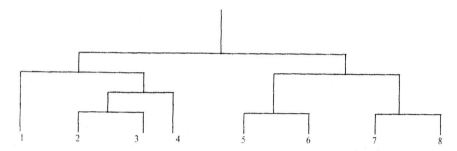

Fig. 38. Classification tree of windows of maps characterized by typology B

Table 7. Composition of window classes into types of landscape units

Typology or class	1	2	3	4	5	6	7	8
Villages	0.09	3.58	0.59	13.12	2.22	0.19	0.83	0.01
Water bodies	0.33	0.30	0.55	0.21	0.19	0.10	0.48	0.52
Crops	96.69	80.76	59.93	33.98	8.45	2.26	23.08	37.70
Pastures	1.85	13.24	18.28	17.52	12.77	6.29	13.72	5.51
Woods	0.57	1.08	12.54	0.82	5.59	1.20	19.45	42.95
Cultivated/abandoned	0.32	0.40	5.14	21.65	28.83	72.83	29.77	8.41
Thyme grasslands	0.15	0.64	2.97	12.70	41.95	17.13	12.68	4.89
Heterogeneity	0.12	0.69	0.96	1.4	1.26	0.4	1.17	1.01

(a) Characterization of classes in terms of composition of landscape units:

Class 1: almost exclusively cropland

Class 2: mostly crops with some pastures

Class 3: dominance of crops with pastures and groves

Class 4: heterogeneous class with crops, mosaic of cropped and abandoned land, and pastures

Class 5: predominance of abandoned land with thyme grasslands, mosaic of cropped and abandoned land, and pastures

Class 6: dominance of mosaic of cropped and abandoned land, with thyme grasslands

Class 7: mosaic of cropped and abandoned land, thyme grasslands, and groves

Class 8: groves and crops

(b) Description:

Maps have been drawn from these typologies of classes of landscape structure. They can be used to analyse the perception of the landscape by different species. The floristic or faunistic surveys can be spatialized and related to the surrounding landscape at different scales. Figure 39 shows some examples of maps.

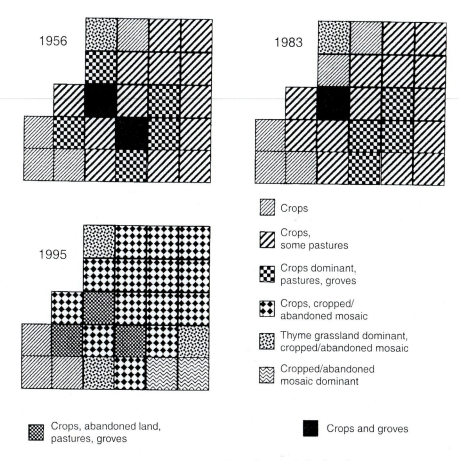

Fig. 39. Mapping of windows of 50 × 50 pixels at three periods of study

To test the link between typologies of windows and periods, the Kullback test, presented in Chapter 4, has been used. The different periods have significantly different types of landscapes.

10.4. Conclusion

The construction of typologies of windows to characterize a landscape can be a valuable aid in comparing landscape structures at different scales.

Indeed, two, three, or four landscapes can be compared. A global characterization of the landscapes by heterogeneity and connectivity would not be enough to measure their similarities and their differences. Burel et al. (1998) used such an approach to compare three bocage zones and a polder. The four landscapes had significantly different structures at all scales of analysis (see Chapters 4 and 7).

The difficulty of implementing the approach arises from the sensitivity of the results to the choice of typology of elements mapped at the outset. The more diversified the initial typology, the more complex the classes of windows will be, and the more difficult they are to characterize. This is part of the differences of variance explained when typology A or typology B is established, in the case developed above. It is lower in the first case and it ends in highly heterogeneous classes. The more information there is at the outset, the more difficult the results will be to interpret. The choice can only be a function of the phenomenon studied, whether it is one of landscape evolution or an ecological phenomenon. There can be no *a priori* solution. Progress can come only from an iteration between ecological and structural studies (see also Suarez-Seoane and Baudry, 2002).

11. GENERAL CONCLUSION

Various methods and techniques are available for the description of landscape structures. Many indexes can be easily calculated and maps constructed. This chapter presents only some of the techniques available; it does not attempt to be exhaustive. Scientific advancement continues to be prolific in this field and will certainly develop subjects such as fractal geometry and landscape simulations, as easy-to-use software becomes widely available.

We have warned, at several stages, of the difficulties of interpreting results of structure analysis. Till now, there has been little joint research on ecological structures and processes. The common approach is to research the explanatory structure after the fact, even if there are structural elements in the definition of protocols for data collection.

These uncertainties and questions are linked to the newness of a discipline such as landscape ecology, which requires descriptive approaches because predictive models do not yet exist. Pickett et al. (1994), for example, remark that if the hypothesis of effects of fragmentation is posed, it remains an exploratory hypothesis because it does not integrate mechanisms linking fragmentation and species richness. In the absence of such a mechanistic model, it is not possible to know what components to integrate in the measurement of fragmentation. Must we take into account the nature of elements between the fragments; if yes, how? Studies on such questions leave open many avenues of research.

Another difficulty of landscape research, and of ecology in general, was highlighted by Allen and Hoekstra (1992): the "evidently" tangible character of landscape elements. "Ecology is a matter of primary human experience," according to these authors (1992: 215.) This could lead to the anthropomorphic kind of approach to landscape that strongly marked the beginnings of the discipline (Pickett et al., 1994), with notions of matrix, island, corridor, and so on. The point of view of other species was taken into account only much later. To practise the precept of Allen and Hoekstra (1992) that "we will try not to be biased in favour of observables in tune with unaided human perception" is a true challenge because, faced with a landscape, we first see the woods, grasslands, hedgerows, and so on, and not the structures detected by the organisms that we study. It is essential for us to conceptualize landscape elements in an ecological perspective before we map and measure anything. There remains a great deal to be done for us to progress in this sense, as proved by the work of Moilanen and Hanski (1998). These authors tested the effects of landscape structure between patches of habitat of a butterfly (*Melitaea cinxia*) without significantly improving the predictions of the presence of the butterfly based only on the size of patches and their isolation. They emphasized that "there are no theoretical reasons to expect that habitat quality or landscape structure have no major effects on metapopulation dynamics". The authors partly attribute the absence of the effect of landscape structure to the available data, which are satellite data with a coarse resolution, in comparison with the resolution of the species studied and the typology of these data.

4

The Dynamics of Landscapes

The concepts and methods of landscape analysis presented in Chapter 3 are only a preliminary way of studying a landscape, whether it is a visible landscape, in the common sense of the term, or an image of a landscape in the form of a map. That chapter gives means of characterizing landscapes and their evolution and means of comparing them. Now come the questions *How do these structures establish themselves? What factors are responsible for their spatial organization and where do they come from?*

The causes of landscape diversity in space and in time are addressed in Chapter 5. The study of trajectories of changes is also included in that chapter. The subject chiefly involves rural landscapes, which are dominant in most regions of the planet. Rural landscapes are largely dependent on human activities and the history of human societies. These landscapes have to be managed and policies are defined for their management, even though they often include the most "natural" part of the landscape. Humans and their agricultural activities are the subject of this chapter. For millennia, agriculture has been the principal factor of transformation and development of landscapes (INRA et al., 1977), whether during a phase of expansion (Bertrand, 1975) or of regression (Baudry and Bunce, 1991). Palynological studies attest to ancient human occupations and their fluctuations (Birks et al., 1988).

Even when landscapes are only slightly altered by humans, they change, sometimes suddenly, under the effect of storms, fires (Romme, 1982), and floods. Studies of natural disturbances (Pickett and White, 1985) have for a long time demonstrated the internal dynamics, the instability, of these landscapes. This dynamics can also be linked to the life and death of existing plant species (Remmert, 1991). Because of such fluctuations, landscapes that are slightly or not modified by humans do not present the peaceful image of evolution towards a stable climax.

Change is an intrinsic characteristic of landscapes: long-term changes linked to geological phenomena, evolution of species, and their migration, or short-term changes linked to physiological rhythms (flowering, leaf drop) and to seasons (cold, heat). We have to study, therefore, not the prevention of change, but the control of rhythms and trajectories.

(a)

(b)

Plate 5. Landscape dynamics, visible information to a reader of landscapes. (a) Succession of beaver ponds, Algonquin Provincial Park, Ontario, Canada. (b) Abandonment in Tuscany, Italy.

Plate 5 (Contd....)

(c)

(d)

Plate 5 (Contd....)
(c) Bocage farm lands in Normandy. (d) Wet grassland and sloped grassland in Pays d'Auge, Normandy.

This chapter draws a brief assessment of change in land use on the planet during the past few centuries. In the second part are presented case studies of localized change on a time scale of a few years. These case studies illustrate the diversity of situations and lead to an analysis of trends to respond primarily to a simple question: Are the trends predictable? Or must we rely on variables external to the landscape structure, on non-ecological factors of organization?

The last part of this chapter addresses natural landscapes and details the case of Yellowstone National Park (USA), which has a well-known history of fires.

It is essential, in the study of transformations and organization of landscapes, to distinguish land cover from land use. In this work we will follow the definitions of Turner and Meyer (1994), which are the most often used. *Land cover* describes the physical state of lands, of the soil surface (type of vegetation, presence of water, presence of rocks). A change in land cover may consist of a conversion (transformation from forest to cultivation) or a modification (density of trees in a forest). *Land use* describes the way in which people use land and the practices followed. This includes agricultural and grazing practices and the type of habitat occupied. A change in land use in a place may consist of a change in use or modification of intensity of use (increase in grazing pressure, suppression of organic or mineral fertilization).

Turner and Meyer (1994) note, rightly, that land cover involves mostly the natural sciences, including ecology, hydrology, and pedology, while land use is an object of study for sciences focusing on humans and their activities (geography, planning, agronomy). Most maps by ecologists are maps of land cover, resulting from satellite data, photographs, or field observations. Using information from farmers and other land users about their practices, agronomists and anthropologists can draw maps of land use.

From the perspective of landscape ecology, the differences are of two kinds. First of all, several uses, several sets of practices, can correspond to a single cover (type of use of a grassland, insecticide treatments of a maize crop, maintenance of a hedgerow) and, consequently, to different local ecological conditions. Examples of mapping are given in the preceding chapter. The other point concerns the mechanisms of landscape organization. The knowledge of uses returns us to the technical systems and the systems of activity that produce them, and therefore to the organizing systems (Berkes and Folke, 1998; Turner and Meyer, 1994).

In this chapter, as in the rest of the work, the landscape will be represented sometimes as a set of land covers, sometimes as a set of uses.

1.   QUESTIONS ON ORGANIZATION AND DYNAMICS OF LANDSCAPES STEMMING FROM OBSERVATION

From the physical appearance of a landscape, the elements of organization and traces of the past can be detected. They can serve as a point of departure for other investigations, to pose hypotheses. From each of the photographs in Plate 5, we can read local evolutions.

The first photograph represents a set of beaver ponds in the Algonquin Park in Ontario (Canada). In the foreground, the pond is recent and full of water. In the middle ground, it is abandoned and colonized by plants; it may later be recolonized by the forest. This photograph indicates that the landscape may change even in the absence of human interventions. We will see other such examples later.

The second photograph shows the dynamics of abandonment in Tuscany (Italy). Trees colonize an abandoned grassland. The landscape will be forested in a few years.

The third photograph shows the conversion of a grassland parcel to tillage in the bocage farm lands of Normandy (France). During this operation, farmers could not restore the ancient limits. Because of relief, they leave the borders untilled. The division of the parcels of land, at least those parcels that are being farmed or grazed, will change.

The fourth photograph is also a representation induced by a variation of agricultural uses in a field in a grazing region, the Pays d'Auge, in Normandy. The farm has adopted a new technique, making hay into rolls. The corners and wet areas of the grassland are left fallow. There is an indication of technical progress and of abandonment. These local evolutions in agriculture may appear minor, but they can lead to profound changes in landscapes in a few years. In non-anthropized landscapes, catastrophic events such as fire can induce landscape transformations much more rapidly.

2.   CHANGES IN LAND USE AT GLOBAL SCALE

The growth of the population, as well as technological progress, have led and always lead rapidly to profound changes in the cover and use of land throughout the world. This is followed by a transformation in the nature and structure of landscapes with many ecological consequences. The transformation of land use is considered one of the causes of present environmental degradation and the loss of biodiversity, along with climatic changes (Lubchenco, 1972). Research in landscape ecology focuses chiefly on consequences on the scale of landscape and small regions, and not on the scale of large bioclimatic regions or entire countries. However, it is in these regional scales that local transformations accumulate, multiply, or are buffered.

Turner et al. (1990) depicted a panorama of transformations on the planet over the past three centuries from the perspective of land use as well as that of population or water resources. Richards (1990) compiled various statistical sources to construct a general model of change in land use, from which Table 1 is drawn.

The decline in wooded formations is widespread, minus 12,000 million ha, or 19% of the area estimated in 1700. The decline of grassland is also significant, minus 560,000 million ha or 8%. Tilling of lands expanded considerably, to an additional 12,000 million ha (an increase of 466%). This

Table 1. Change in land use on the planetary scale since 1700 (Richards, 1990)

Region	Type of vegetation	Years					Change, % (1700 to 1980)
		1700	1850	1920	1950	1980	
Tropical Africa	Forests and woods	1358	1336	1275	1188	1074	−20.9
	Grassy formations	1052	1061	1091	1130	1158	10.1
	Cultivated land	44	57	88	136	322	10.1
North Africa,	Forests and woods	38	34	27	18	14	−63.2
Middle East	Grassy formations	1123	1119	1112	1097	1060	−5.6
	Cultivated land	30	27	43	66	107	435.0
North America	Forests and woods	1016	971	944	939	942	−7.3
	Grassy formations	915	914	811	789	790	−13.7
	Cultivated land	3	50	179	206	203	6666.7
Latin America	Forests and woods	1445	1430	1369	1273	1151	−20.3
	Grassy formations	608	621	646	700	767	26.2
	Cultivated land	7	18	45	87	142	1928.6
China	Forests and woods	135	96	79	69	58	−57.0
	Grassy formations	951	944	941	938	923	−2.9
	Cultivated land	29	75	95	108	134	362.1
South Asia	Forests and woods	335	3317	2289	251	180	−46.3
	Grassy formations	189	189	190	190	187	−1
	Cultivated land	53	71	98	136	210	296.2
Southeast Asia	Forests and woods	253	252	247	242	235	−7.1
	Grassy formations	125	123	114	105	92	−26.4
	Cultivated land	4	7	21	35	55	1275.0
Europe	Forests and woods	230	205	200	199	212	−7.8
	Grassy formations	190	150	139	136	138	−27.4
	Cultivated land	67	132	147	152	137	104.5
Former USSR	Forests and woods	1138	1067	987	952	941	−17.3
	Grassy formations	1068	1078	1074	1070	1065	−0.3
	Cultivated land	33	94	178	216	233	606.1
Pacific	Forests and woods	267	267	261	258	246	−7.9
	Grassy formations	639	638	630	625	608	−4.9
	Cultivated land	5	6	19	28	58	1060.0
Total	Forests and woods	6215	5965	5678	5389	5053	18.7
	Grassy formations	6860	6837	6748	6780	6788	1.0
	Cultivated land	265	537	913	1170	1501	466.4

expansion will continue to accelerate. Cultivated lands have increased by 331 million ha since World War II, while the increase was only 272 million ha between the beginning of the 19th century and 1945.

These transformations are linked to the increase in the human population, which was 425 million in 1500 to 600 million in 1700, 1200 million in 1850, 2000 million in 1920, and 4430 million in 1980 (Richards, 1990).

3.   REGIONAL APPROACHES TO CHANGES IN LAND COVER: VARIATIONS DEPENDING ON MODES OF MEASUREMENT

The preceding section shows that changes in land cover are not homogeneous in space or in time. This brings us to consider the problems of scales of space and time, and the following questions:

(1) How does our perception of changes vary as a function of the area studied? Are the changes additive in the space?

(2) How does our perception of changes vary as a function of the time frame of the analysis? Are the changes observed over 10 years the sum of changes over the course of every one of those 10 years?

(3) What does mapping of land cover or land use contribute? The analyses given in the preceding chapter on fragmentation have demonstrated that the measurements of spatial structures vary non-linearly as a function of the total quantity of landscape elements.

These are the elementary questions to describe changes and their trends, and also to attempt to predict future evolutions. A fourth question thus arises:

(4) From the state of a landscape at a given moment, can we predict its evolution? Will two subsets of a landscape having the same characteristics at a given moment evolve in a similar fashion?

Questions 1 and 2 can be explored with conventional statistical data, but questions 3 and 4 call for landscape mapping and analysis in the sense used in this work.

3.1. Evolution of land cover in France in the 20th century: a variety of situations

The annual census of the Ministry of Agriculture provides data on land cover in each department (an administrative unit), as well as various economic and technical data on agriculture.

Figure 1 shows the evolution of wooded and agricultural areas in France and in a few departments. To make the figure easier to read, only the smooth values are represented. These obscure many interannual fluctuations,

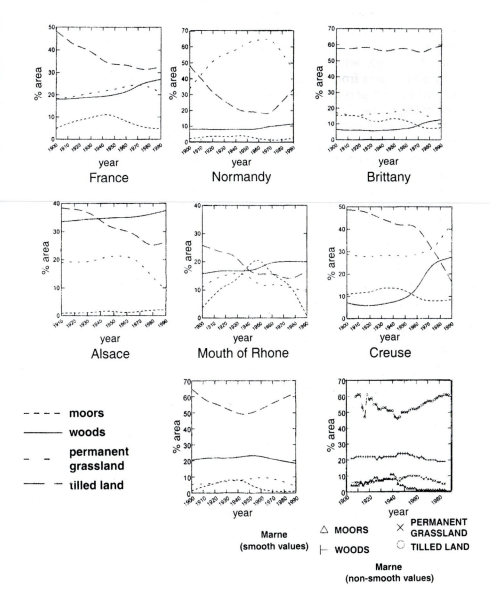

Fig. 1. Evolution of land cover in France since 1880 (from annual statistics of the Ministry of Agriculture)

as shown in the case of Marne. It is reasonable to believe that a considerable decline in cultivated areas between 1910 and 1920 was due to World War I, but the other fluctuations of smaller amplitude could be linked to errors in measurement. The ultimate set of analyses was done from raw data.

We can draw many lessons from this figure: an increase in the wooded surface and evolution varying according to the departments. In certain regions, such as Alsace or Creuse, some land abandonment took place a long time ago, while others, such as the Marne, saw an increase in agricultural area from the middle of the 20th century. Normandy, which is known to be a grassland area, was an extensively tilled department at the beginning of the 20th century. The expansion of grassland took place only later. The land cover in Brittany has changed little, despite a massive intensification of agriculture (Canevet, 1992).

A primary conclusion can be drawn from these few graphs. We must always refer to a particular duration and space when we talk of evolution. Regional diversities and inversion of trends must not be ignored. Moreover, we must question the concept of "traditional land cover" as well as the concept of vocation of lands. Lower Normandy is no more meant to be grassy than Champagne is meant to be verminous.

In comparison, Figure 2 gives the trends in three countries that have experienced different agricultural policies: the Netherlands, Spain, and Algeria. Change is the rule. In the Netherlands, a concomitant reduction of grassy and cultivated areas is noted, undoubtedly partly linked to urbanization.

This dynamics also leads us to question the choice of a reference state from which to judge the ecological characteristics of an environment, especially if we wish to restore that state (Bowles and Whelan, 1994).

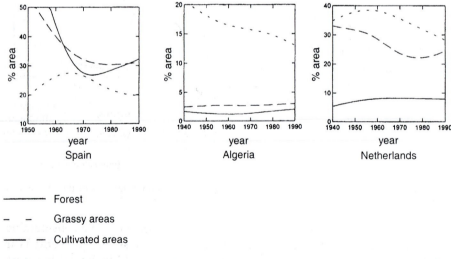

———— Forest

– – Grassy areas

—— — Cultivated areas

Fig. 2. Recent land cover changes in Spain, Algeria, and the Netherlands (from FAO statistics)

3.2. Evaluation of evolution of land cover in western France: methodological assay

3.2.1. Statistical data

Many of the landscape ecology studies presented in this work were carried out in Lower Normandy and Brittany. This section thus serves to focus on evolution of a landscape context. However, it has the primary aim of presenting a method of analysing evolution that takes into account the dependency of observations of change on the spatial and temporal scales of analysis. Two departments of the Loire region (Loire Atlantique and Mayenne) have also been taken into account to widen the range of diversity of change and for their historic links with the regions studied. Figure 1 shows that Brittany and Lower Normandy have experienced very different evolution in the past century. Figure 3 is based on a parallel evolution of three departments of Lower Normandy and a similar evolution in Mayenne. This evolution is characterized by rapid growth of grassy areas until about 1970, then an accelerated return to cultivation. The Breton departments have seen a very different evolution. In Cotes d'Armor, Ille-et-Vilaine, and Loire Atlantique, the tilled areas have always been predominant. In Finistere and Morbihan, the moor areas were dominant at the end of the last century (33 and 37% of the area of these departments in 1886). The increase of grassland remained modest in these departments. As in Normandy, there was a decline in grassland from 1970.

3.2.2. Representation of global trajectories of changes

To take into account the simultaneous evolution of various modes of land cover and to constitute a frame of reference for the analysis, Baudry (1992) carried out a factorial analysis of correspondences on the rates of recovery of various types of land cover (tilled area, permanent grassland, woods, moor) or four variables, for the nine departments and all the years for which data were available, or 1886 individuals. In the factorial space thus constituted, it is possible to place, as a supplementary element, the entities formed from the aggregation of adjacent departments by two or three per region and to aggregate the whole. The principle of distributional equivalence means that each point representing an aggregated unit is at the barycentre of points of elementary unit (Benzecri and Benzecri, 1984).

It is possible to constitute trajectories of change by linking the points representing successive years on the factorial axis.

Results:

The two primary factors of the FAC explain 93.5% of the total inertia, with an eigenvalue of 0.20 on the first factor, an indication of a slight gradient resulting from progressive annual evolutions. The first factor opposes the

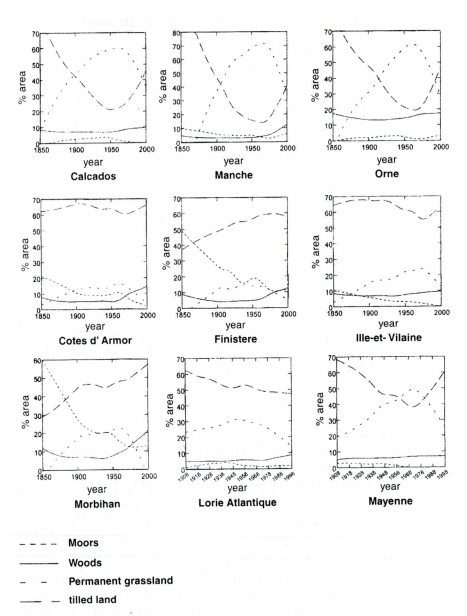

---- Moors

——— Woods

– – Permanent grassland

——— tilled land

Fig. 3. Evolution of land cover in nine departments in western France (from annual statistics of the Ministry of Agriculture)

units with more of permanent grassland (negative part, 58% of the inertia of the axis) to units with a great deal of tilled area (positive part, 27% of the inertia of the axis). The second factor represents the extent of moors (77% of the inertia).

The trajectories of Brittany and Lower Normandy are represented in Fig. 4. They do not cross. These trajectories have many zigzags, which shows that the evolution was not continuous. Their shapes recall fractal figures and therefore a dependence of the measurement on the length of the trajectory, which represents the importance of changes as a function of the time frame of measurement.

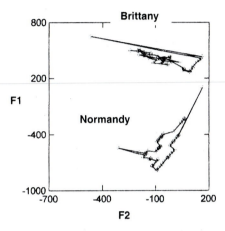

Fig. 4. Evolutionary trend of land cover in Brittany and Lower Normandy over a century

The rates of change have been calculated on time frames varying from 1 to 20 years. The average annual rate of change is the length of the trajectory in the factorial space (the Euclidean distance between the two successive points representing the beginning and end of the time frame considered) divided by the number of years of the time frame (1 to 20 years). Some results are given in Fig. 5. To indicate the scale-dependence relation, the data are transformed into a logarithm.

It may be observed that, in all cases, the average annual change perceived decreases greatly with the elongation of the time frame considered. That is, per year on average, there seem to be fewer changes over a period of 10 years than over a period of 5 years. This strong scale dependence exists for time frames of 1 to 10 years, at most. Beyond that, there are wide fluctuations in the rates of change. Table 2 gives coefficients of correlation between time frames and apparent change for two intervals: 1–10 years and 1–20 years.

From this it is clear that (1) it is possible to extrapolate changes over a period of 10 years (for the entire period studied) and (2) this extrapolation must take into account dependence on the temporal scale. The variations observed over two or three years cannot be extrapolated over 10 years.

Although there is scale dependence in change, i.e., it is not linear or additive, there are inversions and reversions. For example, even if, in the long term, we observe an increase in grasslands in the Lower Normandy

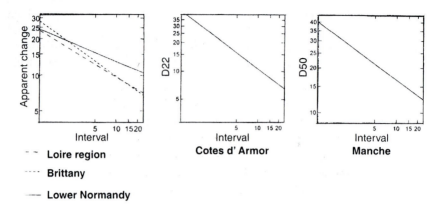

Fig. 5. Average rate of evolution in land cover as a function of time frame (data transformed into logarithm)

Table 2. Correlation between mean annual changes observed and the intervals of observation

Department	Intervals 1–10	p	Intervals 10–20	p
22	0.9	0	0.35	0.3
29	0.95	0	0.81	0.003
35	0.91	0	0.66	0.026
56	0.94	0	0.18	0.6
14	0.98	0	0.02	0.95
50	0.97	0	0.53	0.009
61	0.97	0	0.37	0.27
44	0.92	0	0.79	0.004
Brittany	0.94	0	0.71	0.013
Normandy	0.97	0	0.37	0.26
Loire region	0.84	0	0.56	0.07

Intervals 1–10, intervals of 1 to 10 years. Intervals 10–20, intervals of 10 to 20 years. p = level of significance.

departments from 1900 to 1960, some grasslands appeared and were tilled either within a farm or when a farm was taken over by a new farmer. We must remember that these changes are not concerted. An increase or decrease in the grassland area does not result from decisions taken at the departmental or regional level. Global changes are the result of a multitude of individual, uncoordinated decisions. From these local changes, we can deduce that the changes are also dependent on spatial scales, that they seem more rapid when they are analysed at a fine grain. This is shown in Fig. 6. Changes on the regional level are apparently slower than those at the elementary or departmental level. However, this relationship is not valid when we combine the data for Lower Normandy and Brittany. The changes that occur in these two regions are independent, of different nature, and they add up. They

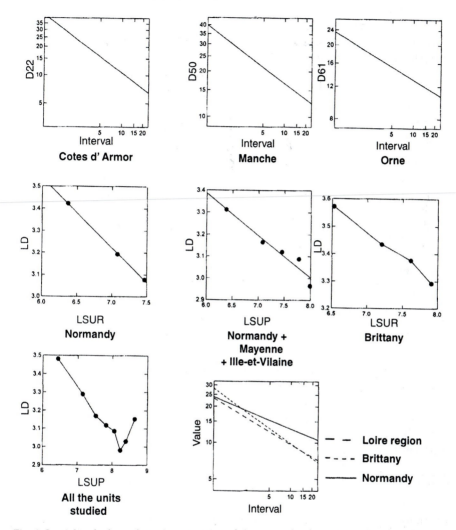

Fig. 6. Spatial scale dependence in perception of changes in land cover in Brittany and Lower Normandy

seem more rapid than those observed in one region or the other. That is, even if no regional planning decisions are made that affect land cover, there are similar behaviours among farmers of each region, and thus the cultivation mechanisms (technological, social, and economic history) give each region a particular quality. We note, moreover, that the three Lower Normandy departments that have seen very similar changes (Fig. 3) have a more marked aggregated result (stronger correlation between grain of analysis and change) than the Brittany departments, which present differences.

The study of the evolution of land cover in the small agricultural regions of Calvados based on data from the annual census on land use (Teruti) between 1977 and 1984 reveals the same phenomenon of scale dependence in space and in time (Fig. 7) (Baudry, 1992).

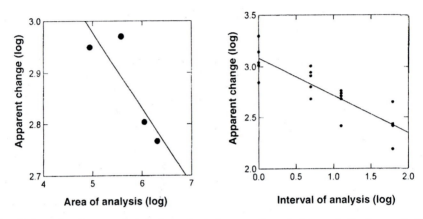

Fig. 7. Spatial and temporal scale dependence in perception of changes in land cover in the Calvados between 1977 and 1984 (from land use census)

Figure 7 shows that there is no possibility of transfer of scale between the small agricultural regions and the department. This tends to show that the mechanisms of change at work at the departmental and regional level are not the same as those at work at more local levels. This is consistent with the analysis of fractal structure of the distribution of permanent grassland areas in Lower Normandy presented in Chapter 3.

3.2.3. Conclusion

This methodological assay shows that it is possible not only to reveal phenomena of scale dependence in the observation of changes in land cover, but also to measure them. The techniques used, which originate from fractal geometry, also are based on levels of organization in space (small region, region), the evolution of which is controlled by different factors.

The results obtained are in agreement with the predictions of the hierarchy theory (Allen and Starr, 1982; O'Neill et al., 1986; May, 1989). The more inclusive temporal and spatial levels have lower rates of change than the levels included, at least when the processes at work are similar. Indeed, the spatial unit resulting from the aggregation of Brittany and Lower Normandy does not follow this law (Fig. 4). Milne (1991) found similar results in landscapes of New Jersey in the United States.

4. LOCAL APPROACHES TO CHANGES IN LAND COVER: IMPORTANCE OF SPATIALIZATION

The analyses presented above are based on statistical tables. Even though they can be mapped, they have no spatial content that characterizes the landscapes. The object of this section is to present case studies based on landscape maps to analyse the evolution of their structure. These studies allow us to distribute the analysis of landscape structures so that it reveals the landscape evolution. The first two landscapes have also been analysed from the biological point of view. The following chapter presents the way in which the dynamics of structure explain the biodiversity.

The first landscape is a Mediterranean terrace landscape in Provence, in southern France. The second is a bocage landscape in Brittany. The third is a rice field landscape in subtropical China. They are all agricultural landscapes that have evolved rapidly under the effect of transformations in human activities.

4.1. Evolution of a terrace landscape in the Mediterranean region

The Mediterranean region is the site of an ancient civilization that was very early transformed by pastoral and agricultural activities. From the 19th century onward, soil erosion, aridity of the climate, and socio-economic changes described later in this chapter led to a massive abandonment of agricultural land (Barbero et al., 1990; Hubert, 1991). In other zones, meanwhile, there was intensification of agriculture.

4.1.1. Basic data

The landscape studied was located in the Maubec municipality (Tatoni, 1992), in the Regional Natural Park of Luberon. It is a small area of about 40 ha on which it was possible to reconstruct the land cover since 1890 using property registers with a time frame of around 20 years. This was a relatively rare situation because, generally, such registers are not up to date.

At present, the vegetation consists of green oak, pubescent or downy oak, and Aleppo pine (Plate 6). From the maps shown in Fig. 8 (Baudry and Tatoni, 1993), the overall evolution of types of land cover can be studied, as well as transitions between types of land cover.

4.1.2. Evolutions

Overall, an increase of 50% of wooded areas and a near disappearance of cultivated areas are noted (Fig. 9). The olive orchards and vineyards spread to some extent, then declined.

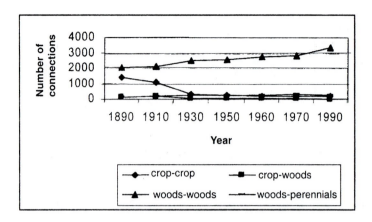

Fig. 11. Evolution of connectivity between the various types of land cover at Maubec

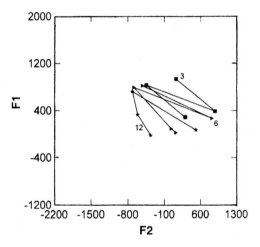

Fig. 12. Evolutionary trends of windows of size 3, 6, and 12 located around point 124, in the factorial axis

stages of perennial cultivation and fallow, which correspond to the complexity of transition matrixes analysed above.

It is possible to characterize overall the evolution of each window by carrying out a principal components analysis on the matrix of windows × factorial coordinates for each year (Baudry and Tatoni, 1993). A classification can also be done on these same factorial coordinates to construct types of evolution.

The space of evolutions is made up of the same two gradients as the factorial space characterizing the windows and visualizing their trajectories: crop lands towards abandonment on axis 1 and crop lands towards orchards on axis 2. Figure 13 represents some windows in this factorial space. We can

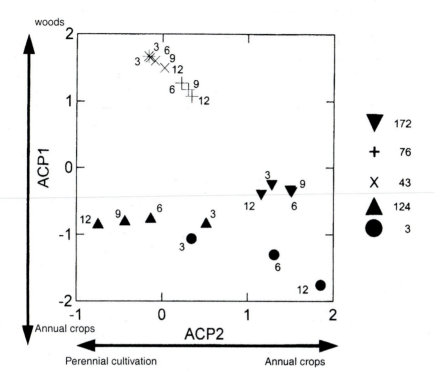

Fig. 13. Positioning of some windows of landscapes centred on different points in space of landscape evolutions. The numbers in the graph represent the size of windows. The symbols correspond to windows around different points.

thus visualize different modes of evolution, according to the various parts of the landscape and differences according to the size of the space considered. Some points have an environment that changes homogeneously (all the points representing the windows of different sizes are similar), others have an environment that changes heterogeneously (dispersed points). The windows around point 124 are much more dispersed than those located around point 76.

4.1.3. Conclusion

This case study relies on the wide range of methods available to study dynamics of landscapes from maps. Thus, we move from concrete space to the map and from the map to mathematical spaces that allow us to characterize and quantify the changes, and then to maps of change.

Baudry and Tatoni (1993) related these evolutions with floristic surveys carried out in 1990 in abandoned fields. They indicated the impact of the local history of the landscape on the state of the vegetation.

4.2. Evolution of a bocage landscape in Lalleu (Ille-et-Vilaine)

Bocage landscapes have long been the subject of research in many disciplines, including geography (Meynier, 1970), ecology, agronomy, and history (INRA et al., 1976). They received fresh attention with the development of landscape ecology (Forman and Baudry, 1984; Baudry, 1988; Burel, 1996). The development of GIS also contributed new forms of representation of landscape states and dynamics (Morant et al., 1995).

A multiplicity of social functions (marking of property, cultural value, aesthetics) and ecological roles (control of water circulation, biodiversity, climate) have made bocage landscapes an important type of landscape to be understood and managed. The subsequent chapters of this book present these functions. In this chapter we discuss the evolution of bocage landscapes. Because of the rarity of documents written during the phases of construction of enclosures and hedgerow planting, often only the reduction in hedgerow length and the uprooting of embankments have been perceived.

4.2.1. The study site

The studies of Burel and Baudry (1990) at Lalleu (a municipality situated about 50 km south of Rennes) addressed the structural changes from 1952, that is, during the period of mechanization of agriculture and up to the consolidations of 1988. This site is the subject of a partial presentation in Chapter 1 (Fig. 6).

The geological substrate of the region is an alternation of sandstone and schist. The sandstone, which is harder, marks the heights in the landscape (Plate 7). The soils are essentially used for farming (grasslands, cereal crops, maize). The size of the farms increased regularly (11.8 ha in 1955, 18.2 ha in 1979, 26 ha in 1988) but remained small, on average. The evolution of agriculture was marked by a reduction in the area of the permanent grassland or an increase in maize cultivation.

During this period, the length of the hedgerows, over the 416 ha studied, fell from 96 to 62 km. The connectivity of the network also diminished, from 4166 to 1768 connections, while the heterogeneity of the parcel size increased and then diminished (Fig. 14). In the absence of a linear relation between these parameters, a global analysis was needed.

4.2.2. Analyses and results

The analysis was carried out over a territory of 416 ha divided into 26 quadrats of 16 ha. The hedgerows were mapped from aerial photographs from the National Geographic Institute at a scale of 1/30,000 for the years 1952, 1961, 1972, and from field studies for 1985 (map represented in Fig. 15). The matrix of 26 quadrats at four periods × their characteristics

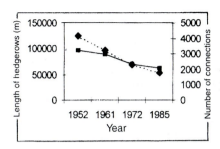 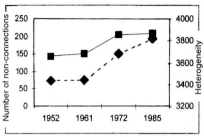

Fig. 14. Overall evolution of principal parameters characterizing the hedgerow networks of Lalleu

(re-coded into four classes of similar number of quadrats) were analysed by a factorial analysis of correspondences followed by an ascending hierarchic classification on the first three factors. Five variables were used (length of hedgerow, number of connections, number of connections with a road, number of non-connections, and heterogeneity). Aggregates of quadrats according to the geographic zones (north, south, village zones, zone of large farms) were counted as supplementary individuals.

The analysis yields a primary gradient opposing the quadrats with long hedgerows and a high number of connections to heterogeneous quadrats with low density of hedgerows. The second axis differentiates the quadrats according to their connectivity.

At each period, there was a wide diversity of situations; types of quadrats present in 1952 were found again in 1985. The quadrats of each year were widely dispersed on the factorial axis. This diversity could be measured by the average Euclidean distance from quadrats to the barycentre of a given year in the factorial axis (Table 3). The more different the quadrats are from each other (dispersed on the axis), the greater this distance. This overall diversity remained stable from 1952 to 1961, then declined in the two subsequent periods. While the quadrats became increasingly similar, their internal heterogeneity increased (Table 3), i.e., locally, the parcel size became

Table 3. Evolution of inter- and intra-quadrat diversity from 1952 to 1985 (Burel and Baudry, 1990)

Year	Distance from quadrats to barycentre		Intra-quadrat heterogeneity
	Average	Standard deviation	
1952	861	301	1.32
1961	895	312	1.32
1972	708	436	1.41
1985	621	337	1.47

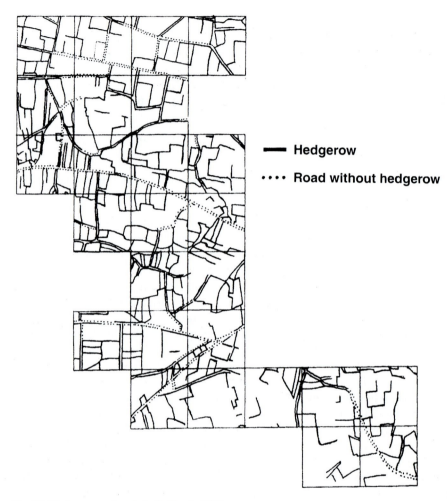

Fig. 15. Hedgerow network in Lalleu in 1985

increasingly different. Here again, the perception of diversity of landscapes changes according to the scale at which we observe them.

The classification produced six types. Their average characteristics are given in Table 4. Types 1, 2, and 3 are characterized by a short hedgerow length. Types 5 and 6 have long hedgerows, low heterogeneity, and high connectivity.

The different types of networks were unequally distributed over the years. The importance of types 5 and 6 diminished, while that of types 1 and 2 increased (Fig. 16). The southwestern part of the study zone evolved most rapidly towards open landscapes.

Table 4. Characteristics of types of hedgerow network

Type	Hedgerow length (m)	Hedgerow-road connections	Hedgerow-hedgerow connections	Non-connections	Heterogeneity (nats*)
1	2187	3.17	77.67	7.08	1.47
2	1947	5.80	55.60	8.73	1.48
3	2400	7.13	41.00	6	1.44
4	2875	6.35	84.60	5.85	1.38
5	3460	10.52	128.69	8	1.33
6	4200	17.50	195.50	4.65	1.30

*A nat is the equivalent of a bit when neperian logarithm is used instead of base 2 logarithm in the computation of heterogeneity.

To better analyse the changes, a typology of trajectories was carried out by hierarchic classification in the table characterizing the quadrats by their factorial coordinates, from the preceding analysis, for each period. Instead of each year, a quadrat represented a different individual; each year intervened as a variable, via the position in the factorial space. Four types of trajectories were thus established (Fig. 17). The first type combined quadrats that have constantly had a long hedgerow length (4460 m on average in 1952, 3360 m in 1985). Inversely, in type 4, the hedgerow length was always less (3155 m to 1778 m). The representation of trajectories of each type in the factorial space better illustrates the differences. Types 1 and 4 were always in different zones. Type 2 moved over the whole plan and evolved very rapidly in the last period. Type 3 evolved mainly in the first period. Thus, the different trajectories are independent and, from knowing the position of a quadrat at a given moment, we cannot predict its evolution (compare types 1 and 2 in 1952).

The difference between the trajectories appears in the analysis of descriptors (Fig. 18). Types that were very different in 1952, such as 1 and 3, grew similar over time. For example, the number of connections became similar. Type 1 lost many connections over time. These two types also ended up having the same number of non-connections (Burel and Baudry, 1990).

Mapping of types of trajectories (Fig. 19) reveals geographic differences. In the south there are mostly types 3 and 4, changing towards open landscapes. The north and especially the west remain among the more closed landscapes.

The rates of change can also be quantified using the distances between points representing the quadrats from one period to another. The Euclidean distance between these points, in the factorial plan divided by the number of years, gives an average rate of change (Smith and Urban, 1988). The average rate of change was greatest between 1961 and 1972; during the other intervals it was three times lower. This period of maximum change corresponds to a period during which high subsidies were given by the

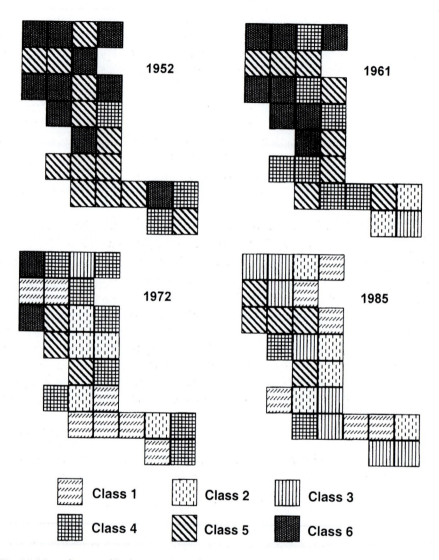

Fig. 16. Map of types of hedgerow networks

French government to enlarge parcels by uprooting hedgerows. It was a time of intensive mechanization of agriculture.

These differences are geographic as well as anthropic. The study zone can be differentiated into the northern zone (old moors), central zone, village zone (in the west), southern zone, and the southeastern zone with large farms (Fig. 20).

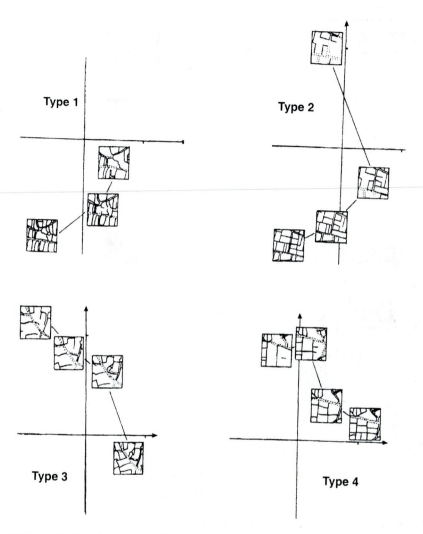

Fig. 17. Examples of evolutionary trend of each type of quadrat, in the factorial axis of gradient of structure

The north and the centre showed heterogeneous evolution, while the village zone remained wooded and the south and the large farm zone opened up.

This differentiation can be used to construct a multi-scale representation of the factorial space (Fig. 21). At the finest scale are found individual quadrats, at an intermediate scale the geographic zones, and at the coarsest scale the entire study zone. The great difference from initial to final state

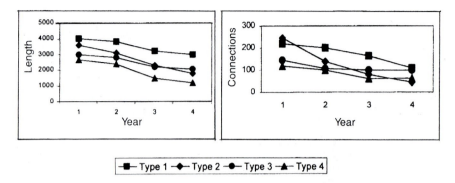

Fig. 18. Descriptors of evolution of networks for the different types of trends

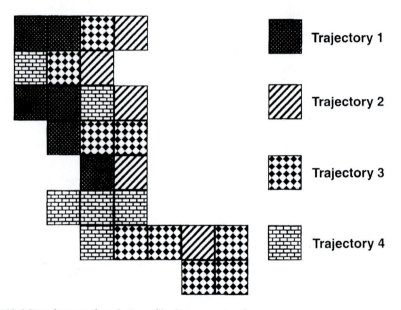

Fig. 19. Map of types of evolution of hedgerow networks

between the village zone and the zone with large farms is clearly illustrated, as is the rapid global evolution between 1961 and 1972. Globally, the landscape evolved less quickly (21, 101, and 18 factorial units per year per time interval) than the average of quadrats taken individually (51, 156, and 50 factorial units per year). We therefore clearly have a scale dependence in the perceived rates of change. We also note that the trajectory of the overall landscape is nearly parallel to the first factorial axis, which clearly represents the strongest constraints of the system.

Plate 6. Terraces of the Montagnette at Maubec (Provence)

Plate 7. Lalleu landscape

(a)

(b)

Plate 8. Villages of Daqiao and Yaunqiao. (a) The wooded village, surrounded by rice fields. (b) Vegetable gardens for household consumption.

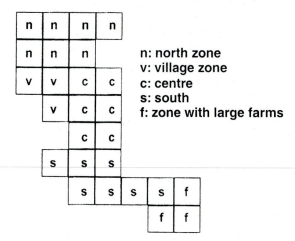

n: north zone
v: village zone
c: centre
s: south
f: zone with large farms

Fig. 20. Differentiation of the study zone

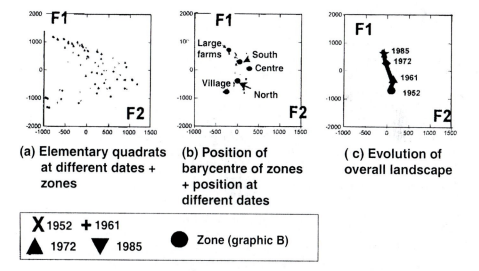

(a) Elementary quadrats at different dates + zones

(b) Position of barycentre of zones + position at different dates

(c) Evolution of overall landscape

X 1952   + 1961
▲ 1972   ▼ 1985   ● Zone (graphic B)

Fig. 21. Hierarchic representation of dynamics of hedgerow networks.

4.2.3. Conclusion

This study had primarily methodological implications. It showed that the evolution of a hedgerow network is more than a simple linear decrease or increase of hedgerows. Connectivity and heterogeneity are also important variables. They explain respectively 25 to 30% of the variance on the primary factorial axis and are not linked linearly to the length.

The hierarchic organization of landscapes here also clearly appears. We may consider that the national policies of the Ministry of Agriculture have an overall impact, while locally the landscape reacts differently according to the type of farm and the type of human habitat.

4.3. Evolution of a rice field landscape in subtropical China

Agricultural landscapes that are hardly or not mechanized may appear stable, ancient, while the techniques of production may seem to evolve only slightly. A study by Baudry and Zhenrong (1999) in two neighbouring villages (Daqiao and Yaunqiao) of the province of Hubei in western central China showed that in fact these landscapes change rapidly (Plate 8). This zone is located in the alluvial plain of the Yangtse, at 35 m altitude. The average annual temperature is 16.1°C, with 254 days without frost. The annual rainfall is 1100 mm, of which 600 mm falls from April to July. The population is dense (475 inhabitants/km²). Flood control has been a constant concern for two millennia of cultivation. A map was drawn from aerial photographs dating from 1955, 1977, 1984, and 1993.

Since 1956, a reduction in dry, non-irrigated crops was noted (Fig. 23), and an increase in rice fields and ponds, for fish farming. The woods, plantations associated with houses, were also increasing. Finally, from 1984 onwards, commercial vegetable crops appeared and were increasing rapidly. Investigations with farmers revealed that respectively 16%, 63%, and 21% of the cultivated land was planted with three, two, and one crop a year (Baudry and Zhenrong, 1999). The rice-wheat succession was common.

Beyond the rates of variation in area of different types of land cover, changes from one type to another increase the rates of landscape dynamics. Between 1955 and 1977, close to 53% of the space changed in terms of land cover. From 1977 to 1984, the rate of change was 62% and between 1984 and 1993 it dropped to 10%. This was represented by a variation of areas and instability of locations of the major modes of land cover (Fig. 22) and by frequent transitions between these modes. Figure 24 gives the example of non-irrigated crops and rice fields. The first seem to have constituted a reserve for new kinds of land cover, including vegetable crops, as well as all sorts of uses between 1977 and 1984. Even though the area of rice fields is increasing, their location became stable only from 1984 onwards.

These trends can be linked to the successive economic reforms that have affected the rights of land use as well as the right to sell products (Tollens, 1993; Baudry and Zhenrong, 1999). They are the result of general

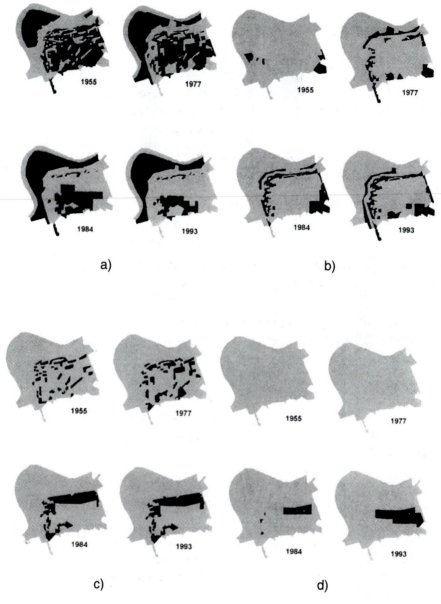

Fig. 22. Evolution of location of some types of land cover (from aerial photographs): (a) dry crops; (b) rice fields; (c) houses and gardens; (d) vegetable crops.

factors external to the landscape that did not even have the objective of changing the landscape.

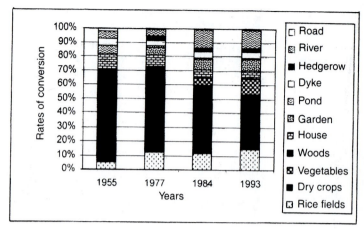

Fig. 23. Evolution of land cover

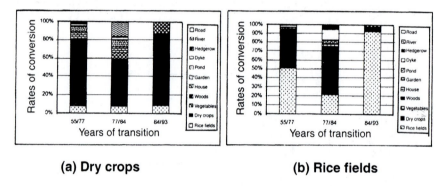

(a) Dry crops   **(b) Rice fields**

Fig. 24. Histograms of transition from dry crops and rice fields. (a) Dry crops. (b) Rice fields.

5.   DYNAMICS OF VALLEY LANDSCAPES: THE WATER COURSE AND ITS CORRIDORS

Water is an essential factor of the organization and evolution of landscapes (Naiman, 1996). Landscape dynamics of the large rivers as well as the relationships between the terrestrial and the aquatic environment are important themes in landscape ecology (Decamps, 1984). The concept of stream corridor was recognized from the beginning of landscape ecology (Forman and Godron, 1986). These studies complement hydrology, pedology, and the study of biogeochemical cycles. Studies of the biogeochemical cycles were the origin of the first long-term experiments on the landscape scale (Likens and Bormann, 1995).

Water courses have always been elements that humans have wanted to control and channel, in the proper sense. The landscape dynamics of large rivers is discussed in this section.

(b)

(a)

Plate 9. Water circulation in landscapes. (a) Dry talweg, Sevilleta Long Term Ecological Research Site, New Mexico, USA. (b) Torrent at the foot of the White and Black Glaciers, National Park of Ecrins.

Plate 9 (*Contd....*)

(c)

(d)

Plate 9. *(Contd....)*
(c) Wet grassland in a talweg, Brittany. (d) Danube Valley, Rumania.

Once people colonized the river valleys, flood control and shaping the bed of the river became objects of constant development. What characterizes a water course in an agricultural zone, and even more in an urban zone, is the uniqueness and stability of the bed. It is a total contrast with a water course that has not been interfered with, which regularly floods its entire valley and has a bed consisting of innumerable, often unstable channels. Plate 9 illustrates the effects of a talweg in the New Mexico desert (United States). The concentration of water and associated minerals allows the growth of some plants. In a mountainous zone, such as the Alps (Plate 9b, White Glacier, France), the torrents have highly irregular flows that disturb their bed. The channels often change. In the plains (Plate 9c, Brittany, France), the streams mark little relief, but the hydromorphy of their valley determines the land cover. A large river (Plate 9d, the Danube, Romania) has wide alluvial valleys, even several stable branches.

In France, the evolution of landscapes of the large rivers was studied, for example the Rhone, the Garonne, the Rhine, and the Lot (Pautou and Decamps, 1985; Decamps et al., 1988; Decamps and Naiman, 1989). The simplification of the water course over time is a constant. This simplification consists first of a reduction in the number of channels, involving suppression of divergent channels and dead branches. The construction of barrages is another mode of development that profoundly disturbs the ecology of water courses and the dynamics of the valleys. The first effect is to regulate floods and to diminish their intensity. If the water is used to produce electricity, it is returned to the river. If it is used for irrigation, the volume of circulating water diminishes considerably.

According to Decamps and Naiman (1989), the Rhine is a particularly interesting example. It is among the most developed and disturbed rivers of the last two centuries. It is flanked by large chemical industries and supplies over 40 million people with water. At the beginning of the 19th century, a canal 200 km long contributed to reduce the Rhine to a single channel in the upper course (Fig. 25).

The first effects were seen on the alluvial forest, which regressed considerably because of the drop in the water table (up to 10 m) and the reduction of floods. These floods play, or played, a major role in landscape dynamics. They are disturbances that lead to a return of the vegetation to the juvenile stages and contribute organic matter and important minerals. These wetlands, ecotones between the terrestrial environment and the water courses, have been studied for their structural dynamics and their functions (Naiman and Decamps, 1990).

The dynamics of ecotones is controlled by a set of factors that intervene over a wide range of spatial and temporal scales (Salo, 1990). Tectonic phenomena govern the forms of the relief at the largest scales. At intermediate scales, it is the hydrological changes within the large watersheds that modify the erosive processes, the form of the valley bottoms. At refined scales,

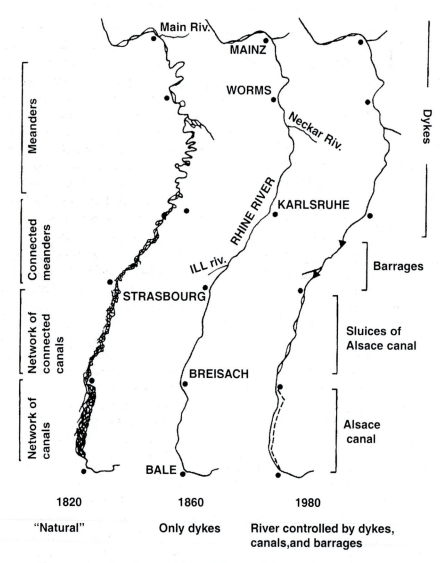

Fig. 25. Rectification and canalization of Rhine between Bale (Switzerland) and Mainz (Germany) since 1820 (Pinay et al., 1990, with permission).

biological processes such as plant successions intervene in the control of the dynamics of ecotones. It is the last two types of processes that are at the usual scales of landscape ecology, even though earth tremors can modify the reliefs in a very short time.

The geomorphology of ecotones varies from the source to the mouth of the water course. The slope of the watershed diminishes and an alluvial plain, subject to flooding, forms. The processes of accumulation gradually lead to processes of erosion. This contributes to the diversity of landscapes. In parallel, the direct interactions between the watershed and the water course diminish, at least in the less disturbed environments, because the ecotone is increasingly wide, but especially because the quantity of water that arrives from the watershed is a smaller and smaller fraction of the water that circulates in the river. We will mention here only the diffuse interactions linked to processes of transport by runoff or flow of the subsurface and not the direct input of pollutants from treatment plants. These last will have a tendency to increase because most large cities are located close to important rivers.

Johnson (1998) presents five key concepts relative to river landscapes and their transformation following development (canalization, construction of barrages): (1) response of the vegetation mosaic to regulation of the water course; (2) hydrogeomorphological controls; (3) changes in equilibrium; (4) the biodiversity; and (5) the management of restoration operations.

The vegetation of alluvial plains is typically a dynamic mosaic, composed of different potential stages of succession. This mosaic is closely controlled by the frequency of floods (Decamps et al., 1988) and processes of sedimentation (Girel and Pautou, 1996). If the alluvial plain of the regulated water courses is not cultivated, which is historically the most common case, the vegetation becomes uniform towards the mature stages.

Johnson (1998) shows that the contrasting response of the Platte and the Missouri rivers to flow reduction following the construction of a barrage is explained by their geomorphology. These two water courses are in the same climatic region, in the western central United States. The Platte is a wide and shallow river (less than 1 m deep), with a relatively high slope (0.00125 m/m), while the Missouri is deep (12 m) and has a very shallow slope (0.00016 m/m). During floods, the Platte overflows and its bed becomes very wide, while the Missouri becomes deeper and overflows only during severe floods. The reduction in flow and floods in the Platte has allowed the rapid extension of poplars, while these trees disappeared along the Missouri in favour of a forest comprising ash, elm, and maple.

The time it took to reach a new equilibrium after construction of barrages may be about 100 years, but it is highly variable.

The effects on biodiversity were overall negative, because of the homogenization of the vegetation. In the uninhabited, uncultivated plains, it can be restored by release of water. In other areas, restoration is impossible.

This overview of the dynamics of river landscapes shows the interactions between human and natural factors. Natural factors are examined in the following section.

6.   DYNAMICS OF NON-ANTHROPOGENIC LANDSCAPES

The study of "natural" landscapes is a long-standing concern of ecology. The debate between Gleason (1926) and Clements (1936) on the origin and stability of plant formations is a major example. The studies of Watt (1947) undoubtedly constitute one of the first theories on landscape dynamics. Focusing on moors as well as beech groves, they showed how phases of vegetation follow one another, not to reach a final state of equilibrium or climax, but to undergo recession and return to the pioneer stages. However, the paradigm of a stable climax continued to dominate the thinking of ecologists for a long time. The reassuring image of nature in equilibrium corresponds to a short-term perspective. For more than 20 years, this representation has increasingly been called into question. Pickett and Thompson (1978) introduced the idea that natural reserves need to reach a sufficient size for all the species present to have a habitat available under any circumstances, especially in case of disturbance (fire, storms). This concept of an equilibrium over a large area in relation to a regime of natural disturbances was developed by Bormann and Likens (1979). They emphasize its practical importance. The vision of Clement attributes only a minor role to fire and other disturbances, which leads to the exclusion of reflection on the management of territories. Bormann and Likens (1979) cite the study of Heinselmann (1973) as pioneering in the field of fire ecology. This study established that in Minnesota, in the central northern United States, fire recurs every 100 years or so. The wind is also an important source of disturbance in the forests of the eastern United States (Reiners and Lang, 1979). The role of disturbances in the ecological dynamics was examined in depth in the works of Pickett and White (1985) and Remmert (1991). Pickett and White (1985), in the conclusion of their work, emphasize the lack of predictive theory on the effect of disturbances, especially the absence of criteria to define a disturbance and its amplitude. Remmert (1991) developed a spatially explicit model of an ecosystem in which all the phases of plant succession are present. He specifies that the succession of different phases does not necessarily lead to a return to the point of departure and that processes of biological evolution may intervene during the cycle. This is an extension of the views of Watt (1947) and Bormann and Likens (1979).

Disturbances produce heterogeneity in landscapes and, consequently, researchers in landscape ecology were very early interested in relationships between heterogeneity and disturbance, to understand how heterogeneity controls the propagation of disturbances (Turner, 1987; Morvan et al., 1995).

Recognition of the dynamic character of natural landscapes has led to intensive study of modelling and theory. Shugart (1998) presents a set of approaches and models ranging from plant successions to simulations of landscapes to the use of these landscapes by fauna.

Case study: Yellowstone National Park (USA)

The first national park established in the world is the site of intensive research in landscape ecology, as well as important subject of debate on management of natural spaces, especially with respect to fire control. In this sense, it is a reference subject of research.

The park was established in 1872, following the expedition of Hayden, who submitted to the US Congress paintings and photographs that testified to the beauty of the region. The landscape of the park is highly diverse because of fires during the 18th century, probably started for the most part by Native Americans (Craighead, 1991). The park occupies an area of 925,900 ha (89 × 97 km) (Knight, 1994). It is part of a region known as the Greater Yellowstone Ecosystem (Keiter and Boyce, 1991), which covers an area of 7.3 million ha, of which 2.5 million ha is included in the national parks (Yellowstone and Grand Teton) and natural reserves. The rest is made up of forests and federal and private land. Figure 26 presents some views of the park. In 26a, the grassy plain is seen with some shrubs and grazing bison at the base of the forest. Part 26b shows a geyser, and in the background is seen part of the forest after a fire. Figure 13 in Chapter 7 shows an elk browsing among charred tree trunks.

The park is located east of the Rocky Mountains, between 110 and 111°S longitude and 41 and 45°N longitude. The mean altitude of Yellowstone is 2300 m. It varies between 1620 and 3333 m. The climate is continental, with long and cold winters (Despain, 1990). The average monthly temperature varies from −11.8°C to 12.9°C. Most of the park receives 760 to 1200 mm water a year. One peculiarity of the park is the large concentration of hot springs and geysers. The vegetation is composed of 80% forests of *Pinus contorta* var. *latifolia*. There are patches of *Picea engelmanii* forest (Engelmann spruce). At high altitudes there are *Pseudotsuga menziesii*. At low altitudes there are *Populus tremuloides* and *Pinus flexilis*. Willows occupy the humid zones. Herbaceous vegetation, shrubs of sagebrush (*Artemisia tridentata*), and lakes occupy the rest. The variety of large mammals (including deer, bison, grizzly bear, black bear, mule deer, antelope, and moose) is remarkable. The largest groups of elk (*Cervus elaphus*) in the world are found here (31,000 individuals in summer and up to 16,000 in winter). Around 2500 bison distributed in three groups are found in the park. The coyote is abundant, and the cougar is found. The wolf was wiped out after the park was established, as in the rest of the United States, but it was reintroduced in 1995.

Romme (1982) conducted one of the first research studies on landscape diversity in this region. He studied a watershed of 73 km², reconstructing

a)

b)

Fig. 26. Landscapes of Yellowstone

the history of fires for the past 350 years by observation of traces of charring on trees. From this information, the study of vegetation, accumulation of litter, and the record of plant successions, he reconstructed the landscape mosaic of about 200 years. Romme then simulated the landscape evolution with three scenarios: (1) spontaneous fires, (2) total exclusion of fire, and (3) controlled fires. He concluded, from these simulations, that fires have maintained a considerable diversity of vegetation over the course of 200 years. A policy of fire control reduces the amplitude of fluctuations of various types of vegetation, while maintaining a greater diversity of the landscape at all times. Major fires cause the return of large areas to herbaceous stages that subsequently become tree populations of a uniform age.

Fire

When the park was established, a policy of fire control was decided on (Romme, 1982). From 1972, park managers no longer attempted to put out fires caused by lightning that did not threaten the welfare or lives of humans. Between 1972 and 1988, 235 fires raged over a total of 13,851 ha (Varley and Schullery, 1991). The largest fire burned over an area of close to 3000 ha. The summer of 1988 was exceptionally dry. In June, precipitation was around 20% of the normal rate. The fires began in May. On July 15, after 3500 ha were burned, the authorities decided to fight all fires when the situation became critical (Varley and Schullery, 1991). In July, August, and September, drought and winds of about 60 to 100 km/h favoured fires. In total, it is estimated that 570,000 ha were burned in the Greater Yellowstone Ecosystem, of which 400,000 ha was in the park itself (Christensen et al., 1989). These authors estimate that the accumulation of inflammable materials was not a significant cause because all types of vegetation were affected. Fires of such magnitude undoubtedly occurred before the arrival of Europeans. In the firefighting operations, 9200 persons were involved and thousands of fire engines were used. The total cost was estimated at US$ 120 million. According to many experts, these interventions had little effect on the extent of the fire but were aimed to protect buildings. It was the 6 mm rainfall and snowfall on September 11, 1988, that put out the fire (Varley and Schuller, 1991).

Research on ecological consequences

About a hundred research projects on the consequences of fire were formulated in 1989. According to Christensen et al. (1989), "The heterogeneity and spatial scale of the GYA fires provide a unique opportunity to determine how features such as site variables, vegetation, climate, and landscape mosaic interact to produce specific responses to disturbance." According to these authors, three major categories of factors will intervene: (1) the heterogeneity of the fire, which has led to major variations in the distribution of plant mortality, ash deposit, and soil temperature; (2) the later climatic conditions that will influence evolutionary trends; and (3) the landscape mosaics before and after the fire that influence seed production and the behaviour of animals, which will have effects even in areas that were not burned. There would be no question of making predictions on evolution other than in terms of possibilities and probabilities (Christensen et al., 1989). One of the primary lessons drawn from this event is that "We have learned enough to know that wilderness landscapes are not predestined to achieve a particular structure." We are very far from tracing the expected trends of vegetation from synchronic studies of small parcels. With respect to the attitude of park managers toward fires, Christensen et al. (1989) emphasize the need for research and recommend prudence. "If prudence argues against aggressive intervention, it also argues against a doctrinaire strategy of laissez-faire."

Three themes of research have emerged: the study of plant successions, effects on large fauna, and effects on aquatic ecosystems.

The vegetation of the park is well adapted to fire, as is shown in the serotiny of *Pinus contorta*. The pine cones are closed and do not open to release seeds until they are heated. The growing plants then benefit from nutrients from the ashes. This characteristic allows a significant accumulation of seeds between two fires. During the fires it appears that few seeds burn (Knight and Wallace, 1989). This leads to the emergence of populations of trees of a similar age. However, all the pines do not have serotinous cones; their proportion is increased by the suppression of fires. Most herbaceous plants are capable of putting out shoots after fires, droughts, or very cold winters. We can thus predict that aspen forests will be rapidly regenerated from shoots of this species. The sagebrush does not put out shoots but produces seeds in abundance. Fires reduce the abundance of seeds. This plant is a winter food resource for some large mammals. Wherever the fire does not burn the roots, the herbaceous vegetation benefits from the input of nutrients.

Singer et al. (1989) evaluated the combined effects of drought and fires on large fauna. They estimate that the combined effects of low snow fall and rain fall, high temperature, and wind reduced around 50% of the production of grass in the summer. The dry grass is often a source of food. The animals that die as a result of the fire are few, often less than 1% of the population. The elk are able to reform their groups, under the effect of panic. The largest group of dead animals was 146 heads (out of a total of 261 carcasses found). Most of the animals died of asphyxiation, in very windy zones in which the speed of the fire reached 4 to 7 km/h. These carcasses were attacked by ravens, bald eagles, bears, and coyotes and at the beginning of December all the carcasses had been consumed. During the winter of 1988–89, the elk mortality was very high. Some groups were reduced by 50% because of the severity of the winter in combination with the scarcity of resources. Surviving individuals lost a great deal of weight and in the following spring the weight of newborn animals was reduced by 17%. In the following summer, the resources were considerably greater in quality and quantity, enabling an improvement in the size of groups. Singer et al. (1989) estimate that the effect of fire on the production of grass would be of short duration (2 to 3 years), but the effect on the production of consumable bark would be 30 years.

The effect of fires on aquatic ecosystems is due to the increase in runoff and the input of nutrients (ash, eroded soil). Minshall et al. (1989) estimate that the variations of runoff must not exceed the normal variations of a hydrological system. Their hypothesis is that the effects would be greater at the heads of watersheds than in water courses of high drainage order. Some basins of drainage order 1 or 2 were entirely burned, which was not the case with any basin of drainage order 4 or higher. With some rare exceptions, the fires hardly increased the water temperature. The primary effect, in the short term, is the stimulation of algal growth, but in proportions that had no negative effect for the other populations. In the medium term, the authors predicted an increase in inputs of nutrients and sediments, as well as increased insolation due to the disappearance of trees. In the long term, water courses evolve in parallel with the development of the forest vegetation. According to Minshall and Brock (1991), over large scales of space and time, fires are a mechanism of regeneration of the productivity and diversity of lotic ecosystems.

There is continuous research and publication on the ecological dynamics of the Yellowstone region. The conjunction of a knowledge of the environment before fire, the magnitude and heterogeneity of the disturbance, the possibility of setting up observations in the long term, and the development of new concepts will render these studies particularly useful to ecology as well as science in general. Debates surrounding nature and management of natural environments (Keiter and Boyce, 1991) are renewed and informed by the contributions of research. It can thus be proposed that the information drawn from research and debates will be useful in the management of anthropized landscapes as well.

7.   LAND COVER AND EVOLVING LANDSCAPES, A GENERAL PHENOMENON

The observations of change described in the preceding sections are found in many articles and books (Turner and Ruscher, 1988; Zonneveld and Forman, 1989). One of the most thorough studies was carried out by Berglund (1991) and fellow researchers, who studied changes in the landscape of southern Sweden over a period of 6000 years.

Here, there is a primary observation: the generality of the phenomenon of landscape dynamics. That is, ecological studies were not done in a stable environment and various parameters of this environment did not change in the same fashion or at the same rate. Moreover, the rates of change appear different depending on the scale of space and time considered. It thus becomes useful, if not essential, to question the nature of the relations between an ecological state (type of species assemblage, biological richness) and the parameters of the landscape. Is there a spontaneous and continuous adjustment or is some delay necessary, and what kind of delay?

In the preceding chapter, the differences of perception of structures according to species were emphasized. Species must also perceive changes differently. One can pose the hypothesis that species with a fine grain perceive changes over small spaces and over short time intervals, and thus more quickly than species of a coarse grain.

Moreover, when changes occur, they may occur in very different directions from one place to another and from one epoch to another. The average changes that characterize the territory of France are the result of contrasting regional and local evolutions.

The existence of a phenomenon of scale dependence in the perception of these changes is another essential point. It has several important consequences for the study and comprehension of landscapes. The first is the existence of regularities, of an organization or at least an order in the changes. This organization can be due partly to phenomena of inertia. Everything cannot change abruptly and the overall changes are an aggregation of local changes. This effect of inertia is also illustrated by the fact that the abrupt changes caused by war in a territory do not prevent the return to the trend preceding the war, as shown in the example of land cover in the department of Marne. The existence of factors of organization that form and cause landscapes to evolve in certain extents of space and time also explains this scale dependence. This harks back to the notion of level of organization in the framework of which the aggregation of Brittany and Lower Normandy does not constitute an entity that functions like either of the regions separately. Scale dependence also signifies that it is possible to make transfers of scale, that the results could be extrapolated in space and time, in the limit of a level of organization.

These levels of organization correspond to spaces (periods) in which the changes have a given direction and intensity.

The last point concerns the predictability of changes. As the extrapolation of past dynamics is not valid over short time periods, in order to predict changes it is necessary to know the mechanisms responsible, to model on the basis of those mechanisms. For non-anthropized landscapes, a general principle may be drawn from the proposition of Reiners and Lang (1979), who state that the diversity of a landscape results from "the superposition of two different vegetation patterns: (1) patterns related to the distribution of species along a gradient of limiting factors, and (2) patterns resulting from portions of landscapes being in different stages of recovery following disturbance." However, that principle cannot hold a place as a predictive model. Predictive models can be drawn from models of growth or senescence, seed dispersal, and reaction to disturbances (Shugart, 1998; Remmert, 1991). However, the example of fires at Yellowstone shows that prediction of effects of fire is very difficult (all types of vegetation are affected) and that the new dynamics is hypothetical.

With respect to anthropized landscapes, interactions between the socio-technological systems and ecological systems are multiple and the uncertainties of prediction are still large. This is the subject of the next chapter.

5

Organization of Landscapes

The organization and evolution of landscapes have always been and still are the essential subjects of geography. The landscape is the regional landscape, defined according to the major types found in it (bocage, open field) (Lebeau, 1979). In turn, the determinants of the environment or the possibilities of human societies are invoked to explain landscape states and transformations (Bertrand, 1975). This type of work pertains mainly to historic or prehistoric periods and to large areas (Bertrand, 1975; Berglund, 1991). There are few works addressing the mechanisms of change on small scales. Still, it is the local modifications, at the level of individual actions, that generate the transformations of the landscape, even if there is no doubt about the fact that these actions are, most of the time, responses to information coming from larger levels in the form of laws, regulations, technological innovations, and market forces.

In the preceding chapter, the possibility of making extrapolations to anticipate changes in the landscape was discussed. Beyond that, can models be constructed to foresee these changes? The objective of this chapter is to present some studies that aim to explain the way in which anthropized landscapes, essentially agricultural, are organized. These studies seek explanatory factors in characteristics of the physical environment and technical and socio-economic characteristics of land use systems. Related studies address the explanation of the diversity of landscape elements as well as their relative localization.

Serious efforts are being made to construct models in landscape ecology (Harms et al., 1992; Fresco et al., 1994; Schoute et al., 1995). They aim to help public bodies define policies to regulate changes in these landscapes. Such policies are designed to control environmental problems in rural landscapes. Studies on spontaneous changes are more rare but have begun to be developed in the context of changes affecting the earth as a whole (Meyer and Turner, 1994), or in interactions between agricultural activities and risks of water pollution (Benoit et al., 1997) or soil pollution (Papy and Boiffin, 1988), as well as in a perspective of management of biodiversity (Baudry, 1993; Sauget and Balent, 1993; Thenail, 1996). The historical approaches, in which the activities of humans are essential, are related to social and technical mechanisms of changes and, in the same way, provide a framework of reflection for the present.

1.   CATEGORIES OF MODELS

Geographers very early on developed theories and quantitative methods relating to the organization of geographic areas (Pinchemel and Pinchemel, 1988). The oldest theory is that of Von Thunen, published in 1826, looking at an organization from a single centre (Ciceri et al., 1977). This theory explains how differentiations in land use operate in physically homogeneous regions such as the great plains of Central Europe. Agricultural production was located as a function of cost of transport to a market centre (small town). In the model, the fresh products, which were difficult to transport at the time, were cultivated in the immediate vicinity of the town. Cereals, which could tolerate transportation that was longer and less frequent, were cultivated further away. Extensive livestock farming was carried on furthest from the town. The model is not necessarily drawn in circles. Navigable routes may cause deformations, because they facilitate transport and reduce the geographic distances. The Von Thunen model also allows an explanation of local differentiations as well as differentiations in the distribution of crops on the national or continental scale. For example, in France, the Pays d'Auge was the centre of meat production for Paris, till the development of the railway, which allowed animals to be transported over long distances. Until then, the animals were herded from Vilette to the Paris slaughterhouses.

In the field of research on interactions between landscape and environment, which is covered by the ecologist, we can distinguish, independent of modelling techniques, four major categories of models:

1. Models based on transition matrixes. The inputs are maps and probabilities of transition from one mode of land use to another, and the outputs are maps.

2. Models seeking correlations to explain landscape status and changes. The inputs are maps of land cover or use as well as maps of the physical environment or spatialization of types of farms. The outputs are correlations that can be used to simulate the effects of changes in agricultural activities. In the case of non-anthropized landscapes, the biological successions in various environmental conditions constitute the outputs of the model.

3. Models using correlations or cause-effect relationships between the socio-economic structures and land use. The inputs are proportions of land use in various socio-economic situations at different periods. The outputs are correlations or functional models giving proportions of types of land use under various hypotheses of change in the socio-economic context.

4. Models based on internal functions of farms to simulate small-scale changes. They are then aggregated to constitute new landscapes. The inputs are models of spatial distribution of uses within farms. The outputs are maps.

The first category of maps do not use mechanisms at work in landscape dynamics. Only past trends are taken into account. The use of transition matrixes, based on the extrapolation of probabilities of change from one type of cover to another over a given period, is not a realistic solution, as attested by the differences between transition matrixes from one period to another in the case of Maubec or the Chinese villages studied in Chapter 4. Moreover, by definition they cannot predict the appearance of a new type of cover, such as the vegetable crops in China.

Turner (1987) furthered this development by comparing three spatially explicit simulation models with the real evolutions of a county in the piedmont of Georgia state (USA). The first of these models is random and the other two incorporate the effects of vicinity. The author was able to correctly simulate the evolution of cultivated areas, in area and number of patches, but not the other types of land cover such as forest, urban zones, or abandoned areas. A final model was developed incorporating the type of owners and the limits of ownership (Turner et al., 1996). It gave the best results. In doing this, the authors introduced, albeit hesitantly, a socio-economic variable, a factor of the influence of human activities, without the knowledge of which it is practically impossible to make any prediction of change in landscapes dominated by humans. The authors are working in the direction of other categories of models, on the basis of researches of correlation between land cover and geographic position, as along a gradient of distance from a town (Wear et al., 1998).

Categories 2, 3, and 4 necessitate the knowledge of mechanisms or, at least, of correlations between the changes and independent variables that could be the cause of them. The factors of organization and change in agricultural landscapes can be physical (soil, climate, topography), social (type of agricultural household, collective organization), economic (market forces, type of revenue), technological (equipment, animal breeds, plant varieties, fertilizers, pesticides), or cultural (family tradition, personal or collective goals). These models attempt to respond to a deficiency highlighted by Baker (1989), who wrote: "The most important present limit to the development of better models of landscape change may be the lack of knowledge of how and why the landscape changes, and how to incorporate such knowledge in useful models, rather than a lack of technology to develop and operate models of landscape change."

Categories 1, 2, and 3 consider the landscape as a whole and the entire modelled space changes at each stage of the dynamics. The essential differences are that categories 1 and 2 are spatially explicit, while category 3 essentially produces histograms of distribution. Category 4 provides results on the scale of a farm or a type of farm. In this category, the parcels of land are the basic units of the map. In categories 1 and 2, the basic units are often pixels. From a functional point of view, the field is a significant unit, because it is the scene of action of a farmer or any other land user. It is the place in which human activities directly modify the physical or biological environment (Turner and Meyer, 1994). It is a unit of management (Baudry and Thenail, 1999). According to Deffontaines (1993), it is the scene of articulation of various disciplinary approaches necessary to the understanding of interactions between human activities and ecological processes. Agronomists and ecologists have a different view on what a field is. The former consider it a portion of a system used with an objective of production (cattle fodder, wheat production, cider manufacture). The latter consider it a piece of a mosaic with associated borders that may constitute ecological habitats.

Models of type 3 serve in scenarios of regional evolution. The numbers obtained (area of modes of land cover) are mapped on the small scale (aggregated statistical unit: region, country) (Rabbinge et al., 1994).

Multi-scale approaches, used to analyse the landscape structures and their dynamics, are also essential to explain the phenomena of landscape differentiation and evolution. This is an important point because the causes of changes, the factors of organization, may differ from one level of organization to another, in the sense that they have been defined above. The lack of attention to problems of scale is often a source of fruitless discussions (Allen and Starr, 1982). Generally, although changes in landscapes are linked to local actions, the latter are nearly always influenced by larger regional, national, or international factors. According to Richards (1990), the gradual integration of all areas into the market economy explains the rapid transformation of land use since the 19th century. Moreover, links must be established between the scales at which transformations occur and the scales of decision-making that controls these transformations. For example, European agricultural policy has a major influence on landscapes. Sometimes this influence is not desirable, and its effect on all the regions of Europe may not be uniform (Meeus et al., 1990).

The factors of landscape organization are thus analysed from a spatial perspective (distance from an organizing centre) and from the perspective of the hierarchical structuration of these factors into functional sets (interlocking of systems of decision and incentives).

These two points of view constitute the frame of reference for the analysis of case studies presented in this chapter and some theoretical developments that are now possible.

2. THE CONCEPT OF ORGANIZATION

This chapter focuses on organization, a central term in any research activity, but one that must be defined particularly for each approach. What is the organization of a landscape? When can we say that a landscape is organized? Just as changes are a function of a frame of reference (scales of space and time, measurement method), the organization depends on the position of the observer, the frame from which the observer evaluates and measures the organization.

In any case, an organized structure is demarcated from randomness. A landscape is organized to the extent that it differs from a random structure in the eyes of an observer. The formalism of the information theory offers concepts and tools to define organization.

2.1. Spatial organization of the landscape mosaic

That formalism has already been used in this book to measure the heterogeneity of a landscape. For a given number of elements, in particular proportions, a landscape is spatially less organized to the extent that the heterogeneity is high. That is, in the sense of concepts defined in section 4 of Chapter 3, when one moves from pixel to pixel, the landscape is less organized to the extent that the predictability of the nature of the following pixel is low. A fragmented landscape is less organized than a landscape in which the various types of elements are in large patches. Thus, we have a preliminary measurement of organization that is the observer's viewpoint, which looks at the spatial arrangement of landscape elements.

2.2. Organization of a landscape mosaic vis-à-vis other factors

The first definition of organization considers the horizontal aspect of a landscape. A second definition considers the "vertical" aspects. The landscape is thus represented by a set of layers of information (plant cover, land use, soil type, slope, ownership). The question is: what information do we have on one layer (land cover) when we know one or several other layers? Phipps (1981) developed studies on this theme following the concepts of Hill (1975). Plate 10 shows some examples visible in the landscape.

In Plate 10a, the vegetation along the Nile (Egypt) is strongly structured by the presence of water and alluvia in a desert climate. The transitions from the water in the river to the vegetation on the flood plain and then to the desert are strongly marked.

In 10b, the physical gradient is more gradual. In this landscape on the shores of the Henry Pittier National Park (Venezuela), there is a shift from the dry vegetation along the sea, with palms, to the rain forest, which is marked by the constant presence of clouds that keep the humidity high. This sharp gradient is the cause of a great richness of animal and plant species.

In 10c, the shrub vegetation of the mountains (Picos de Europas, Spain) is clearly delimited from the herbaceous vegetation by human intervention.

This human intervention is an important factor of the organization of landscapes that is superimposed on the physical gradients. In the Mont Saint Michel bay (France), shown in 10d, the polderization leads to a great difference between the polder cultivation and the salt marshes that are regularly subjected to tides.

The interactions between human activities and the environment may be more gradual, as on this slope in the Pays d'Auge (Plate 10e). The summit is covered with a wood on thin soil over clay with flint. The base of the slope, on marl, is ploughed. The intermediate fringe, on a steep slope, is covered by undergrazed grasslands colonized by bracken fern, which makes the texture of the vegetation irregular.

The principles of the process are the following: we calculate the co-occurrence between each form of the landscape mosaic (types of land cover, for example) and each form of variables describing the environment. These could be the characteristics of the physical environment (soil, slope, humidity), the social environment (ownership, type of farm, municipality), or technology (fertilization, uses). This process gives links variable by variable. As landscapes are complex systems in which the variables interact, we need to know the relationships between the forms of the mosaic and the combinations of variables.

The Shannon and Weaver (1949) information theory and Atlan (1992) provide concepts and methods that are specially adapted to measure these relationships and establish combinations of variables (Phipps, 1985). According to these authors, the landscape can be seen as a channel of information between a set of messages defined from abiotic descriptors of the environment and a set of messages defined from biotic descriptors of the environment.

2.2.1. Basis for measures of information

The Shannon formula is expressed as follows:

$$H = - \Sigma_{i(1,n)} \, p_i \log p_i$$

where n = number of results possible and p_i = probability of occurrence of result i. This formula constitutes the basis of the formalism.

2.2.2. *Various types of information*

In the case of the landscape, it is desirable to establish a relationship between a set U of units of the mosaic and a set E of independent variables describing the environment. We carry out some calculations described below:

Given a matrix U × E in which U has l possible states {1, 2, ... i, ... l} and E has m states {1, ... j, ... m}; and nij the number of elements in the state Ui and Ej, ni. the number of elements in the state Ui, n.j the number of elements in the state Ej, and Snij = N.

$$H(U) = - \Sigma_{i(1,\lambda)} \text{ ni.}/N \log \text{ ni.}/N$$

is the heterogeneity of states of U, or entropy of U, or quantity of information of U

$$H(U)max = \log(\lambda)$$

$$H(E) = - \Sigma_{j(1,\mu)} \text{ n.j}/N \log \text{ n.j}/N$$

is the heterogeneity of states of E, or entropy of E or quantity of information of E

$$H(E)max = \log(\mu)$$

$$H(E, U) = - \Sigma_{i(1,\lambda)} \Sigma_{j(1,\mu)} \text{ nij}/N \log \text{ nij}/N$$

is the heterogeneity of the matrix U × E.

The mutual information or neguentropy

$$T(E, U) = H(E) + H(U) - H(E, U)$$

is the mutual information between E and U. It is a measurement of the knowledge of the state of U when one knows E and vice versa.

The **redundancy** R = T(E, U)/H(U) × 100 is a **measurement of the organization of the system**. It is the rate of heterogeneity of U linked to the heterogeneity of E. It varies between 0%, if there is no link between U and E, to 100%, if U is totally explained by E.

The **significance test** is the criterion of Kullback. This criterion is **K = 2 N T (E, U)**. It follows a χ^2 law at $(\lambda - 1)*(\mu - 1)$ degrees of freedom, λ being the number of modalities of U and μ the number of modalities of E. The test is done by using the ratio of significance R. It is the ratio between the Kullback criterion K and the value of χ^2 for the same number of degrees of freedom. If this ratio is greater than 1, the relation is significant at a threshold of 5%. For a set of variables, the variable that provides the greatest information is that for which R is the highest.

The Pegase procedure

The procedure used was proposed by Phipps (1981). It allows the constitution of combinations of variables (E1, E2, ... Ee) that maximize the redundancy of the relationship between a set of modalities of states to explain {U} and {E1 ... Ee}.

The program is constructed to work in successive stages. In the first stage, the procedure tests the significance of links between the modalities of the variable U and each of the explanatory variables {E1 ... Ee}. For the next stage, the variable having the greatest ratio of significance is retained and a distribution is done of the set of individuals according to the modalities of the explanatory variable retained. At the second stage, the procedure tests first of all, for the first subset of the distribution (first modality of the explanatory variable retained), the significance of links between "U" and the other explanatory variables. Again, the variable having the greatest ratio of significance is retained. The procedure continues in the same way for the other subsets defined at the first stage and new distributions are done. The procedure continues thus till it faces one of the rules for stopping. A stage is a solution (a solution subset is constituted) when the heterogeneity of the subset is no longer significant, that is, when the number of individuals that constitute it is less than 10 and/or when the mutual entropy is less than 0.5 and/or when there is no longer an explanatory variable having a significant value with the variable to be explained.

The results of the information tests are of three kinds: (1) a measurement of rates of information between "U" and each of the explanatory variables and, at the end of the test, the rates of information between "U" and the set of explanatory variables; (2) a hierarchy of contributions of different explanatory variables to the explanation of modalities of "U"; and (3) a solution matrix giving the numbers of each solution subset for each of the modalities of the explanatory variables that can be used in order to better describe the modalities of "U", in relation to the modalities of explanatory variables.

2.3. Example

A simple model can be used to illustrate the two types of organization. Suppose an agricultural landscape used by several farmers in which the soil cover is strongly dependent on the age of the farmers, an invisible variable that can only be ascertained on inquiry. In this case, the ratio of ploughed fields in relation to permanent grasslands diminishes with the age of the farmers, according to a rule defined in Fig. 1a. This ratio increases over time for the young farmers who take over the farms. Figure 1b represents the state of land cover at three successive periods. In the evolutionary scenario, the link between age and rates of ploughing becomes increasingly strong and the overall rate of ploughed land increases (younger farmers establish themselves). The landscape becomes increasingly heterogeneous (Fig. 1c). The spatial distribution of types of land cover is close to a random distribution, as shown in the great diminution of redundancy. In parallel,

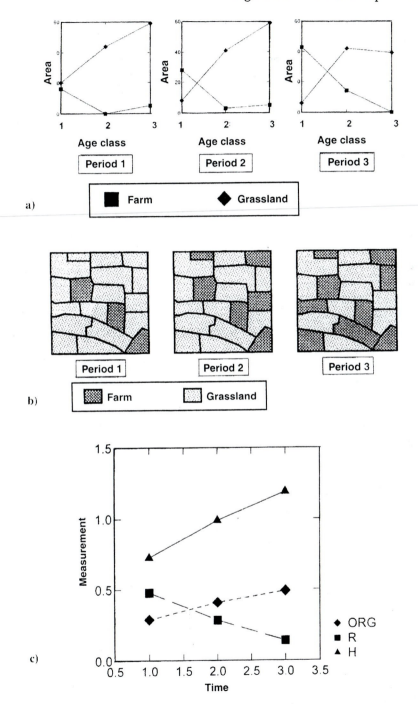

Fig. 1. Simulation of evolution of land use in a landscape. (a) Rules of allocation of land use as a function of type (age) of farmer. (b) Evolution of land use. (c) Evolution of parameters of evaluation of different forms of landscape organization.

the organization of the landscape (variable ORG) in relation to the type of farmer increases. In consequence, even though the cover of each field may be increasingly determined, the spatial structure is decreasingly so.

Thus, there are two points of view on landscape organization. This type of process of spatial disorganization was demonstrated by Di Pietro (1996) in a valley of the Central Pyrenees. The collective control of land use disappeared in favour of individual control. The consequences were an increase in the spatial heterogeneity of land use and a stronger link between land use and type of farm. The authors conclude that the landscape was disorganized, while in fact one type of organization was replaced by another.

3.   ECOLOGICAL ORGANIZATION OF LANDSCAPES

From the ecologist's point of view, the hypothesis of an organization of land cover and uses, which are combinations of species, as a function of the physical environment, was the first to be formulated. The landscape mosaic is the product of interactions between abiotic components (climate, soil) and biotic components (spontaneous vegetation, crops, type of agricultural production) in particular relationships. The studies presented in this section address the correlation between the characteristics of a place and its mode of cover or use. The "place" may be a point or a pixel or have a certain extent. It may be a landscape window. The variation of the extent allows us to analyse the effects of scale.

3.1.   Spatial organization of farming systems in Ottawa (Canada)

Phipps et al. (1986) analysed the relationships between land use systems and the physical environment in two municipalities (Nepean and Gloucester, 27,000 ha) south of the green belt of Ottawa. The objective was to establish a relationship between land use and the physical environment or the pressure of urbanization, and to analyse the evolution of such relationships between 1956 and 1978.

The relief of the study zone results from the alternation of a marine plain and morainic deposits with chalky rocky ridges (Marshall et al., 1979). The average duration of the frost-free period is 142 d; the temperature is higher than 5°C for 200 to 240 d. The average temperature in July is 20.7°C and that in January is –10.9°C. In this recently colonized region, the territory was divided into regular squares that were assigned to colonists and constitute the basis of the present division of property. It is a peri-urban zone in which various housing developments have been built. The sector includes many woods and fallow land. The average size of farms was 82 ha in 1978 (source: *Statistics Canada*).

(a)

(b)

(c)

Plate 10. Examples of organization of the landscape mosaic vis-à-vis factors of the physical and anthropic environment. (a) Nile valley, lower Egypt. (b) Shore of the Baya de Cata, Henry Pitier National Park, Venezuela. (c) Grasslands, Picos de Europas, Spain.

Plate 10 (Contd....)

(d)

(e)

Plate 10 (Contd....)
(d) Border between polders and salt marshes, Mont Saint Michel bay. (e)
Abandoned landscape in Pays d'Auge.

3.1.1. Basic data

The Ministry of Agriculture established maps of agricultural land use systems based on land cover in 1978 (Huffman and Dumanski, 1983) (Table 1). These are land use systems in that they represent a combination of modes of land use.

Table 1. Definition of farm systems (source: Ministry of Agriculture, Canada)

Agricultural system	Definitions
	1956–1965
Mixed	Agricultural system with at least 25% ploughed land
Hay	Agricultural system with less than 25% ploughed land
Pasture	Extensive system of permanent pastures
Fallow	Abandoned land
Woods	Wooded land
Urban	Land used for non-agricultural purposes
	1978
Monoculture	System of monoculture (generally maize without rotation)
Maize	Intensive mixed system (more than 30% area with maize)
Mixed	Intensive mixed system (less than 30% area with maize)
Hay	Semi-intensive system of cereals and hay (temporary grasslands)
Hay-pasture	Hay + pasture system (generally permanent grasslands)
Pasture	Extensive system of permanent pastures
Fallow	Abandoned land
Woods	Wooded land
Urban	Land used for non-agricultural purposes
Quarries	

Phipps et al. (1986) reconstituted the land use systems for 1965 from corrected minutes of the *Inventaire des Terres du Canada,* and those for 1956 from aerial photographs. These maps were sampled with a grid of 1108 spaced at 500 m. The same grid was used to sample the map of land use and soil capability (Marshall et al., 1979). In the analysis of the land use map, the base variables were noted (Table 2). To these descriptive variables of the physical environment, the authors added a variable describing the urban pressure (distance and density of urban zones for each point).

Table 2. Descriptive variables of physical environment

Parent material	Type of substrate and origin (marine, glacial, fluvial)
Texture of parental material	Sand, silt, clay … in combination
Surface texture	From coarse (A) to fine (clay = E) and organic
Natural drainage	From excessive to very poor
Rockiness	Presence of stones, from nil to significant
pH	From highly acid to highly alkaline
Slope	From flat to steep and depression
Agronomic constraints	Nil to great + type of constraint (fertility, variation of humidity)
Urban pressure	From no urban centre within 1 km to large city within 1 km
Agricultural pressure	Proportion of intensive agricultural systems within 1 km

The analyses concerned partly the entire zone, and partly particular sectors differentiated according to the urban pressure, the UR1 sector.

The relationship between the land use system (dependent variable) and the variables descriptive of the environment (explanatory variables) was established from calculations of mutual information, done according to the Pegase procedure (Phipps, 1981). The authors thus established landscape units characterized by a specific combination of environmental variables linked in a preferential manner to land use systems.

3.1.2. Results

Between 50 and 65 landscape units have been known at different times. The nature of the substrate is always the principal factor of land division. During the period 1956–1978, it was the substrates of marine origin that were the site of the greatest intensification of agriculture. Inversely, peat lands, occupied partly by grasslands in 1956, were totally wooded in 1978. Similarly, agriculture was practically non-existent in the fluvio-glacial substrates that became quarrying sites during this period (Fig. 2).

The texture of the surface horizon was also an important variable. Lands with medium texture (silt) supported intensive culture, while those with coarse texture (sand) were gradually abandoned.

Overall, a reduction in landscape organization was observed. The theoretical redundancy went from 45% in 1956 to 38% and then to less than 33% in 1978. In the last period, this was explained partly by the diversification of land use systems.

Table 3 gives the contribution of different variables to the organization of a landscape. The substrate is the dominant variable, then the ease with which the land can be cultivated or drained (the texture). The agronomic variables, marked out on maps of soil capability, are only a secondary consideration, no doubt because they are correlated with the substrate and texture of the soil surface.

Table 3. Hierarchic contribution of factors of organization to total mutual information

Rank	Descriptor	Contribution (%)	Cumulative
1	Substrate (parent material)	24.49	24.49
2	Texture of the surface horizon	16.57	41.06
3	Urban pressure	16.52	57.58
4	Effect of agricultural contiguity	10.05	67.63
5	Texture at depth	6.21	73.84
6	Drainage	4.82	78.66
7	Slope	4.72	83.38
8	Permeability limit	4.60	87.98
9	Rockiness	4.50	92.48
	Other	7.52	100.00

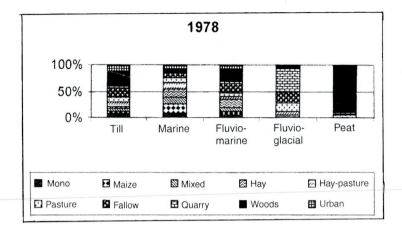

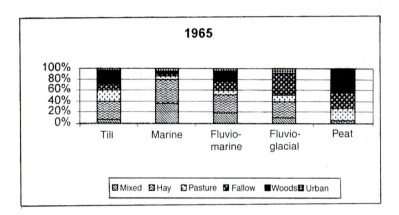

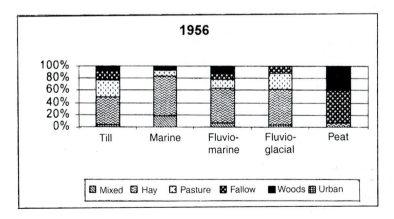

Fig. 2. Frequency of various land use systems in the types of substrate

3.1.3.   Concept of landscape niche

By analogy with the ecological niche, Phipps et al. (1986) proposed to define the landscape niche of a land use system as the combination of environmental factors in which a land use system is most frequent. To the extent of the process of division of land use systems into homogeneous subsets, combinations of relevant environmental variables are formed. Thus, the division into classes of parental material is the most relevant division in the first stage of the procedure. At the second stage, the class "material of marine origin" is divided according to the urban pressure, and at the subsequent stage the class "no urban pressure" is divided according to the surface texture. At each stage, the entropy of each subset diminishes. It is used in a more homogeneous manner. For the progression of division considered, the entropy goes from 2.22 to 2.18 and to 1.96. The most intensive system, milk production based on maize, is most frequent on soils of marine origin (18% of the area occupied against 11% on average). Within these, the frequency is maximal far from the urban zones (25% of the area), then on intermediate silty soils (35%) (Fig. 3). This example shows how the method allows us to acquire knowledge about the location of land use systems in the landscape.

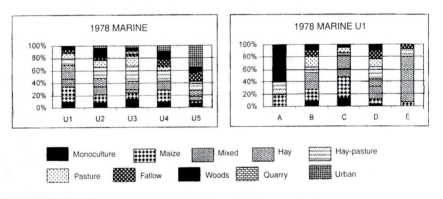

Fig. 3. Example of distribution of land use systems along a sequence of the Pegase procedure (Phipps, Baudry and Burel, unpublished)

At the end of the process of division, we obtain a matrix of land use system × combination of variables. This matrix is analysed by a factorial analysis of correspondences (FAC). In the space thus obtained, the land use systems are marked out in relation to the set of environmental variables. In effect, each point "combination of variables" can be represented by one of the variables included in the combination.

The analysis for the year 1978 is based on four significant factors. The first opposes the forest to cultivated areas, the second and the third use urban systems and quarries, and the fourth is made up of a gradient of agricultural systems, from "maize" to "hay-pasture". The factorial plan 1–4 thus represents the biologically productive systems (Baudry, 1985). From this plan, the following can be defined: (1) the resources used for the different land use systems and (2) the similarity between the systems from the viewpoint of resource use.

The landscape units on marine alluvia are located on the right half of the factorial plan and are linked to the most productive systems (maize, mixed, monoculture). In contrast, the units on glacial till are in the left part, with a few exceptions. The fluvio-marine materials, highly heterogeneous, are located on the entire plan. They have no specific use. The peat and fluvio-glacial materials, on the other hand, have very specific uses, forests for the former, quarries for the latter.

In this factorial space, the place of the land use systems point defines the centre of its landscape niche. The specificity of the niche (its size) is the greater or lesser aptitude of land use systems to use varied environments. It is defined as the difference between the overall environmental entropy of the landscape (H(E)) and the entropy of the land use system in relation to the set of environmental or landscape units (HUI(E)). The more a land use system is present in numerous units, the less specific it is.

Table 4 gives the specificity of niches of various land use systems at the three periods studied.

Table 4. Specificity of land use systems in the Ottawa region

		Mono-culture	Maize	Mixed	Hay	Hay-pasture	Pasture	Fallow	Woods	Quarry	Bldg
1956	value	–	–	1.13	0.21	–	0.22	0.61	0.54	–	–
	rank	–	-	1	5	–	4	2	3	–	–
1965	value	–	–	0.8	0.18	–	0.36	0.53	0.39	–	0.83
	rank	–	–	1	6	–	5	3	4	–	2
1978	value	0.59	0.72	0.67	0.42	0.59	0.47	0.33	1.36	0.35	0.9
	rank	5	3	4	6	5	7	10	1	9	2

This process provides a descriptive model of the landscape organization and evolution. It is remarked that the less demanding systems such as fallow or pasture have low specificity in 1978, which was not the case earlier. The specificity of woods tends in the opposite direction. The high specificity of built-up sites may be due to the fact that the density of urbanization is a variable that is included in the definition of niches. The relative specificity of agricultural systems evolves little. Concerning fallows, the reach of urbanization may explain the evolution of the specificity. Before urban planning, the least arable soils were left fallow. In 1978, it could be the reach of urban development that determined the fallows and no longer physical criteria. The woods occupy zones that cannot be farmed and cannot be built on, and thus they are limited to particular sites.

The fact that in 1978 the specificity of various agricultural uses was highly correlated to gross margin (Phipps et al., 1986) shows that the choice of production systems is not independent of the physical environment, despite the evolution of drainage and other technologies. On the contrary, it is probably that in intensive systems, it is the soils responding best to the increase of chemical and energy inputs that support the most productive systems. The absence of agronomic analysis of farms prevents us from drawing further interpretations.

3.2. Organization of agricultural landscape in the Pays d'Auge (Normandy, France)

Baudry (1992) took up a similar process to analyse the organization of a landscape in the Pays d'Auge. First, he analysed a landscape transect, then the entire municipality of Sainte Marguerite de Viette in the Pays d'Auge. However, instead of sampling the landscape at points, he used segments for the transect and windows for the entire territory. Each segment or window was characterized from a landscape point of view (combination of elements present) and from a physical and agricultural point of view (characterization of farms using the segment or the window). Apart from these technical differences, the essential difference in relation to the studies conducted in the Ottawa region was that the landscape organization was analysed in relation to spatial and physical factors and in relation to socio-technological factors (the diversity of farms). Thus, it was a contribution to the ancient debate on physical determinism and the potential of technology. The objective was not to take one side or the other but to show the importance of the perspective adopted and the method of analysis in the results obtained.

3.2.1. Factors of landscape organization along a transect

—Material and methods

The study transect was made up of 72 segments represented by a square of 50 × 50 m (Baudry, 1992). The elementary segments were aggregated two by two for a multi-scale analysis on segments of 100, 150, 200, ... 1000 m length by 50 m width. The grouping was done according to the method of sliding windows. This simultaneously compensated for the arbitrary nature of the division into segments and multiplied the data for the coarsest scales. For the base transect, the author obtained, for a group of 20 segments, 20 transects that were 3 segments long. The set of information contained in the aggregated segments was conserved to be characterized, unlike in studies by Turner (1990) and others, which conserved only the spatially dominant information.

Each segment (square of 50 × 50 m) was characterized by a set of variables involving the landscape elements, the physical environment, and the type of farm (Table 5).

Table 5. Descriptors used to characterize the elementary units of analysis

Landscape		Physical environment		Farms	
Grasslands	Yes/no	Arable land	quality a, b, c	Farmer age	< 35, 35–55, > 55
Cultivated	Yes/no	Non-arable grassland	quality d, e, f, g	Farm area (ha)	< 10, 10–20, 20–50, > 50
Degradation	Presence	Relief	1 to 4	Ploughed area (ha)	0, 1–10, 10–30, 30–60, > 60
Hedgerow	Yes/no			No. of cattle	0, 1–10, 10–40, > 40
Hedgerow intersections	No.			No. of dairy cows	0, 1–10, 10–30, > 30
Roads with hedgerows	Presence				
Water course	Presence				

So as to make the widest possible reference ecological space, the elementary segments of the transect were replaced over the set of 2492 squares of 50 × 50 m covering the municipality. The characterization was done by FAC, an analysis by category of variables. The aggregated segments were placed in supplementary elements in these analyses. By virtue of the principle of distributional equivalence (Benzecri and Benzecri, 1984), each aggregated segment is placed at the barycentre of elementary segments that it is made up of in the factorial space.

From these analysis, two types of results can be obtained:

(1) A measurement of the diversity of the landscape of the transect that is the length of a transect in the factorial space (sum of Euclidean distances between two successive segments). This measurement can be obtained at different scales of analysis.

(2) A measurement of the relationship between the different groups of descriptors. One such measurement is the coefficient of determination of the linear relation between the position of segments on the axes of different factorial spaces. Baudry (1992) retained the calculation for the position of segments on the first landscape gradient and the other types of gradient.

—Results

The first landscape gradient opposes annually cultivated segments to segments dominated by permanent grasslands and/or signs of abandonment of agriculture. In the last areas, the hedgerows are dense. The second gradient represents a variation of the hedgerow density.

The first gradient of physical environment opposes arable land to non-arable land on the moderated relief. The second gradient opposes land with moderate constraints to land with severe constraints (grasslands on slopes, often inaccessible to mechanized farming) and to arable land.

The first gradient of agricultural structure opposes major farms with a large herd of dairy cows to small, old farmers with no tilled land. The second gradient distinguishes young farmers with large ploughed areas and the third opposes small farms without animals that produce only cash crops to medium-size farms.

—Phenomena of scale dependence

The apparent landscape diversity of the transect varies with the scale of analysis. At the finest scale (segments of 50 m), the transect describes many zigzags in the factorial space (Fig. 4). Thus, segments of grassland follow on segments containing a hedgerow. All the fine variations appear. To the extent that segments are aggregated, they become more similar. They contain grasslands as well as hedgerows. The inter-segment variation declines, while the intra-segment variation increases (Levin, 1989). The grouping into segments of 1000 m (20 elementary segments) relies on a grassland zone (positive position on the factorial axis), in opposition to the rest of the transect, which is poorly differentiated (position close to zero).

The length of the transect (diversity of the landscape) in the factorial space is closely dependent on the scale of analysis (Fig. 5). Two points of inflexion are noted on the diversity-scale curve. The first point is located at the level of aggregation 5 (segments of 250 m). The second, very marked, is located at the level of aggregation 17 (segments of 850 m). From this point, the perceived diversity no longer varies at any aggregation. The two points of inflexion can be considered to correspond to changes in the level of organization. Between these points, the landscape has a particular fractal dimension that allows transfers of scale. The curve allows us to infer the diversity at scale 15 from the diversity at scale 9. Beyond the second point of inflexion, the landscape no longer has a fractal character. Changes of scale are no longer possible.

3.2.2. Factors of landscape organization

The coefficient of determination of the relationship between the first landscape gradient and the gradients formed by the explanatory variables was calculated for all the scales of analysis. Figure 6 gives the most significant examples. The coefficients of determination vary widely according to the scale of analysis, and this variation is regular, with two points of inflexion notable at scales 5 and 17, which are also points of inflexion of the diversity-scale relation. Below scale 5, there is no significant relation. Between scales 5 and 17, the relation between the landscape and the agricultural gradient

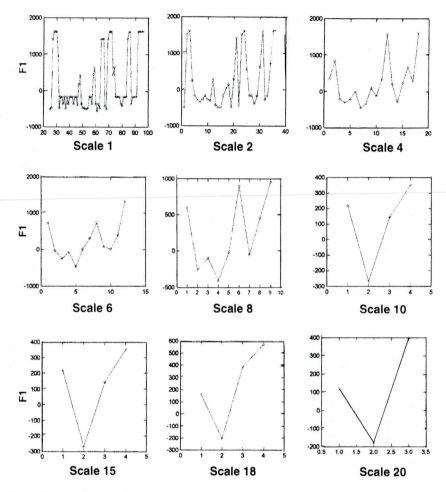

Fig. 4. Representation of the apparent diversity of the landscape according to the scale of analysis. The graphs give the successive position of the transect segments on the first landscape gradient (adapted from Baudry, 1992).

A1 is significant and dominant. Beyond that, the physical environment is better correlated to the landscape.

Baudry (1992) concludes from this that the sets marked out as being levels of landscape organization by the analysis of diversity correspond closely to the spatial sets for which the controlling variables change. This conforms to predictions of the hierarchy theory (Sugihara and May, 1990). A methodological remark is warranted: as the variables relating to landscape, farms, or physical environment are not of a common scale, it is logical to collect their value at fine scales, finer than the size of the smallest patch formed by one of these variables. However, the most significant relations

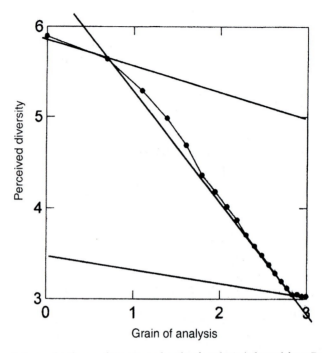

Fig. 5. Relation between landscape diversity and scale of analysis (adapted from Baudry, 1992)

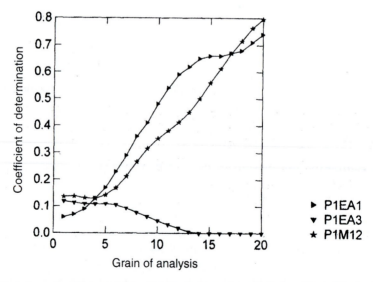

Fig. 6. Variations of coefficient of determination between the principal gradient of the landscape (P1) and the agricultural gradients (A1 and A3) and the gradient combining the first two gradients of the physical environment (M12) (adapted from Baudry, 1992).

between the combinations of these variables are found at average or coarse scales.

The results obtained can be interpreted in the following manner. Scale 5, from the first point of inflexion, corresponds to the size of fields (3 to 4 ha). These fields constitute the first pattern of the landscape. The intermediate scales correspond to the choice of the system of production, particularly the importance of ploughing in the farm. Since many arable lands are not ploughed, this may explain the preponderance of the set of agricultural variables over those of the physical environment. Beyond that, the second point of inflexion appears around 60 ha, or the largest farm size. Environmental factors, particularly the opposition of slope to plateau, become the preponderant factors of the landscape organization.

4.   FROM FARMING SYSTEMS TO LANDSCAPE DIVERSITY

The studies presented in the preceding section consider agricultural activities as an important factor of landscape organization, but they do not address how these activities themselves are organized. They do not refer to technological, social, and economic mechanisms within the farms or to any other system of activity intervening in the landscape. This section focuses on the newly emerging studies that address those mechanisms.

These studies require particular representations of landscapes, drawn from agronomy and not ecology. On the one hand, the diversity of types of land cover must be linked to the diversity between farms and within farms. On the other hand, an agronomic representation must be drawn that spatializes the activities. This is therefore a different process of agronomy that studies the potential of soils, the potential of environments, to predict their uses, often drawing on a cartographic presentation (Vink, 1983).

It is a developing field of research. In the framework of this book, only a few examples and basic concepts are presented.

The diversity of land covers and uses and landscape elements in relation to farms constitutes a first set to be explored to understand where this diversity comes from. Subsequently, the spatial arrangement of these various elements is presented, always in relation to the functioning of farms. It is an analysis of agriculture as a spatialized activity.

4.1. From farms to a diversity of landscape elements

In a very general sense, the diversity of elements in an agricultural landscape is primarily due to the diversity of production (cereals, milk, etc.). This manifests itself in the landscape in the form of specific types of plant cover, which may even identify the landscape, as with the Normandy grassland, the wheat fields of Beauce, or the olive groves of Provence. The increasing specialization of regions may thus be read in the landscapes. In the case of annual crops, agronomic rules of soil conservation and protection of the health of crops have given rise to the practice of crop rotation. This adds to the diversification of landscapes, even in places where monoculture is common.

The technical choices and individual practices of farmers are another source of diversity. Cattle farms are good examples. Cows may be fed on grass, in pastures in summer and in stalls in winter. Thus, there will be a landscape of often permanent grasslands. The intensification of production techniques and increase in potential productivity have often led cattle farmers to grow maize as a source of fodder, at least in winter, and that has caused rapid and profound changes in the landscape. During these changes, not only did maize replace the grasslands, but also mechanization led to the increase of parcel size.

Plate 11 and Table 6 give an example of this landscape diversity due to different systems of production (Deffontaines et al., 1995). The example involves two neighbouring farms in the Pays d'Auge, Normandy. Even though both have herds of dairy cows, one (belonging to L.S.) is more specialized in cereal cultivation. This is visible in a more open landscape, with fewer grasslands and increased signs of abandonment. In the other farm (belonging to L.F.), the milk production is more intensive, the cows are more productive, but the grass remains the basis of production, and hence there is a more wooded landscape with only a few maize fields.

Table 6. Characteristics of farms owned by L.F. and L.S. (Deffontaines et al., 1995)

Farmer	Age	Area (ha)	Total milk quota	Milk quota/ ha	No. cows	No. dairy cows	Prod./cow
L.F.	40	44	187,000	4250	72	28	7400
L.S.	35	117	268,000	2291	151	55	5300

Beyond these diversities of land cover, variations in use of a single area may have major ecological consequences. The study of permanent grasslands and the borders of fields provides examples that indicate the link between the diversity of exploitation and farms, and the diversity of landscape.

4.1.1. Diversity of permanent grasslands in the Pays d'Auge, Normandy

Some 99 floristic surveys have been conducted in a zone located between Livarot and Saint Pierre sur Dives (Baudry et al., 1997). Each survey corresponds to about 50 quadrats of 1 m^2 distributed over the entire field or zone with a homogeneous environment in the case of heterogeneous

(a)

(b)

Plate 11. Landscape diversity due to different systems of production. Landscapes of farms belonging to (a) L.S. and (b) L.F.

fields. In total, 1085 quadrats were studied. The results pertained to the presence of species. Two hundred fifty species were counted, but only 114 were included in the analysis, the others being too infrequent. An FAC on the matrix quadrats × species was based on two gradients. On the first factor (eigenvalue 0.325; 4.98% of the inertia), mesotrophic to eutrophic environments (fertilized grasslands, characterized by *Lolium perenne, Trifolium repens, Festuca arundinacea*) were opposed to heterotrophic environments (unfertilized, even abandoned grasslands, characterized by *Potentilla repens, Stellaria graminea, Sucisa pratensis*). On the second factor (eigenvalue 0.282; 4.31% of the inertia), limestone chalky environments with a xerophilic tendency (*Genista tinctoria, Circium acaule, Brachypodium pinatum*) were opposed to humid environments (*Angelica ssylvestris, Mentha aquatica, Equisetum palustre*).

The study of 150 farms of the Pays d'Auge and modes of managing grasslands (around 1700 parcels of land) allowed the establishment of a link between the overall level of intensification or production of the farm and the level of intensification of the grasslands. The classes of practices in the grasslands were established by a classification on the basis of factors resulting from an FAC. They range from class 1, extensive practices without fertilization, to class 5, more intensive practices with a nitrogenous fertilization of the order of 150 kg of nitrogen a year (Baudry and Denis, unpublished data).

When the first floristic gradient is related to the gradient of intensification (Fig. 7), it is confirmed that the fertilized grasslands are located in the negative part.

The floristic richness varies quite abruptly on axis 1 of the gradient of intensification from 15 to 45 species when we go from fertilized mesotrophic grasslands to oligotrophic grasslands. It stabilizes after the centre of inertia, which corresponds to the separation between fertilized and non-fertilized grasslands. The floristic richness diminishes in the abandoned grasslands located at the negative extremity of the gradient of intensification (Baudry et al., 1997).

From the perspective of agro-ecological analysis, we find the classic causes of diversity of grasslands in a landscape, relating physical factors and agricultural practices. This raises the problem of the distribution of various types of grassland between the farms. In other words, do the most productive farms only have intensified grasslands?

To test the link between exploitation, agricultural practices, and flora, 170 floristic surveys were done in 15 farms. These farms were divided into two groups according to their dairy quota. The quantity of milk the farmer was authorized to produce reflected the overall level of intensification. Group A had a quota less than 4000 l of milk per cow per year. Group B had a higher quota. Figure 8 shows that the entire gradient of practices of grassland use were found in the two types of farms, in different proportions.

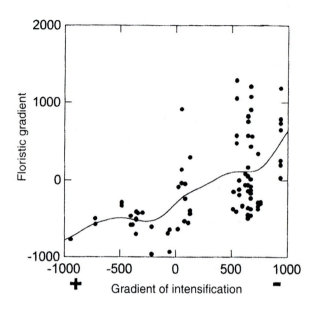

Fig. 7. Floristic and intensification gradient

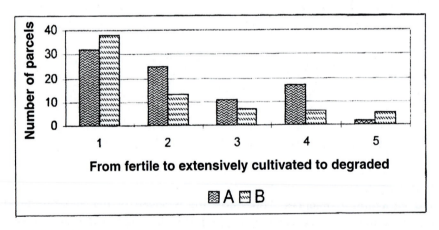

Fig. 8. Relationships between the type of farm and type of use of permanent grassland (Baudry and Denis, unpublished data). Farm: A, low quota; B, higher quota. Types of grassland use: increase in intensification from 1 (no fertilization) to 5 (more than 150 kg of nitrogen/ha/year).

Figure 9 shows that a real floristic diversity existed in the two types of farms, even if the most productive farms (type B) had the majority of surveys concentrated in the negative part of the axis. It can also be seen that they had fewer dry grasslands.

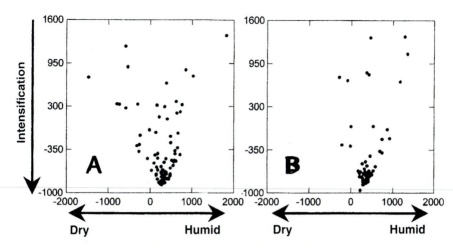

Fig. 9. Inventories of grassland flora in the floristic gradient graph for the two groups of farms (A and B)

The maintenance of the grassland diversity within the farms can be explained by their technical functioning (Thenail, 1992). The major explanatory factor is the diversity of herds and lots of animals having different needs. The second is the diversity of the physical environment, which, in the case of steep slopes and hydromorphy, prevents a uniform management. The diversity of herds allows the farmer to avoid these "constrained" zones; for example, the slopes are put to use for grazing beef cattle. Thus, the intensification of one part of the farm, which allows the farmer to improve his or her revenues, can favour the extensive use of another part. Similar observations were made in the Cotentin and in Brittany (Thenail and Baudry, 1996).

4.1.2. Diversity of field boundaries in the Armorican bocage farm land

Basic data
The uncultivated elements of the landscape, such as groves, hedgerows, fallow land, or ditches, have an important ecological function as a habitat for species that cannot survive in cultivated land and as a corridor. Ecologists have developed the concept of *field boundary*, which is the entire uncultivated space extending between two adjacent cultivated fields (Fig. 10) (Way and Greig-Smith, 1987; Marshall and Arnold, 1995).

Field boundaries may be highly complex (comprising ditches, an embankment, and other elements) or may consist of a simple grassy strip. The complexity comes particularly from the fact that the boundary comprises two sides, each abutting on a different field (Fig. 10). Each side may present

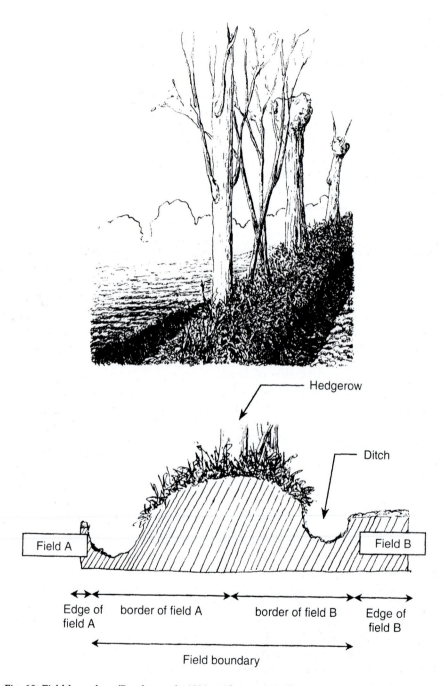

Fig. 10. Field boundary (Baudry et al., 1991, with permission)

a very different structure linked to the history of the field and is the result of individual management practices. Field boundaries are in this sense associated with the cultivated fields.

Baudry et al. (1998) studied the diversity of the structure of vegetation in field boundaries in a given bocage landscape (south of the Mont Saint Michel bay) in the context of a larger programme on the ecology and agronomy of bocage landscapes. The primary objective of the study was to propose a method of *in situ* evaluation of the structure of the vegetation through its physiognomy, using a series of descriptors. The secondary objective was to explore the relationships between the physiognomic diversity of the field boundary vegetation and a certain number of factors associated with (1) characteristics of the hedgerow in the field boundary, (2) the use of adjacent fields, and (3) farmers and their modes of managing hedgerows and field boundaries.

In order to do this, Baudry et al. (1998) analysed all the field boundaries of 10 farms, chosen after a series of investigations out of a total of 69 farms present in three study sites (Thenail, 1996; Cassegrain and Barmoy, 1995). The major characteristics of system of production, type of household, and modes of maintaining hedgerows and field boundaries are given in Table 7.

All the hedgerows were indexed and described from these three sites. For their study, the authors selected the following information: the height of the hedgerow, the tree cover, the shrub cover, the width of the canopy, the number of holes in the tree and shrub strata, the total permeability to wind (the denser a hedgerow, the lesser its permeability), and finally, the width of the field boundary at the foot of the hedgerow.

Characterization of physiognomic types of field boundaries
From a multivariant analysis out of 1168 boundaries described, the researchers constituted seven types of field boundaries (Table 8).

Hierarchy of explanatory factors of the physiognomy of field boundaries
Various tests, based on the Pegase procedure, were successively carried out in order to hierarchize the factors that could explain the physiognomy of the field boundaries. The first tests were done each from a single explanatory variable, then on the combination of all the variables (Table 9).

When all the explanatory variables are combined in a single test, the understanding of the physiognomy of the field boundaries is clearly better than when a single variable is used. This shows that this physiognomy depends on all the factors. The diversity of hedgerows or fields cannot be reduced to that of the farms.

The total redundancy obtained by this complete test (36.7%) is still lower than that obtained (45.4%) when we artificially add the redundancies resulting from tests on explanatory variables taken one at a time. This indicates a certain correlation between these variables. Nevertheless, there is each time a perceptible preponderance of descriptive variables of

Table 7. Characteristics of 10 farms studied

Farm	Area (ha)	Farming system (EU classi-fication of farms	Source of household revenue	Working days/year		Mode of managing boundaries	Estimated hedgerow length (km)
				Mechanical clearing	Chemical clearing		
P08	42	81	9	0	3	2	5.7
P14	75	43	8	0	2	2	9.7
P24	49	81	1	1	0	4	4.9
T05	17	41	7	24	0	1	6.9
T08	35	41	7	0	1	4	10.3
T09	19	41	7	5	0	3	7.5
T13	29	41	3	0	6	2	5.3
V04	36	41	7	20	4	1	13.7
V07	43	41	8	0	4	2	12.4
V13	8	11	3	4	0	3	1.6

Farming system, according to the European Union classification of farms
11: major crops, oriented towards cereal production
41: dairy cows
43: dairy and meat cows, breeding
81: major crops and herbivores

Source of household revenue:
1: cooperative society, non-family farm, salaried workers
The other classifications are all family farms.
3: Part-time farm (< 1 annual worker unit), household with revenues from external activities and without pension
7: Full-time farm (≥ 1 annual worker unit), household with revenues from external activities and without pension
8: Full-time farm, household with no revenues from external activities but with pension
9: Full-time farm, household with no revenues from external activities and no pension

Mode of management:
1: Mechanical clearing dominant, highly significant
2: Chemical clearing only, significant
3: Mechanical clearing only, rather significant
4: Mechanical or chemical clearing, very little

hedgerows (redundancy of 17.6% in the single test) and structure of household revenues (redundancy of 11.4% in the single test) over the variables of crop succession of the field (redundancy of 7.9% in the single test) and mode of management of the field boundaries (redundancy of 8.5% in the single test). The type of crop succession in the adjacent field gives a better picture of the physiognomy of the field boundary than the land cover, that is, the plant cover observed during the given year in the field. In this last case, the redundancy is only 4.4%.

Relationship between physiognomy and explanatory variables
Following the process of mutual information test, the authors conducted a multiple correspondence factorial analysis on the matrix made up of 34 solution subsets corresponding to 884 field boundaries and physiognomic

Table 8. Characteristics of field boundaries of 15 farms in the north of Ille-et-Vilaine (Baudry et al., 1998)

Type	No. of boundaries	Characteristics
1. Bare soil	2	Borders made up nearly entirely of bare soil.
2. Grassy	331	Visually homogeneous. Brambles (base) and ferns are rare, grasses are dominant, dicotyledons are poorly represented.
3. Grassy with brambles	205	Physiognomic type fairly close to type 2. The boundaries are relatively homogeneous. Brambles and ferns are well developed, as are grasses and dicotyledons.
4. Herbicidal	52	Highly heterogeneous appearance due to the quantity of dry grass and plant debris after pulverization of a herbicide. The charred grasses and dicotyledons are few. Brambles are few, and ferns develop.
5. Sparse vegetation	13	Mostly bare soil. The rate of cover with grasses, dicotyledons, brambles, ferns, dry grasses, and plant debris is very low.
6. Forest with dead leaves	276	The most heterogeneous boundaries. There is a mix of grasses, dicotyledons, mosses and lichens, dead leaves, plant debris, ferns, and brambles with, however, some bare soil. In this group ivy cover is the most significant.
7. Forest with brambles	285	Heterogeneous. Brambles and ferns, grown quite high. Grasses and dicotyledons are equally important. Little plant debris, bare soil. Dead leaves are rare.

Table 9. Redundancy between the diversity of field boundary types and some variables (Baudry et al., 1998)

Variable	Redundancy (%)
Crop succession	7.9
Mode of boundary management	8.5
Source of household revenue	11.4
Hedgerow structure	17.6
All together	36.7

types. From this analysis, the relationships between the physiognomy of field boundaries and the explanatory variables as they were hierarchized by means of the mutual information tests were explained. The physiognomic types of field boundaries were positioned in this manner, in relation to the solution subsets that correspond to a combination of hierarchized explanatory variables. During a preliminary analysis, the "herbicide" type seemed to be located outside the gradient. It was excluded from the analysis and designated a supplementary variable.

The relationships between the physiognomy of field boundaries and hierarchies of explanatory variables as they were analysed by Baudry et al. (1998) are described in the following paragraphs.

Physiognomy of field boundaries and characteristics of farms

—Overall management of field boundaries in the farms
Physiognomic type 4, "herbicide", was characterized by a peculiar mode of management. It was associated with the exclusive use of chemical clearing (96% of field boundaries of type 4 belong to farms that manage them in this way). Physiognomic type 2, "grassland", was also associated with chemical maintenance of field boundaries. Type 7, "forest with brambles", was associated with a very low level of management. Type 3, "grassland with brambles", and type 6, "forest with dead leaves", were not correlated to a mode of management.

—Structure of household income
Boundaries of type 7, "forest with brambles", were found in cooperative farms or full time farms, with income from external activities. Those of type 3, "grassland with brambles", belonged only to full time farms, with external revenue. Type 2, "grassland", was linked to full time farms, without external income but with a pension. Type 4, "herbicide", was associated with full time farms, the resources of which come solely from agricultural activities. Type 6, "forest with dead leaves", was not linked to a particular household income structure.

Physiognomy of field boundaries and hedgerow and field boundary structure
The boundaries of physiognomic type 7, "forest with brambles", were associated with dense hedgerows, fairly impermeable, having substantial tree and shrub cover. Type 6, "forest with dead leaves", was linked chiefly to relatively permeable hedgerows having significant tree cover but few shrubs. Type 2, "grassland", was associated with boundaries without trees and with rare shrubs. Type 4, "herbicide", was associated with boundaries without trees or shrubs. Type 3, "grassland with brambles", was again the least homogeneous with respect to the structure of the associated hedgerow.

Physiognomy of field boundaries and characteristics of fields
Field boundaries of type 4, "herbicide", were adjacent to fields used for cash crops. Types 2, "grassland", and 3, "grassland with brambles", more often bordered fields in which the crop rotation was maize and wheat or maize, wheat, and ryegrass. The other physiognomic types seemed to be less correlated to one type of crop succession. Nevertheless, boundaries adjacent to permanent grasslands were for the most part of type 7, "forest with brambles", or 6, "forest with dead leaves".

Conclusion
The field boundary is a more complex object than it ordinarily appears. It is not a simple, autonomous band of vegetation. Its characteristics are linked to the structure of the hedgerow that overhangs it as well as the use to which the adjacent field is put and characteristics of the family that farms the field.

4.1.3. Landscape elements as parts of farming systems

Grasslands, field boundaries, as well as all the fields, groves, riverbanks, and all the elements that constitute the agricultural landscape are also elements of systems other than farms and their technological systems. The relationship between ecology and man here is particularly clear and strong. The way in which these elements are assembled depends also, in large part, on the functioning of farms. This is the focus of the following sections.

4.2. Organization of land use in an Armorican bocage farm land

4.2.1. General principles of land use patterns in a livestock farm

The proposed process (Thenail, 1996) starts from mechanisms within the farm towards the landscape. In the framework of a work on landscape ecology, the analysis of the effect of spatial variables will be sufficient, without going into mechanisms of livestock feeding or the raising of crops. The size of fields, their distance from the farm centre, and their place in the landscape are the essential variables, apart from the type of farmer and farm.

In a cattle farm, it is possible to enumerate a certain number of uses and characterize them by the number of times it is necessary to come to the field and by the time spent in each place after each displacement (Fig. 11a). For example, dairy cows must be led to pasture and brought back twice a day (if they are grazing). Each time, the time spent in the field is short. Inversely, when the field is being ploughed, the farm worker goes out to the field only once but stays there the whole day. From this can be drawn probable thresholds limiting the land use as a function of the distance or the area of the fields (Fig. 11b). The dairy cows consume energy in moving around and thus have a lower production. Thus, the more productive herds remain close to the centre of the farm. The silage of maize necessitates the transport of the produce to the silo, which is always close to the farm buildings. A large number of tractors may compensate for the distance, but the cost would be higher. In some cases, the area can compensate for the distance. For example, when fodder is grown over an area of a few hectares, the displacements are profitable.

From these general elements, we can draw a general model of land use in dairy farms (Fig. 12). In this diagram, the dairy cows use the fields closest to the buildings, then come to fields reserved for winter forage production, such as maize. In practice, the crop successions mean that scattered pasture grasslands and maize alternate over time. Further out, cereals are planted. The young cows use the furthest fields, which are difficult to reach.

This type of analysis leads to new representations of the landscape (Fig. 13), such as maps showing field access (13a), hydromorphy of fields (13b),

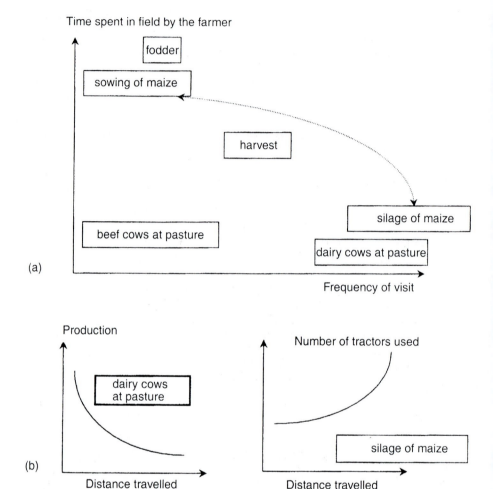

Fig. 11. Model of relationships between agricultural activities and land. (a) Displacements and uses. (b) Effects of distance on use.

structure of farm fields (13c), and strategy of hedgerow management (13d). Some of these maps necessitate a knowledge of farms and their territory, which is not always possible. Such data cannot be collected directly in the field, but only by investigation. Unanswered questions lead to gaps sin the maps.

Fields can also be represented by more inclusive categories such as farm types. The fields of grouped farms, for example, can be distinguished from those of scattered farms. Thenail (1996) distinguishes scattering from fragmentation in the following way: fragmentation corresponds to the fact

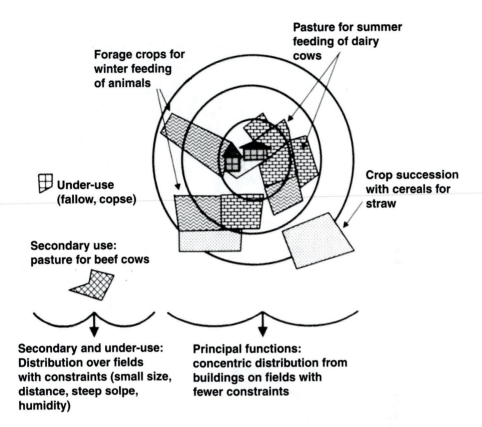

Pasture for summer feeding of dairy cows

Forage crops for winter feeding of animals

Under-use (fallow, copse)

Crop succession with cereals for straw

Secondary use: pasture for beef cows

Secondary and under-use: Distribution over fields with constraints (small size, distance, steep solpe, humidity)

Principal functions: concentric distribution from buildings on fields with fewer constraints

Fig. 12. Land use in a dairy farm

that the fields are not grouped but are in a number of distinct blocks, while scattering corresponds to the fact that the fields are far from each other. Thus, a territory may be scattered, but not fragmented. From this Thenail (1996) proposes the division of a farm territory into aggregates of contiguous fields (Fig. 14).

4.2.2. Case study

Available data

The study involved three bocage territories of almost 500 ha each (Baudry and Thenail, 1999; Burel et al., 1998; Thenail, 1996). They differ in the density of hedgerows, which decreases from site A to site B to site C. They also differ in their land cover and the types of farms present (Fig. 15). Thenail (1996) conducted surveys of 69 farmers. The questionnaire focused on the family, the production orientation, and the use of each field whether or not it was within a study site.

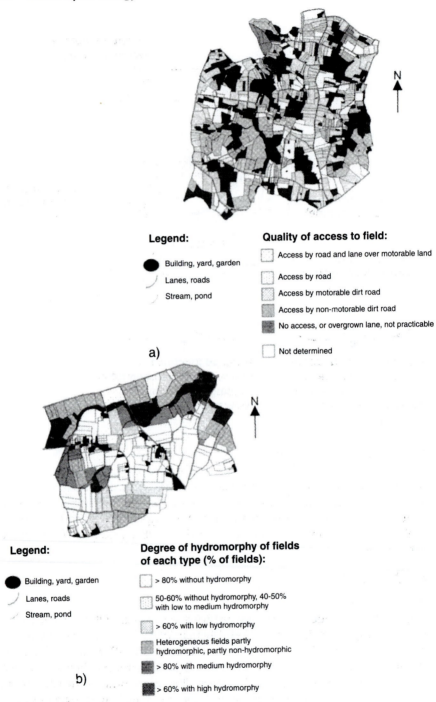

Fig. 13. Maps giving information on field structure

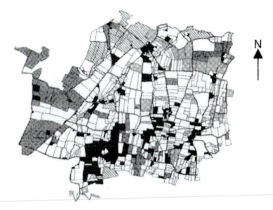

Legend:

● Building, yard, garden

⟋ Lanes, roads

⟋ Stream, pond

Farm type	Total area (ha)	Fragmentation (number of fields)	Scattering (relative distance of fields)
	< 30 or 30-50	++	+
	30-50	+++	++
	< 30 or 30-50	++	++
	< 30 or 30-50	+	+
	30-50 or > 50	+	+++

c)

Legend:

● Building, yard,garden

⟋ Lanes, roads

⟋ Stream, pond

Hedgerow management strategy

Farm type	Pruning	Mechanical clearing	Chemical clearing
	Very high	High	High
	High	Very high	Nil
	Medium	Medium	Medium
	Medium	Nil	High
	Low	High	Nil
	Non-continuous		

d)

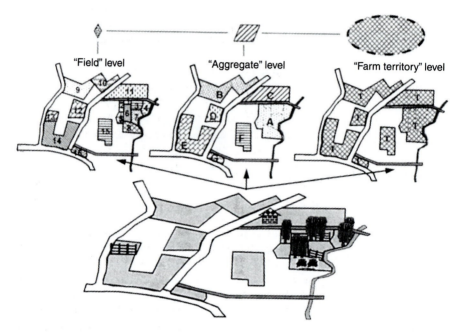

Fig. 14. Sections of a farm territory

The map of land cover (Fig. 15) shows a structural diversity among the sites, which is expressed by a different overall heterogeneity (Fig. 16).

From the study of Thenail (1996), we present only a few results relating to the explanation of land cover, depending on whether the data from the surveys were available or not. The analysis can be refined by including the field use, particularly the crop succession, in each farm.

The results

Two types of results were obtained: (1) the information contributed by each variable, in each case, and (2) the overall level of organization of landscapes, measured by the redundancy between the types of land cover and the explanatory variables. The value of the information obtained by survey can thus be measured.

Figure 17 gives the overall links between the types of land cover and the area of the fields, their level of hydromorphy, and the distance from the farm centre. These three variables have significant influence on the land cover. The proportion of fields occupied by maize increases with their size, while the proportion of grasslands diminishes. Maize benefits from a slight increase in soil humidity, at the cost of cereals (essentially winter wheat). The grasslands are the main cover of hydromorphous soils. The land closest to the farms is essentially grassland. Maize is grown further away, but its extent reduces beyond 3 km. This is essentially due to the cost of harvesting for ensilage. Studies by the Federation Nationale des Cooperatives d'Utilisation de Materiel Agricole on farm labour time confirm the significant excess costs associated with the distance of fields from the farm centre (Fig. 17).

Table 10 gives the redundancy values for the four variables taken into account (area, hydromorphy, and absolute and relative distance) and their combinations for the three sites. Two sets of fields were analysed: the set of fields on the site and the fields surveyed that were the only ones for which the distance from the centre was known. It emerges from this that the organization of the mosaic increased from site A to site C, for almost all the cases considered. As the explanatory variables were not independent, the level of redundancy of combinations was less than the sum of redundancies. In all the cases, the area of the fields was the primary explanatory factor of the land cover, followed by hydromorphy and, finally, the distance.

The comparison of two extreme sites in terms of the structure of the parcelling (sites A and C) helped to refine the analysis (Fig. 18). The characteristics of the fields varied only in a single criterion, the area. Eighty per cent of the fields in site A were smaller than 1 ha, as opposed to 30% of fields in site C. In this site, the parcels were also less scattered (Fig. 15). When we look at the factors of organization of the mosaic, we observe that although the trends are overall the same, some differences appear. For example, in site C, no field located further than 3 km from the centre of the farm was planted with maize, while the corresponding figure was close to 25% in site A. In this site, large numbers of the small fields were also planted with maize. Beyond the differences between the farms, the constraints linked to the parcelling (small fields, scattered fields) led farmers to use land in different ways, even though the basic patterns were similar.

Table 10. Redundancy between diverse variables and combinations of variables in the three sites, for all the fields (All) and the surveyed fields (Thenail and Baudry, unpublished data)

Variables	A	B	C
All A + H	19.40	20.50	28.00
All A	12.50	12.30	16.70
All H	6.10	7.90	16.20
Survey S	12.20	12.30	17.20
Survey H	8.90	9.60	16.20
Survey absolute D	2.80	6.10	12.30
Survey relative D	2.04	3.60	7.50
Survey S + H	21.90	20.70	27.50
Survey H + D	15.50	19.90	30.80
Survey S + D	14.20	23.50	33.40
Survey S + H + D	21.90	31.00	36.10

A, area of field. H, hydromorphy. Absolute D, absolute distance between field and farm centre. Relative D, relative distance from farm centre.

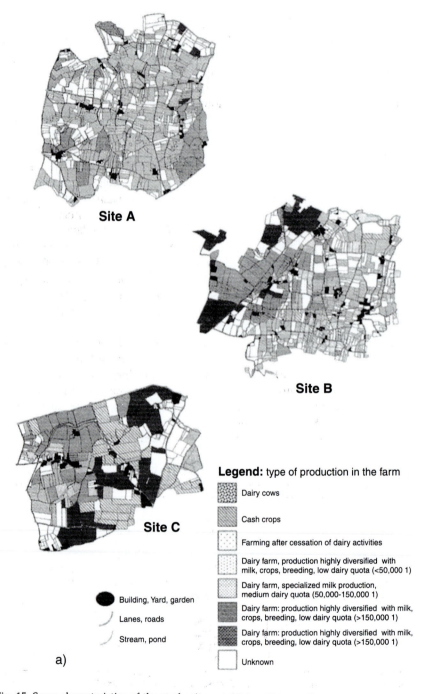

Site A

Site B

Site C

Legend: type of production in the farm

Dairy cows

Cash crops

Farming after cessation of dairy activities

Dairy farm, production highly diversified with milk, crops, breeding, low dairy quota (<50,000 1)

Dairy farm, specialized milk production, medium dairy quota (50,000-150,000 1)

Dairy farm: production highly diversified with milk, crops, breeding, low dairy quota (>150,000 1)

Dairy farm: production highly diversified with milk, crops, breeding, low dairy quota (>150,000 1)

Unknown

Building, Yard, garden

Lanes, roads

Stream, pond

a)

Fig. 15. Some characteristics of the study site

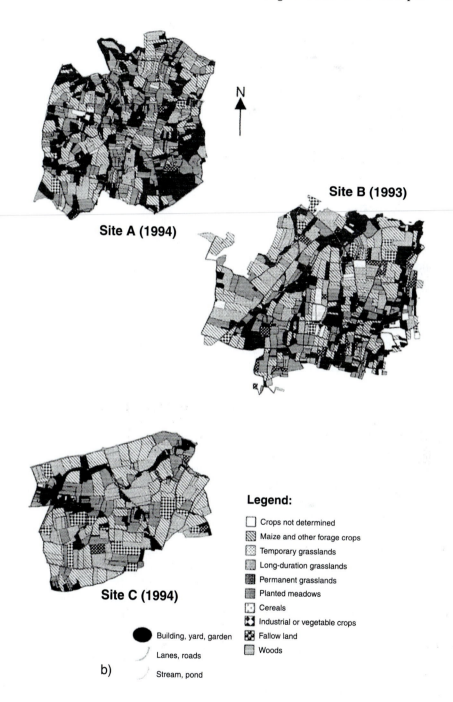

N

Site A (1994)

Site B (1993)

Site C (1994)

Legend:

☐ Crops not determined
▨ Maize and other forage crops
▧ Temporary grasslands
▦ Long-duration grasslands
▩ Permanent grasslands
▨ Planted meadows
⬚ Cereals
✦ Industrial or vegetable crops
▨ Fallow land
▨ Woods

⬤ Building, yard, garden

╱ Lanes, roads

b)

╱ Stream, pond

Fig. 15. (contd.)

This resulted in a structuration different from the mosaic in the landscape. For example, maize fields and permanent grasslands were more fragmented and dispersed in site A, whence the greater heterogeneity observed in Fig. 16.

Conclusion

This case study shows that the articulation of agronomic and ecological approaches is productive in analysing the way in which a landscape is organized. It also allows us to understand why the redundancy levels remain relatively low. Farmers must grow certain crops in particular proportions, even if the farm is affected by constraints. They must achieve an equilibrium between needs and constraints. Size is always a constraint, for example, but the perception of the size of a field varies as a function of the size of other fields in the farm.

The agronomic approach comprises an aspatial phase, centred on the farms, their diversity, and their functioning. This phase allows us to explain the diversity of landscape elements. Then there is a spatial phase to explain the location of these elements. This process supposes a knowledge of agriculture and the use of concepts outside the framework of ecology (Baudry, 1997).

5.   GENERAL APPROACH OF DYNAMICS AND ORGANIZATION OF AGRARIAN LANDSCAPES

The case studies presented clarify the concepts of organization in space and in relation to systems of activities. Although the structure of landscapes may change, it always does so within the framework of a given physical and socio-technological environment. This environment determines, at a given moment, the types of elements present and their relationships in space. The organization of a rural, agrarian landscape varies as a function of the evolution of user systems. It is also perceived differently according to the spatial scale of analysis.

There are some keys here to link the several studies on landscape evolution that are most often centred around changes in structure rather than the mechanisms of change. Sometimes, the mechanisms are taken on at coarser scales than the changes. There are exceptions to this rule.

In France, the landscapes of the middle mountains have been the focus of many studies on the abandonment of land (INRA et al., 1977; Bazin et al., 1983; Auricoste et al., 1983; Hubert, 1991).

The studies of Deffontaines and colleagues in the Vosges (INRA et al., 1977) provide an example of the spatial arrangement of modules of land use on a slope that is particularly easy to read in the landscape. In chapter 2 (Fig. 6), there is an illustration of this sector. The basic module of a landscape is made up of mown grasslands, below farm buildings. These

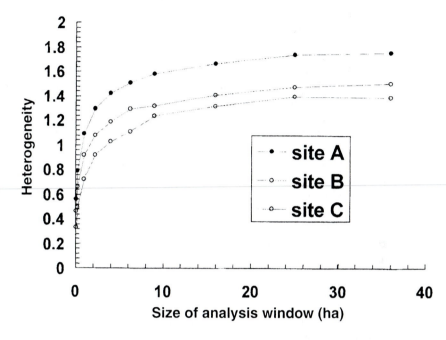

Fig. 16. Comparison of overall heterogeneity of the three sites (Morvan, 1996)

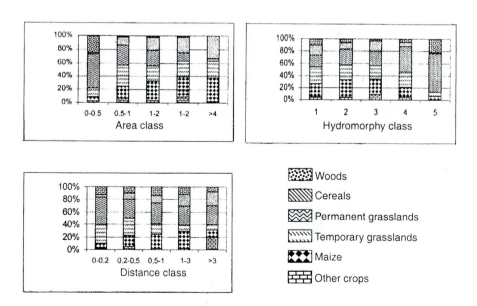

Fig. 17. Overall relationships between use of fields and their physical and spatial characteristics

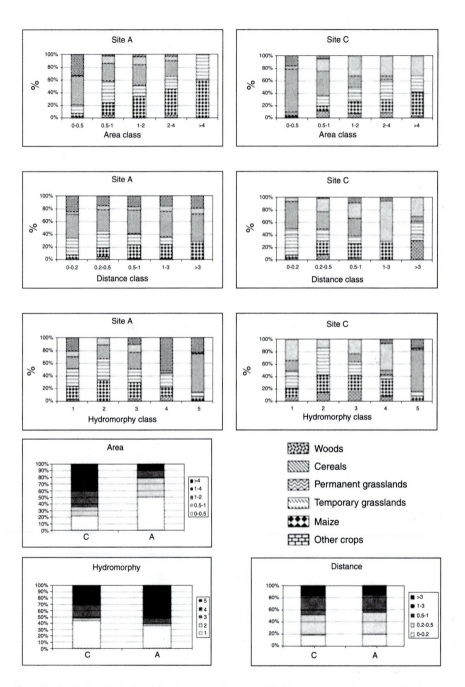

Fig. 18. Analysis of relationships between the use of fields and their characteristics in two different bocage sites. A: fragmented and scattered fields. C: large fields.

grasslands are irrigated by the liquid manure from the stables. Above the buildings there are cultivated fields and, further above, the pastures. Since the farms are aligned along roads perpendicular to the slope, this module is repeated all along the mountain side. At present, the irrigation canals of the mown grasslands are no longer maintained and the crops have disappeared.

Bazin et al. (1983), in the Dome, Massif Central, proposed a historical approach to the evolution of land cover. The relief is greater than in the Vosges and, consequently, the climatic variations are greater on the terroir (land used by a community). Until the end of the 19th century, each village and each community had to produce cereal crops. These crops were raised on land that was under a system of collective rotation. This made available a pasture outside the period of cultivation (of rye, most often). When the fields are fallow between two crops, they benefit from the nightly penning of sheep, which contribute manure after feeding on the grazing lands. This system led to significant transfer of minerals from pastures to the ploughed fields, thus reorganizing the distribution of fertilizers in the landscape. The ploughed fields received around 92 kg of nitrogen, 12 kg of phosphorus, and 68 kg of potassium per hectare per year. The system was common in many regions and countries (Balent and Barrue-Pastor, 1986) and significantly modified the flora of pasture zones, which could go as far as highly oligotrophic forms (Auricoste et al., 1983). These practices were the object of very strict social regulations assigning each household a certain number of nights.

World War I brought about great social transformations in Europe. The village economy was opened up to the region and then to the whole country. The shift to a market economy led to a profound reorganization of landscapes. It was possible to buy wheat in the valley, where the production was better, and to produce grass in the mountains for the purpose of milk and cheese production. The land cover no longer depended on primary needs but on what could be sold in the best conditions (Bazin et al., 1983). Bazin et al. add that, taking into account the great diversity of local situations, the farmers chose, among the possible systems of cultivation and livestock farming, those that were appropriate to the size of their holdings, to the capacities of their soil, and to the technologies that they possessed or could acquire. Integration in the market economy thus corresponded to a diversification of landscapes. It is regrettable that the authors did not make a detailed analysis of the process of landscape transformation, but then that was not their objective.

The Mediterranean mountain region soon became the site of considerable agricultural abandonment. This process was analysed by Bazin, as reported by Hubert (1991). Here also, inclusion in the national market economy caused abandonment of the cultivation of cereals, which became available at low prices because they could be easily transported from the plains, where they were grown. This led to concentration and specialization of land cover. Agriculture was carried out on the most fertile land, while the forest overran the sloped lands that were sensitive to erosion. This resulted in the present landscapes in which a highly productive, high value added agriculture occupies the irrigated valleys, while extensive livestock farming is carried out on those slopes that have not been abandoned.

Fernandez Ales et al. (1992) also indicated this phenomenon of concentration in southern Spain, especially in the region of the Donana dunes. Generally, between 1956 and 1977, the landscape changed from a highly heterogeneous mosaic comprising herbaceous crops (cereals) and ligneous cultivation (olive) to a landscape in which the former were concentrated on the most fertile soils and the latter on the poor soils. An ancient organization based on polyculture was replaced by a new organization based on intensification, with larger farms and greater sensitivity to factors of the physical environment.

In an entirely different context, in Ohio (USA), Simpson et al. (1994) also observed a differentiation in the evolution of two neighbouring zones with different geomorphological conditions. The fertile plain was the site of an increasingly intensive agriculture, while the morainic hills were either covered by young woods or built on.

In Portugal, the studies of Pinto Correia (1993) in a municipality also indicated rapid and significant changes. The traditional agro-pastoral system of the montado (32% of the area in 1958) nearly disappeared with the beginning of the European Community. A large part became fallow (28% of the area of the municipality in 1990). There was also an increase in the area planted with eucalyptus.

6.   LANDSCAPE DYNAMICS AND (RE)ORGANIZATION: MULTI-SCALE AND MULTIDISCIPLINARY APPROACH

Local and synchronous approaches to landscape organization can be used to make a refined analysis of the processes and relations between land cover and the physical environment. The examples of dynamics involve regional phenomena, covering a large number of farms, and differentiations appear at the medium and coarse scale. Phenomena of spatial autocorrelation can be recognized within the geomorphological entities. Spatially continuous factors such as relief and geological substrate determine the land cover. This form of organization is emerging in the landscapes of Ottawa and the Pays d'Auge. The size of the latter highlights the essential weight of technological or socio-economic differentiations.

These socio-economic factors and technical differentiations are also present in larger areas, but they are of another nature and are determined by other factors. Market forces and regional politics also play a role. Moreover, all the dynamics, or nearly all, presented in the preceding section are linked to the integration of a local economy with an autarkic tendency into an encompassing market economy. The example of the Dome mountains is particularly illuminating. It illustrates what Allen and Starr (1982) call a collapse of the system because of over-connection with the more inclusive levels (Baudry and Bunce, 1991). These more inclusive levels lead to a modification of local interactions between systems of agricultural production and the physical environment and field divisions.

Breuning-Madsen et al. (1990) provide a particularly detailed example of these multi-scale processes in their analysis of risks of abandonment of land in Denmark. At the regional level, excess water or slope are the chief factors of abandonment. At the infra-regional level, on sandy soils of western Jutland, the authors note that farmers react slowly to the continuous drop in cereal prices. In an arid zone, farming is made possible by the availability of irrigation water. In a wet zone, it is made possible by drainage. Farming may be discontinued, especially in peat soils, if the farmers are not authorized to replace drainage networks that no longer function. Thus, abandonment of cultivation is caused by physical, economic, and regulatory factors.

This complexity of factors and their interactions are now widely acknowledged. In a situation of agricultural overproduction, the availability of land for environmental uses becomes a stake in development. This is often expressed by a desire to maintain agricultural activity to manage landscapes or habitats for fauna and flora. The total abandonment of agriculture must be avoided in such a situation, apart from the fact that it has social consequences (loss of employment) that are equally harmful. How can abandonment be avoided? In their studies on this subject, Baldock et al. (1996) defined marginalization (fragilization, risk of abandonment) as being a process steered by a combination of social, economic, political, and environmental factors, which render agricultural activity unviable in the existing conditions of land use. They subsequently defined the conditions of fragilization at various scales.

On the regional scale: In the European context, a region may be at risk in physical and socio-economic terms. Agriculture is no longer globally competitive because of the poor yield or distance of markets.

On the infra-regional scale: Some types of uses may become marginal because of socio-economic changes, as has happened in the Massif Central and the Mediterranean mountain region.

On the individual farm scale: Small size, absence of machinery, and unfavourable land division are causes of marginalization. In this case, the land may be taken over by other farmers.

Within a farm: Some fields may have conditions of humidity, slope, size, and distance that are incompatible with the pursuit of agricultural activity.

The landscape is at a given moment a fluid, transitory product of a hierarchy of factors of organization, from international agreements on commerce to the location of agro-food industries, local non-agricultural employment, and local land transactions. At the same time, it has perennial structures that make up its cultural identity.

Information circulates constantly between these levels of organization. Reactions at refined scales are more or less directly linked to more global evolutions. Inversely, the accumulation of local changes affects decisions at higher levels. There is no model that allows us to link these levels within a perspective of analysis of evolutions of the landscape from an ecological point of view. The links between land cover, systems of production, and mode of maintaining marginal zones of farms must be explored.

Some factors such as the socio-technological history of regions are still insufficiently studied (Laurent and Bowler, 1997). It seems nevertheless that this history must be an important reason for the continuing diversity of modes of production, as for example the acceptance of maize in place of permanent grasslands in livestock farming, which contrasts Normandy and Brittany.

To predict the evolution of landscapes is a hazardous enterprise, even if it is necessary to simulate the repercussions of policy decisions affecting agriculture and land use. In the framework of a paradigm combining physical, socio-economic, technological, and biological environments, the advantage of simple approaches based on agronomic potentialities of land is limited (Flaherty and Smit, 1982). The concept of potential of production, however, can be useful in the protection of land resources in development operations, especially those linked to urbanization and infrastructure. Giles and Koeln (1983) and Rabbinge and van Ittersum (1994) propose a new, useful distinction between different approaches to land use: (1) a descriptive and comparative approach, (2) an exploratory approach, and (3) planning studies. They emphasize that studies attempting to identify and explore technological possibilities and limitations are generally narrow and require only a biophysical knowledge, whereas exploratory studies having socio-economic objectives and constraints are included in a wide interdisciplinary approach.

The evolution of the European common agricultural policy has given rise to many studies in the possible evolution of landscapes. Land use planning and the construction of scenarios are fields of intense research activity. The ecological consequences of scenarios are rarely addressed, a notable exception being the studies of Harms et al. (1992) in Holland.

Schoute et al. (1995) focus on scenarios in the rural environment. In that work, Schoonenboom (1995) clarifies that scenarios are different from predictions. Predictions aim to describe the most probable future, while scenarios aim to explore possible futures or the feasibility of desirable futures. He thus emphasizes that a work such as *Ground for Choice* (Rabbinge et al.,

1994), which proposes various land use scenarios within the European Union, deliberately ignores the national and European "inertia" of agriculture. This "inertia" has historical and cultural causes.

According to Schoonenboom (1995), scenarios comprise three elements: (1) a description of the present situation, (2) a certain number of alternative futures, and (3) the possible routes between the present and these images of the future.

Scenarios are analysed by means of simulation tools, spatially explicit in the case of landscapes, that make it possible to see the effects of particular trends. Farjon et al. (1995) present the future of potentially natural zones (unused by humans) in the Netherlands by using only biophysical variables. The result is a map of the entire country. Scenarios of landscape evolution can also lead to illustrations visualizing the result (Harms et al., 1995).

Scenarios involving agricultural landscapes on the refined scale, that is, taking into account mechanisms within farms, remain rare. The studies of Thenail et al. (1997) rely more on expert knowledge than on an actual simulation. Works in progress (Baudry et al., in press) involve the implementation of simulators on the field scale, allowing us to test the effect of changes in production systems or redevelopment of parcels of land.

Normative approaches, which attempt to propose ideal landscapes, are outside the scope of this field of research, and social and economic stakes are excluded. The planner relies on scientific knowledge in organizing the landscape. This type of process has been experimented with in various agro-environmental measures based on land management contracts including specifications for technical charges. Their ecological efficiency is difficult to evaluate (Primdahl and Hansen, 1993) and their inclusion in production systems is often problematic (Perrichon, 1994).

The protection of water and soil resources and biodiversity and the search for sustainable modes of development have led to a new field of research in the relationships between socio-technological systems and ecological functions (Berkes and Folke, 1998). In designating space as an essential factor of these ecological functions, landscape ecology leads to a questioning of the organization of landscapes, their establishment, and their evolution. This implies an analysis of spatialization of human activities. (What are we doing, and where? Who is doing what, and where?) It also implies an analysis of the spatial constraints on these activities. (Why must one carry out this activity here?) It is a field that is still to be developed, even though geographers have several answers to the first type of question. Agronomists are increasingly exploring this field (Deffontaines and Lardon, 1994; Deffontaines, 1996; Thenail, 1996) in order to answer the second type of question. Agricultural and rural landscapes are not merely scenes of production with particular ecological functions. They are also spaces with a large historical and cultural dimension. The input of archaeologists, historians (Berglund, 1991), and anthropologists (Collectif, 1995) is also essential in understanding landscapes.

Part III

Ecological Processes within Landscapes

Introduction

During the last two decades, landscape ecology has developed three major themes of research in the general context of environmental studies and issues at the international level. The first is population dynamics in a fragmented environment, a theme initiated principally in the United States in response to forest fragmentation and the associated risks of species extinction (Burgess and Sharpe, 1981). The second theme, especially since the Rio conference in 1992, involves the maintenance of biodiversity at the landscape level, whether the landscape has been anthropized severely and from ancient times, as in Western Europe, or more recently subjected to the pressure of human activities, as in the Amazon forest (Huston, 1995). The third theme is the control of water and nutrient flows in landscapes. It is developed most in the anthropized landscapes, where the maintenance of water quality as well as availability is an increasingly acute problem (Naiman, 1996).

6

The Functioning of Populations at the Landscape Level

The landscape is essentially heterogeneous and dynamic, as we have seen in the introductory part of this work. Whether under the influence of human activities or natural disturbances, it is a mosaic of habitats, and a large number of species use several elements of this mosaic during their life cycle. The complexity and heterogeneity of structures and their dynamics are often recognized but they have been simplified by empirical ecologists or theoreticians. Until the 1970s, researchers restricted their domain of investigation to zones that were homogeneous or considered to be so (Lefeuvre and Barnaud, 1988). Since then, the development of patch theory (Pickett and White, 1985; Chapter 4 of this book) has offered theoreticians a simplified view of heterogeneity defining space as a set of patches arranged in an ecologically neutral matrix (Wiens, 1995). Such simplification of spatial heterogeneity has led to many developments in the ecology of fragmented populations, based mostly on models of metapopulations (Gilpin and Hanski, 1991; Hanski and Gilpin, 1997). In this context, the specialist species, that is, those using only one type of landscape element, have been abundantly studied because they are sensitive to fragmentation (Farina, 1998).

The gradual complication of this spatially explicit model, by recognition of the different nature of patches and the presence of corridors (Fahrig and Merriam, 1985) and consideration of the permeability of landscapes (Verboom and van Apeldoorn, 1990), has brought researchers to consider spatial heterogeneity in its continuity and its complexity (Fig. 1). The movements between landscape elements and exchanges between ecological systems are therefore considered key processes in our comprehension of ecological processes, at the individual as well as the population level (Wiens, 1997).

Multi-habitat species use different types of landscape elements to complete their life cycle. These elements together form functional elements whose spatial arrangement determines the conditions of survival of populations.

All these relationships between populations and landscapes are dynamic. In this chapter, we discuss the frequent asynchrony between change in

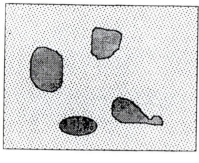 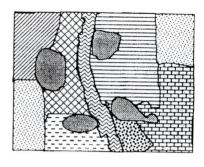

A   B

Fig. 1. Complication of the spatial representation. (A) Patches of elements of similar nature but varying size, form, and isolation are inserted in a neutral and uniform matrix. (B) The landscape is represented as a mosaic of elements of different nature and a set of networks of linear elements.

landscapes and population dynamics. Finally, we present the modelling process, which is essential to an understanding of processes in complex systems.

1.   PATCH THEORY AND FUNCTIONING OF METAPOPULATIONS

The first studies on fragmentation of forest habitats referred to the island biogeography theory. The size of wooded islands is said to be a good indicator of the species richness of birds (Burgess and Sharpe, 1981; Harris, 1984; Genard and Lescourret, 1985; Freemark and Merriam, 1986; Lauga and Joachim, 1992). First of all, the model considers the number of species present but not their density. Moreover, the effect of the cover in itself has rarely been isolated from the internal heterogeneity of the island, which increases with the area. Even though references to the island biogeography theory have declined significantly since the 1980s, the concept of metapopulation enunciated by Levins (1970) has served as a basis for many theoretical and empirical studies on the effects of habitat fragmentation on populations (Gilpin and Hanski, 1991; Hanski and Gilpin, 1997).

1.1. The concept of metapopulation

1.1.1. Definitions

Levins, in 1970, introduced the concept of metapopulation as a population made up of local populations that die out and are recolonized locally.

A metapopulation can survive in a region only if the average rate of extinction is less than the rate of migration. The local populations are established in particular habitat patches, which could be occupied or vacant at any given time. Some dispersing individuals may leave a patch to go and colonize an empty element or reinforce a small population. The populations established in a patch may disappear following environmental accidents (fire, felling) or demographic accidents (epidemic, ageing). Levins' model (Fig. 2) is very simple. It gives each patch the same value as a source of dispersing individuals and the same probability of extinction, and it considers the probabilities of successful dispersal identical in all the patches.

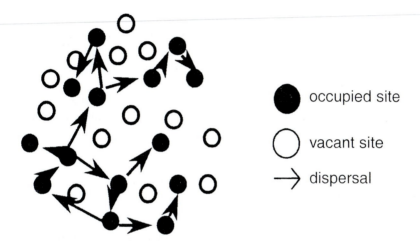

occupied site

vacant site

→ dispersal

Fig. 2. The metapopulation model of Levins

Levins individualizes the dynamics of each local population and that of the set of local populations, which he describes by means of a variable p(t), which is the fraction of patches occupied at a time t. He quantifies the processes linked to individuals and to local populations by two parameters, e and m, which represent respectively the rates of local extinction and colonization in empty patches. The equation of metapopulation dynamics is thus:

$$dp/dt = mp(1 - p) - ep$$

Although the model was inapplicable to field data, it was a great innovation in population ecology and served as the basis of theoretical and empirical research in this domain.

The three major components of the dynamics of metapopulations are: (1) processes of local extinction, (2) movements between patches, and (3) processes of colonization. All three are dependent on the landscape structure and dynamics. From this premise, Hanski and Gilpin (1991) wrote: "the fusion of metapopulation studies and landscape ecology should make for an exciting scientific synthesis." During the past few years, worthwhile studies have linked these two disciplines and have led to applications in development projects and conservation biology (see Chapter 9).

The terminology used in metapopulation studies is particular and will be used repeatedly in the rest of this work. Some definitions, on the basis of Hanski and Simberloff's studies (1997), are presented below:

—*patch* (or habitat patch, island, site, landscape element): a continuous space in which a local population finds all the resources needed for its survival and that is separated from other favourable patches by an unfavourable space

—*local population* (or population, sub-population, deme): a set of individuals that live in the same patch and are thus in interaction

—*metapopulation* (or multipartite population, species assemblage): a set of local populations among which the migration of a local population towards at least a certain number of others is possible

—*turnover* (or ratio of colonization to extinction): extinction of local populations and establishment of new local populations in vacant habitats by individuals migrating from existing local populations

—*persistence of a metapopulation* (or survival time of a metapopulation): duration needed for the extinction of all the local populations

1.1.2. Conceptual models

Levins' model served as a base for the elaboration of various models in which the function of different patches is variable: as source, habitat, or accessibility. These models, even though they are mostly theoretical, aim to represent the different scenarios observed during their field studies. Ouborg (1993) showed, for example, in studying a series of data over a long term, that certain vascular plants of the Rhine watershed are organized into metapopulations of the Levins type, but the spatial isolation of patches has an effect on the processes of colonization.

1.1.2.1. The Boorman and Levitt model

The Boorman and Levitt model is made up of a large central fragment supporting a permanent population that feeds many peripheral sub-populations (Boorman and Levitt, 1973) (Fig. 3). This model is similar to the island biogeography theory in that it supposes that a landscape element functions as a permanent source of dispersers, just as a continent serves as a source for islands in the ocean. Murphy et al. (1990) showed that the butterfly *Euphydryas bayensis* forms a metapopulation of the Boorman and Levitt type for a set of patches located in a radius of 7 km around the source of individual dispersers.

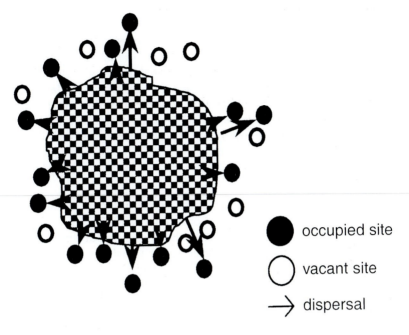

Fig. 3. The metapopulation model of Boorman and Levitt

1.1.2.2. The source-sink model

The source-sink model is that of a metapopulation in which the growth rate, at a low density and in the absence of immigration, is negative for some patches, called sinks, and positive for other patches, called sources (Pulliam, 1988). The different sub-populations generally occupy habitats of different quality: the rates of increase are greater than 1 in the source population and less than 1 in the sink population, so that the survival of populations in the sinks requires a constant restocking from nearby sources. Blondel et al. (1992) give an example of such a system in the blue tits (*Parus caeruleus*) in the mosaics of Mediterranean habitats.

1.1.2.3. Metapopulations in a state of non-equilibrium

In a metapopulation in a state of non-equilibrium, the rates of extinction over the long term are greater than the rates of colonization, or the contrary. In extreme cases, the local populations are so far from one another that there cannot be migration among them and thus no possibility of recolonization. An example of this is the series of extinctions that took place among mammals during the climatic changes of the post-Pleistocene, which led to an isolation of mountain peaks (Brown, 1971). Many other examples exist in habitats fragmented by humans, for example, salamanders (Welsh, 1990) or woodpeckers (Stangel et al., 1992) in the remaining groves of mature forests in the United States.

1.2. Metapopulations and landscape

The landscape structure affects several components of the structure of metapopulations: the size, form, and quality of patches determine the collection capacity of each patch and are thus linked to the probabilities of extinction. The nature of the ecotone, the heterogeneity of the land between the patches, and the nature of the neighbourhood influence the intensity and nature of individual movements between the patches, and thus partly determine the processes of immigration. These two processes are dependent on the species considered.

1.2.1. Patch size

The process of extinction is linked partly to the internal dynamics of the sub-population, to its demographic performance. The size of the fragment determines the size of the population. As the population declines, it becomes more vulnerable to demographic stochasticity (Hastings and Wolin, 1989; Hanski, 1989). A three-years study of the presence of the nuthatch (*Sitta cothraustes*) in all the groves of a rural landscape in the Netherlands showed that the frequency of local extinction was strongly correlated to the size of groves (Verboom et al., 1991).

Large fragments allow the survival of large populations (Hastings and Wolin, 1989; van Dorp and Opdam, 1987). The size of fragments may be much more important for the persistence of a metapopulation than are overall rates of extinction and recolonization (Harrison, 1991).

Turchin (1986) showed that the emigration rate of a coleopteran pest on bean was inversely proportionate to the area of the source fragment. The shape of a fragment, which can be defined by the ratio of the perimeter over the area, also influences the rate of immigration. Between two fragments of the same area, the one with the longer perimeter and less compact shape will have the greatest proportion of individuals reaching the edge and thus susceptible of leaving the fragment (Kareiva, 1985; Stamps et al., 1987).

1.2.2. Isolation of patches

The distance between patches determines the probability of arrival of colonizers in the patches. Fahrig and Merriam (1985) showed a reduction in the growth rates of the most isolated local populations. A simulation model of a metapopulation of nuthatch was used to indicate the importance of isolation and proximity on the survival time of the metapopulation (Opdam et al., 1993). When the patches are relatively numerous and close together, the probability of extinction decreases, because the probability of the arrival of immigrants in each grove increases. The survival time of the metapopulation increases considerably when a large forest that is considered a permanent source of individuals and not subject to extinction is integrated in a landscape (Fig. 4) (Verboom et al., 1991 in Opdam et al., 1993). The probability of presence in a grove increases from 76% at a distance of 1 km from the source to 64% at 18 km from the source. Field results obtained in landscapes studied by a team from Holland confirmed this effect of distance for squirrels (Verboom and van Apeldoorn, 1990), birds in mature forests (Opdam et al., 1985), and russet voles (van Apeldoorn et al., 1992).

The use of relic groves disseminated in an agricultural matrix in New Zealand by kiwis (*Apteryx australis*) is explained by the size of the patches as well as by their isolation (Potter, 1990 in Wiens, 1997). Kiwis do not fly. They must walk between isolated fragments. All the fragments located at less than 80 m from the forest fragments, no matter what their size, were occupied by the birds. The movements over a distance greater than 1 km occurred gradually and the birds used small groves as relays. In this case, the spatial arrangement of habitat patches is a critical factor that determines the effect of the isolation and consequently the dynamics of the metapopulation.

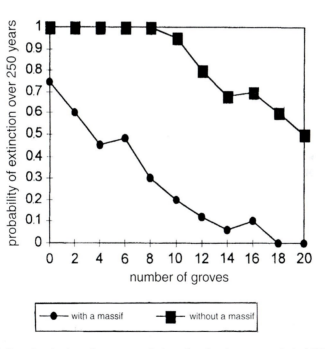

Fig. 4. Probability of extinction of a metapopulation of nuthatch over a period of 250 years. The graph represents the probability of extinction as a function of the number of groves present in a given area. The two lines represent a situation without a large forest (the number of groves varies from 0 to 20) and a situation in which a forest comprising at most 20 territories of nuthatches is added (Opdam et al., 1993).

1.2.3. Ecotones and their configuration

The edges of patches play a role in the movement of disperser individuals. The edges are more or less permeable depending on their immediate environment, their structure, and the species considered. They can be perceived as discontinuities or as marked gradients. A large ecotone may be perceived as a transition zone by a very mobile organism that crosses in a short time, while another less mobile organism may perceive it as a patch with narrow borders. Relatively sedentary organisms perceive differences between patches as more marked ecotones than do the more mobile organisms because they encounter few ecotones in their travels (Fig. 5).

When an individual reaches a permeable ecotone, it can choose whether or not to cross it. If it has a marked preference for one of the habitats bordering the ecotone, it will probably cross easily from the less favourable to the more favourable habitat. This response has two consequences: when

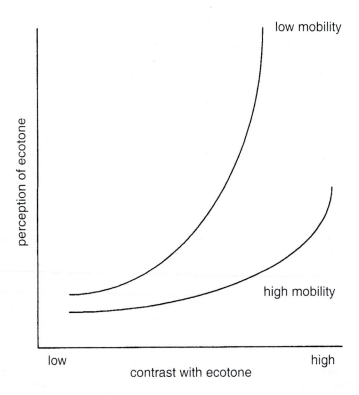

Fig. 5. Hypothesis on the perception of ecotones by organisms, as a function of the contrast between the habitats. When the contrast increases, all the organisms perceive the ecotone more clearly, and, for a similar degree of contrast, perception of the ecotone is greater in less mobile species than in the highly mobile species (Wiens, 1992).

the preference for one habitat increases, directed movements increase, and individuals gradually aggregate in the better patches. For example, the immigration rates of the butterfly *Melitaea cinxia* increase with the proportion of favourable habitat patches bordered by open field zones (Kuussaari et al., 1996).

1.2.4. The role of corridors

Corridors, which are linear landscape elements, play a particular role in the flows on the landscape level. They can conduct the flows (as corridors) or hamper them (as filters), or even prevent them (as barriers). In the framework of metapopulation studies, corridors have been mostly studied for their role as conduit, facilitating the passage of disperser individuals from one patch to another.

This idea was developed initially by the team of H.G. Merriam in Canada (Fahrig and Merriam, 1985). Their studies were done on small forest mammals *Peromyscus leucopus* and *Tamia striatus* in the agricultural landscapes of the Ottawa region (Figs. 6 and 7). It is a sector with a relief resulting from the alternation of a marine plain and morainic deposits with chalky rock boundaries. These lands were cleared about 200 years ago. There are remnant groves of the deciduous forests on rocky outcrops, and a network of spontaneous or relic hedgerows borders the cultivated fields

Fig. 6. Aerial view of part of the Ottawa region, Canada

Fig. 7. *Tamia striatus*

(Baudry, 1985). The two species of small mammals live in the groves during the summer. The winters are severe (average temperature in January is −10.9°C), and often there are only females in the groves at the end of winter (Middleton and Merriam, 1981; Henderson et al., 1985). The local extinctions during this period are frequent. The vacant groves are recolonized in spring, from groves that remain occupied in which individuals have found refuge in the farm buildings.

Behavioural studies have shown that these two species move around farm land, preferentially along the boundaries of the hedged fields, which offer them protection from predators (Merriam, 1989). The processes of recolonization of vacant habitats are thus favoured by the presence of hedgerows that serve as corridors allowing individuals to cross the farm land more easily when going from one wood to another. Merriam (1984) defined connectivity as the set of characteristics of a landscape that facilitate movement between patches.

A series of simulations made on this model demonstrated the importance of connectivity for the survival of the metapopulation on the landscape level (Fahrig and Merriam, 1985) (Fig. 8). Without an exchange of individuals between groves, local extinctions are more frequent. The increase in the

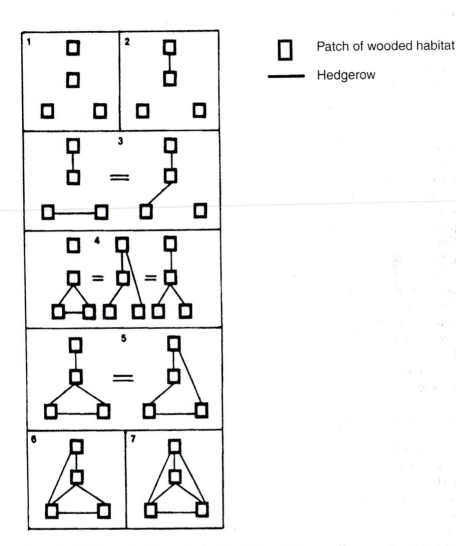

Patch of wooded habitat

Hedgerow

Fig. 8. Increase in survival time of a metapopulation with four patches, as a function of the increase in the number of corridors, determined according to the model of Fahrig and Merriam (1985). The survival time increases from the top of the diagram to the bottom (Merriam, 1991).

number of corridors between the patches increases the survival time of the metapopulation.

Verboom and van Apeldoorn (1990) showed in the Netherlands that the presence of squirrels in a grove was positively correlated with the length of hedgerows nearby. Carabid coleopterans of the forest use the hedgerows of bocage landscapes to move from one local population to another (Petit and Burel, 1998).

1.2.5. *Gene flows*

The flows of individuals associated with the functioning of metapopulations generate genetic variability at the landscape level, which can be used as an indirect means to understand the functioning and delimitation of metapopulations. The subdivisions and dispersal have an impact on the dynamics of local populations and on the genetic diversity. Genetic and ecological approaches can be integrated to enable an interpretation of the structure of natural populations. It is no doubt unrealistic to expect that the turnover of populations or gene flows can be determined from genetic data, but frequency data can also be used to identify the appropriate scale at which to study the metapopulations (Hastings and Harrison, 1994).

Small, isolated populations of russet voles compared among themselves diverge, while no difference is observed between strongly connected populations (Bauchau and Le Boulange, 1991). Kozakiewicz (1993) put forth the hypothesis that in a metapopulation, divergence between local populations observed in the spring disappear in summer under the effect of transfer of individuals. Fluctuations of divergence from one year to another are observed (Bauchau and Le Boulange, 1991). Corbet (1975) remarked that a population of russet voles newly established in a recently formed habitat (a plantation) diverge greatly from the surrounding population. After several years, the two populations become similar. According to this author, the new habitat is colonized by the founding individuals, then, as the vegetation develops, the exchanges between metapopulations increase, making them uniform again.

In Canada, Merriam et al. (1989) sought to prove the existence of frontiers of metapopulations of *Peromyscus leucopus*. From the analysis of salivary amylase polymorphism, the authors concluded the non-existence of divergence between populations on the regional scale. Even a very slight flow between populations suffices to homogenize them. Merriam (1991) proposed a schematic view of metapopulations illustrating the types of flows between the populations (Fig. 9). The corridors affect the demographic and genetic processes, not only at critical thresholds of fragmentation, but at all levels of heterogeneity. The scale and relevance of various landscape elements depend on the species considered. For example, the absence of wooded cover on a few tens of metres may constitute a total barrier to the crossing of individuals of one species (Mader, 1988), while for other species only rivers or highways are real limits to the crossing of individuals (Forman and Godron, 1986).

2.   MULTI-HABITAT SPECIES

Multi-habitat species use several types of landscape elements in the course of their life cycle. The changes in habitat can be daily for species such as the

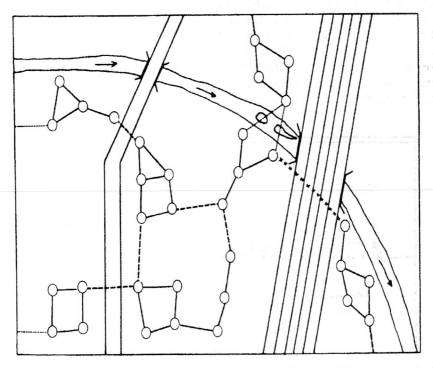

Fig. 9. Various flows that may exist within metapopulations. The scale of elements is adapted to the species considered. For example, a river is perceived as a barrier by some species while it is easily crossed by others. Flows supplementing the numbers of local populations are indicated: the broken line indicates recolonization of local extinction, the dotted line indicates gene flow, and the line of x's indicates complete isolation (Merriam, 1991).

squirrel, which feed in cultivated fields and rest in a covert (Wetzel et al., 1975), or the wild rabbit (Papillon and Godron, 1997), or seasonal as a function of spatial variations in food resources or various phases of the annual life cycle.

Forman and Godron (1986) reviewed 10 animal species whose movements were studied on the landscape level. They defined the hospitality of various landscape elements, expressed by their use or avoidance by animals. Elements inhospitable to terrestrial species, such as lakes, wetlands, and cities, are generally avoided by species such as the skunk, fox, or wolf. Zones that are homogeneous on the large scale are also avoided, more because of their isolated character than their homogeneity.

Many species need several elements of different kinds that are close together, as with the skunk, whose vital area is about a square kilometre around its burrow and which feeds mainly on small mammals and other small animals of the hedgerows. In Illinois (USA), it digs its burrow in cultivated fields and prospects the surrounding land for food. The spatial distribution of burrows suggests an avoidance of large forest areas, marsh land, and large corn fields. The skunks preferentially use zones that are heterogeneous on the scale of their territory. Within their territories, they search for food particularly in the hedgerows and move around in the corn fields: the mature crops offer an effective cover against predators and the bare soil is convenient to move around in and offers abundant food (arthropods) (Fig. 10).

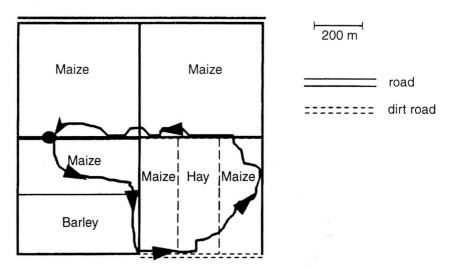

Fig. 10. Movement of a skunk during one night. The plain lines are the hedgerows, the broken lines indicate a change in cultivation within a single parcel, a double line represents a road, and a double broken line indicates a dirt road (B.J. Verts, *The Biology of the Striped Skunk*, cited in Forman and Godron, 1986).

Similarly, Papillon and Godron (1997) showed that the wild rabbit needs a set of resources available within a limited radius, and that its spatial distribution is linked to the simultaneous presence of several landscape elements in which it finds food and shelter and reproduces. The organization of these different landscape elements, on the scale of the rabbit population (280 ha), varies with the structure of the surrounding landscape: the perception of the space depends on what the authors call landscape families. These landscape families represent a particular combination of topographical, pedological, and climatic parameters and land cover on the scale of the department.

2.1. Daily movement between landscape elements

Daily data on the use of land by animals are scarce. On this time scale, only the mobile species are concerned with the heterogeneity of the landscape.

Among the mammals, the skunk, rabbit, fox, and boar use various landscape elements in the course of a day, to satisfy all their needs. Birds, during the nesting season, use areas near their nests to feed in.

The linnet nests in the moor and feeds on the resources there, as well as on resources in nearby agricultural areas. A population established in a moor of 7 ha in the middle of a hedged landscape was followed for 10 years, before and after a consolidation of land parcels (Eybert, 1985). This, in facilitating the access to parcels and in increasing their average size, led to a modification in the cultivation systems. Grasslands and fallow lands declined in favour of cultivated land. The food preferences of the linnet, however, did not change (Table 1), despite the smaller representation in frequency of prairies and fallow land around the moor. This means that the individuals had to prospect in a larger radius around the nest to meet their needs. Before land consolidation, the linnets found enough food to satisfy their needs in a radius between 200 and 500 m. After consolidation, most of the population exploited the grasslands and fallow lands at a distance of 900 to 1000 m away from the nest. This example clearly highlights that the linnet, which is inevitably found in this type of agrarian landscape, takes the landscape heterogeneity into account in its behaviour. The grain of heterogeneity perceived by a species depends on the characteristics of its life history, but with some plasticity depending on the organization of a landscape. Here, the scale of perception of land varied as a function of the landscape change.

$$D = (r - p)/(r + p - 2rp)$$

where r is the proportion of individuals in a feeding site, and p is the proportion of a feeding site out of the entire study zone; D varies from -1 to $+1$ and indicates a negative selection for values between -1 and 0 and a positive selection for values between 0 and $+1$. This index has the advantage of being insensitive to variations of relative abundance of different types of crops.

Mobile insects such as butterflies, bumblebees, and dragonflies can also use several elements of the landscape mosaic during their search for food (Taylor, 1997).

Table 1. Food preferences of the linnet (Eybert, 1985), calculated according to the index of D. de Ivlev, reviewed by Jacobs

	Other crops	Rapeseed	Permanent grasslands	Fallow
Before consolidation	-0.59	$+0.68$	$+0.24$	$+0.68$
After consolidation	-0.86	$+0.75$	$+0.47$	$+0.79$

2.2. Seasonal movements between landscape elements

Many species of invertebrates use various landscape elements during their annual cycle depending on the stages of their life cycle (larval development, hibernation, mating, egg-laying). Fry (1995) adopted a classification proposed by Duelli on the use of agrarian landscapes by insects. He distinguished the following:

1) specialist species, which spend their entire life cycle in a given habitat;

2) species that, from non-cultivated elements, exploit the surrounding crops;

3) species that hibernate in field boundaries and spend the rest of the year in the fields;

4) species that disperse over long distances during part of the year and use all landscape elements while they are moving.

Many species of carabid Coleoptera use cultivated land during the period of vegetation and hibernate in the field boundaries in the unploughed areas. The grassy zones serve as a refuge in the agro-ecosystems (Desender, 1982; Sotherton, 1985; Thomas, 1990). The plant structure of these uncultivated areas determines the abundance of individuals that hibernate there. In Hampshire (England), various species of predatory carabids are more numerous on planted embankments and at the foot of windbreaks than on grassy embankments and belts of grass. The combination of embankment, ditch, and field boundary is the most favourable habitat for these insects (Sotherton, 1985).

The empidid Diptera are part of the fourth group described by Fry. In this family, there is a large variety of species as a function of their biological characteristics: food regime, mating behaviour, or larval development. Nearly all the species with edaphic larvae develop in an undisturbed soil. The adults are either predators or flower-users. The adult empidids can be observed when they are in the sun or under the leaves of bushes. This phase corresponds to a hunting or resting phase.

During the reproductive period, several species of empidids execute original nuptial parades. *Hilara* spp. are easily observed above the water, where they swarm. The males hunt prey while flying above the water level, surround the prey, envelop it in silk, and offer it to females (Fig. 11). The empidids therefore require different habitats in a given landscape to complete their life cycle: (1) sites on undisturbed soil for larval development; (2) leaves of trees or shrubs for resting periods; (3) areas rich in flowers or insects for feeding on; and (4) sites favourable for the formation of mating swarms for reproduction (Fig. 12). These different habitats must be present within a given area in a landscape to allow individuals to complete their life cycle. This area varies with the dispersal capacities of various species (Morvan et al., 1994).

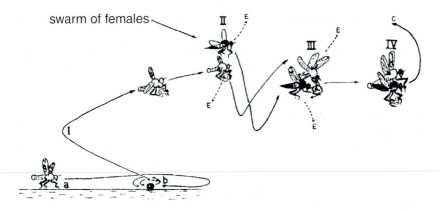

swarm of females

Fig. 11. Empidid of the genus *Hilara*: reproductive behaviour, hunting, spinning of a cocoon, offering, mating, and dispersal of females (drawing by Nathalie Morvan)

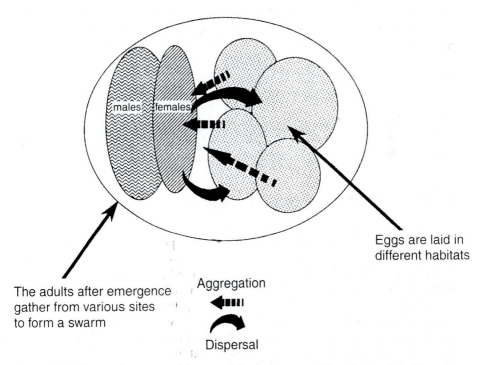

males   females

Eggs are laid in different habitats

The adults after emergence gather from various sites to form a swarm

Aggregation

Dispersal

Fig. 12. Landscape elements used by empidids of the genus *Hilara* during their life cycle (Morvan, 1996)

Some vertebrate species use different habitats at different seasons and return to the same places in the following year. This could be considered a complex form of migration between three, four, or more habitats (Forman, 1995).

The black bear (*Ursus americanus*), in the coastal plains of North Carolina (USA), avoids five types of landscape elements (residential areas, farm land, pine plantations, sparsely planted dunes, and lakes). It uses three key habitats. It feeds on sparsely wooded ridges at the end of autumn; it makes its den and finds a rich and abundant nourishment in spring in the dense, damp woods of the Carolina bay; and it hides from hunters and their dogs in the tidal zones (Landers et al., 1979) (Fig. 13). Some individuals can no doubt survive if one of the first two habitats disappears, but the zone of refuge against predation and disturbances is absolutely critical to the maintenance of the bear population.

2.3. Functional units

The landscape units used by a species during its life cycle define what is called the functional unit (Merriam, 1984). The functional unit may be made up of elements of the same nature for specialist species, fragmented or otherwise, including corridors. For multi-habitat species, it takes into account the heterogeneity of the landscape mosaic. The biological characteristics of the species define the grain of the landscape, and the heterogeneity of that landscape, the nature of elements that must be considered, needs to be understood.

The survival of individuals or populations depends on the integrity of the functional unit, measured by the connectivity, heterogeneity, composition, and spatial arrangement of the landscape mosaic.

3.   MOVEMENT IN LANDSCAPES

The two preceding sections emphasize the importance of movement of individuals for the survival of populations in heterogeneous landscapes. Whether it is movement of disperser individuals in the case of structures of metapopulations or daily or seasonal movements of individuals using several habitats, the movement and its interactions with the structure and dynamics of landscapes form a key process in the survival of populations in the heterogeneous environment. As we will see in the following chapters, it is also a central point in plans for development and land management.

In systems including mobile organisms, many ecological processes are determined by two parameters of individual behaviour: movement and the choice of the landscape element used (habitat). The intensity and trajectory of an organism's movement determine the probability with which it will

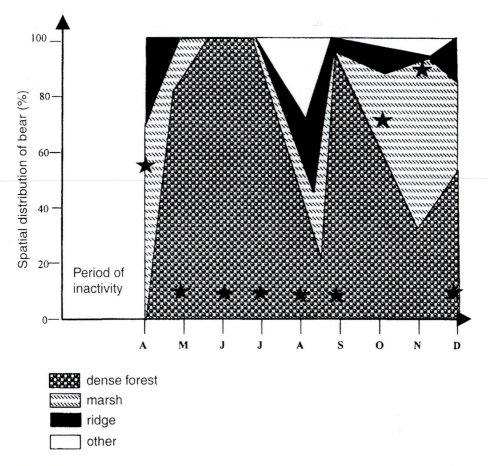

Fig. 13. Location of resting zones and feeding zones of the bear during the year. The graph represents the percentage of individuals tagged with a transmitter found in each habitat in the coastal plains of North Carolina (USA). "Other" indicates the five habitats avoided by the bear. The stars indicate the habitats in which the majority of food resources are consumed each month (> 33%).

move from one place to another in the mosaic during a given period. Trajectories of individual movements have been analysed with simple models of diffusion (Okubo, 1980), percolation networks (Gardner et al., 1989), or models of random movement that are correlated or not correlated (Turchin, 1991). The choice of the patch, or the habitat, expresses the response of an individual encountering an ecotone and thus determines the probability of its entering an adjacent patch.

These two processes are influenced by factors whose effects vary in space and in time (Fig. 14). For example, the movement may be more difficult

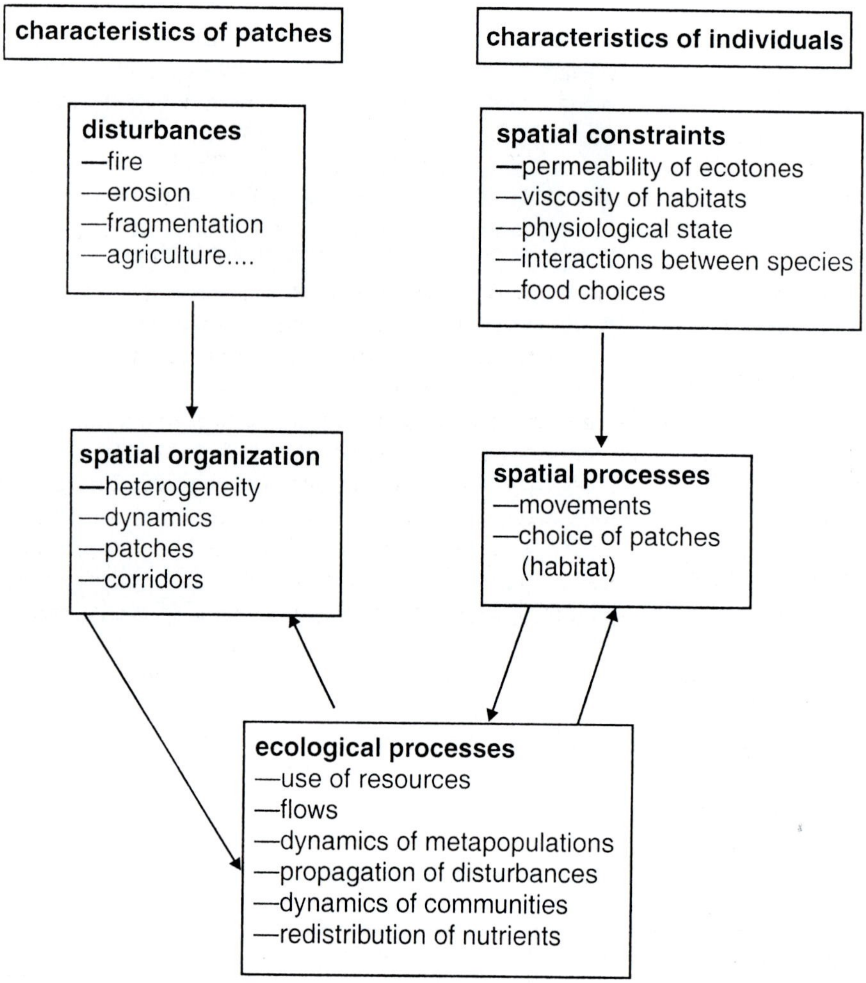

Fig. 14. A methodological framework for the study of spatial structures from the spatial processes (Wiens et al., 1993)

between some habitats, which is expressed as the viscosity or the rugosity of landscape elements, and the ecotones are more or less permeable depending on their structure and the nature of adjacent elements (Stamps et al., 1987; Wiens et al., 1985). Highly mobile organisms perceive the heterogeneity of spatial structures at different scales from those perceived by more sedentary individuals (Kolasa and Pickett, 1991). The movements thus differ according to the spatial structures and the organisms studied.

3.1. A hierarchical approach to movements

The movements of an individual are influenced by the structure of the landscape at several spatial levels, depending on its biological activities. For example, Senft et al. (1987) showed that the movements of large herbivores are influenced by spatial heterogeneity from the local level to the regional level, but for different kinds of activities (Table 2). At the local level, the choice of plant species or plant parts (buds, leaves) ingested depends on the feeding behaviour of the herbivores. At the patch level, the herbivores seek to maximize the efficiency of ingestion of nutrients either by selecting particularly rich elements that may be widely dispersed or by less selectively taking the forage that is available. The movements depend on the distribution of plant species in the patch, the size of the patch, obstructions that may exist on the refined scale (streams, rocks), and more or less specialized regimes of herbivores. On the landscape level, the animals choose patches on the basis of their food resources. The patch is chosen as a function of its richness (ratio between the percentage of total time spent on feeding and the percentage of the area of the vital domain), whether for domestic species (sheep, cows) or for a large number of wild species (wapitis, bison, kangaroo, Camargue horse) (Duncan, 1983). The response of animals is linear in relation to the proportion of landscape occupied by the favourable patches. At the regional level, animals use the different landscapes as a function of variations in availability of resources, in a manner that is predictable (transhumance) or unpredictable (nomadism).

At each hierarchical level considered, the movements are linked to the spatial structure of that level, as well as of adjacent levels (Kotliar and Wiens, 1990).

Table 2. Types of animal movements and their relation to spatial structures: a hierarchical approach (Ims, 1995)

Spatial scale	Type of movement	Spatial structure
Patch of resource	Selection of food	Distribution of food Size and form of patches Obstacles on the refined scale
Patch of habitat	Search for feeding areas, surveillance of territory	Configuration of patches of resources Shelter Topography and abiotic factors
Landscape mosaic	Dispersal	Size, form, isolation of patches Connectivity, permeability of landscape
Region	Migration	Geomorphology Barriers on the regional scale

3.2. Quantification of movement: intensity and nature

Individual movements, their intensity, and their form as a function of the spatio-temporal scale and the spatial structure have mostly been studied in invertebrates, particularly insects.

Simulations of correlated random movements have been used to analyse the movement of Coleoptera. Wiens and Milne (1989), Johnson et al. (1992), and Crist et al. (1992) compared the real movements of ants and tenebrionid coleopterans with the results of simulations of a correlated random march: the change in direction between two intervals of time was not totally random but linked to the direction of the preceding movement. The results showed that the real movements of insects generally differ from a correlated random movement of the first order. Most of the time, the real movements are longer and less complex than the simulated movements. The displacements of insects are linked to the structure of the vegetation and are dependent on the species studied.

In the bocage landscapes of northern Ille-et-Vilaine, Charrier et al. (1997) studied the movements of a carabid coleopteran, *Abax parallelepipedus*, for a summer. This species walks; local populations are established in groves or the intersections of hedgerow networks and the coleopterans disperse through the hedgerows (Petit and Burel, 1993). Individuals equipped with diodes linked to a copper antenna were observed every 48 hours by means of a harmonic radar emitting waves reflected by the diode (Mascanzoni and Wallin, 1986). The analysis of trajectories (Fig. 15) showed that changes of direction between two periods occurred at random and that the average distance travelled in two days was less than 50 cm. The fractal dimension of trajectories was strongly correlated to the structure of the vegetation in the habitats crossed. It decreased progressively from the groves to the least dense hedgerows, indicating that movements are more complex in the most favourable habitats than in the others.

These mechanistic approaches to movement constitute just one of the elements of a theoretical process of studying landscape-population interactions. They can be used to implement conceptual models (Petit and Burel, 1998) or mathematical models.

3.3. Connectivity or permeability of landscapes

Movement between patches of a landscape mosaic is analysed on the basis of the landscape composition, its configuration (spatial arrangement of landscape elements), and adaptation of the behaviour of organisms to these two variables. This is what defines the connectivity or permeability of landscapes (Merriam, 1984; Taylor et al., 1993). This notion is dependent on the species considered and is distinguished from connectedness (Baudry

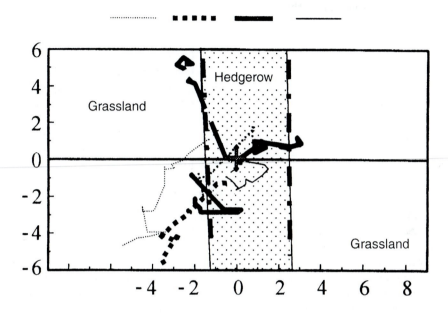

Fig. 15. Trajectories of individual movements of *Abax parallelepipedus* in a grove. Each individual is represented by a trajectory; all the individuals were released from the origin of the graph, 0, 0.

and Merriam, 1988), which considers only the spatial arrangement of the landscape elements.

Connectedness and connectivity may be similar, for example when we consider the specialist and barely mobile species that move only when there is a direct connection between elements of the same nature (Fig. 16).

3.3.1. *The spatial arrangement of patches and percolation theory*

The parameter of connectivity, or permeability, is vital in understanding the population dynamics in the landscapes. Fahrig and Merriam (1985), Hanski (1991), and Andren (1994) showed that the spatial organization of patches is the key factor of the dynamics of populations subdivided in the case of an intermediate fragmentation or a low recovery rate of favourable patches (less than 30%).

The effect of the spatial arrangement of fragments is often compared to the percolation charts (see introduction), which serve as a neutral model (Gardner et al., 1987) for evaluating the probability of passage of flows of individuals in the landscapes (Wiens et al., 1997). For example, around the

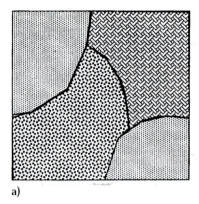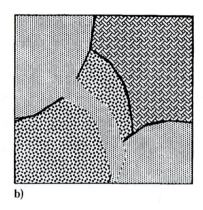

a)   b)

Fig. 16. Connectedness and connectivity for specialist species of habitat A. In landscape (a), two populations are isolated in two landscape fragments, and in landscape (b) the addition of a corridor of a similar nature links the two patches and thus ensures the exchanges of species. Here, the connectedness and connectivity are identical.

percolation threshold, the propagation of disturbances such as forest fires and epidemics is very rapid, while beyond that threshold it is difficult (Turner, 1987). In the case of individual movements, the percolation theory allows us to measure the probability of passage from one place to another, using the resource patches. This passage is possible only when the proportion of such patches is greater than the percolation threshold. Beyond the critical threshold, the habitat forms a continuous cluster from one end of the map to the other. An organism located on this cluster may cross or "percolate" in the landscape (O'Neill, 1988; With and Crist, 1995). These maps of spatial contagion of habitats are not sufficient to determine the connectivity of the landscape, because they are established *a priori*, without taking into account the preferences of species and their mobility. They are, however, used in ecology to determine the time at which a landscape becomes disconnected and thus to identify when the fragmentation of a landscape may have consequences for the population dynamics. They serve as a null model for understanding ecological processes in a heterogeneous environment, offering general models of habitat distribution.

By modelling the population dynamics in random landscapes (percolation maps), we can put forth hypotheses on the behaviour of these populations as a function of the landscape structure. With and Crist (1995) simulated the behaviour of species using different habitats and having varied dispersal capacities, on random maps. For specialist species with an average dispersal capacity (which cross at least 3% of the landscape studied), the connectivity of the landscape is less important than the abundance of the optimal habitat. For generalist species, connectivity varies as a function of the spatial arrangement of various habitats and the average distance of dispersal.

These simulated landscapes can be generated in a more or less complex fashion from a fractal distribution of pixels (With et al., 1997) or a hierarchization of random maps (Lavorel et al., 1993). They determine a theoretical framework that must be linked to field research or to experiments to be validated, because the real landscapes, as we have seen in the preceding chapters of this book, have a social and environmental organization that differentiates them from the random ones.

3.3.2. *Permeability of land between patches of favourable habitat (matrix)*

To go from one patch of favourable habitat to another, individuals often cross several elements of the landscape mosaic. Those elements are more or less favourable to movement, and their spatial organization determines the accessibility to patches. This has been illustrated by Taylor et al. (1993) in their models of complementation and supplementation of landscapes. These two notions were defined by Dunning et al. (1992): complementation is the use of different kinds of landscape elements that are necessary to meet the needs of a species, and supplementation is the use of fragments of similar nature, when each is too small to provide the resources needed by a population. According to Dunning et al. (1992), the structures of favourable landscapes for these two processes are a function of the distance between fragments, while Taylor et al. (1993) state that the nature of the surrounding matrix critically influences the exchanges between patches (Fig. 17).

Foxes (*Canis vulpes*) live in underground dens and are nocturnal, and they feed almost exclusively on small and medium-size vertebrates. Storm et al. (1976) studied foxes over several years in the rural landscapes of Minnesota, Iowa, and Illinois (USA). The landscapes were mosaics of farm fields, urban zones, lakes, and groves. The dens, located on the heights, are more than 200 m away from houses. In the autumn, the sub-adults and some adults leave their territory and disperse in all directions, over an average distance of 31 km for the males and 11 km for the females. The residential areas and highways are obstacles on the routes of dispersal and divert the individuals from their initial trajectory. Small or medium-size streams slow the advance of foxes but do not modify its principal direction, while rivers are barriers that cannot be crossed. The other landscape elements (groves, hedgerows, some crops) are perceived as homogeneous by foxes, whose movement is not altered by the crossing from one element of the mosaic to another.

The agricultural mosaics are perceived as heterogeneous by other species of vertebrates or invertebrates. *Abax parallelepipedus*, a carabid coleopteran, occasionally uses permanent grasslands, but its mortality rate is 100% if it

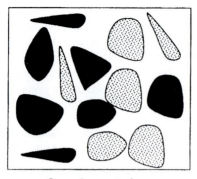   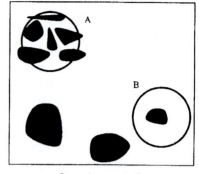

Complementation   Supplementation

Fig. 17. Processes of complementation and supplementation of landscape (Dunning et al., 1992). In processes of complementation, individuals use different kinds of elements in the landscape, and only metric distance affects the measurement of accessibility to different resources. In processes of supplementation, the patches of favourable habitat are too small to allow the populations to survive. In site A, if the exchanges are made within the circle, there is a possibility of survival, while in site B there is no such possibility.

enters maize fields (Charrier et al., 1997). Field mice (*Apodemus sylvaticus*) are present in the woods and hedgerows in autumn and winter. Their density declines in spring and the population breaks down completely during the summer. This corresponds to a dilution of populations in the cultivated fields during the period of crop growth. The mice do not disperse homogeneously in the cultivated space. Ouin (1997) has studied the spatial distribution of field mice in the vegetable crop zone of the Mont Saint Michel bay, from one part to another of the wooded dykes that shelter the populations during the winter (Paillat and Butet, 1997). The field mice are capable of travelling long distances (Szacki and Liro, 1991). Their abundance in a site is positively correlated with the ratio of wooded area in a radius of 500 m, which is explained by their winter dependence on wooded habitats. During the summer, the field mice spread out in the fields depending on the nature and phenological state of the crops. A large number are captured in areas planted with wheat and peas in May and June and the numbers decrease and fall to zero in August; the numbers remain small in May, June, and July in maize and carrot fields, and they increase in August in maize fields (Fig. 18). The capture rate is correlated with the closure of the vegetation, a parameter that combines the height and density of the plant cover. This could be explained by the fact that the closure of the vegetation offers field mice protection against predators (Tew, 1989). After the harvest, the field mice leave the fields. About half are eaten by predators and half find shelter in the hedgerows (Tew and Macdonald, 1993).

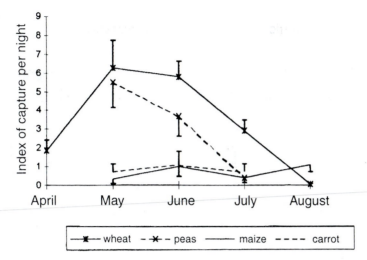

Fig. 18. Seasonal variations in abundance of field mice (plus or minus standard deviation) in the cultivated parcels. The monthly values of the index of capture per night (number of field mice captured by traps per night) in the crops and adjacent hedgerows, with the standard deviation (Hansson, 1967). Different scales were used for the indexes, in order to facilitate the reading of standard deviations and differences in the amplitude of variations.

3.4. Corridors

Corridors are linear landscape elements whose physiognomy differs from the adjacent environment. They may be natural (rivers, ridges, animal trails) or man-made (roads, high tension wires, ditches, hedgerows). They are mostly organized in networks and their linearity gives them a special function in the circulation of flows of matter or organisms. Forman (1995) attributes five principal functions to corridors: habitat, conduit, filter, source, and sink (Fig. 19).

As with the concept of metapopulation, the corridor concept has given rise to the development of many models, but field data are still scarce (Simberloff et al., 1992). The function served by the corridor depends on its structure, its place in the landscape, and biological characteristics of the species considered. In western France and in England, hedgerow networks have been studied extensively with respect to a large number of animal and plant species. Hedgerows are habitats for many animal and plant species of the forests (Pollard et al., 1974; INRA et al., 1976; Forman and Baudry, 1984). Some reptile species are even exclusive to the hedgerows in western France (Saint-Girons and Duguy, 1976). Hedgerows serve as a seed source for recolonization of abandoned fields (Baudry and Acx, 1993), as a filter for the movement of carabid Coleoptera of the field, which are slowed down

Corridors are linear elements that serve as:

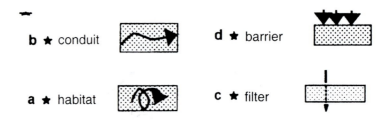

Fig. 19. Roles and efficiency of corridors: (a) linear elements may serve as a habitat for species with a small vital domain, (b) movement of species may be facilitated in or along the corridor, (c and d) flows in the matrix may be filtered, or even interrupted, by the corridor.

as they move from the hedgerow to the field boundary (Fry, 1994; Frampton et al., 1995; Mauremooto et al., 1995), as a seasonal refuge for many insects (Lefeuvre et al., 1976), and as a barrier for matter flows (Burel et al., 1993) or insects dispersed by the wind (Brunel and Cancela Da Fonseca, 1979). Their conduit function has been demonstrated for carabid Coleoptera of the forest (Burel and Baudry, 1989) and small mammals (Paillat and Butet, 1997), as well as flows of water and nutrients.

3.4.1. Structure of the corridor

The structure of the corridor, that is, its more or less regular form, its width, and the vertical structure of the vegetation, partly determines its role.

In his study on the forest flora of the hedgerows of New Jersey (Chapter 1, Figs. 7), Baudry (1988) demonstrates that the role of hedgerows in the diffusion of species depends on their width. The decrease in the number of forest species as a function of the distance from a wood is perceptible mostly for the wide hedgerows (more than 8 m between the last ploughed rows of the fields located on either side). For hedgerows that are 4 to 8 m wide, the relationship between the number of species and the distance from woods is not statistically significant (Fig. 20).

The russet vole (*Clethrionomys glareolus*) is found only in hedgerows that are dense and continuous, with a bushy and well-stratified shrub stratum and an abundant herbaceous stratum (Saint Girons et al., 1987). Zhang and Usher (1991) show that this animal uses the hedgerow as a dispersal corridor and as a habitat for reproduction, if the structure of the hedgerow is sufficiently complex. Merriam and Lanoue (1990) used radio-telemetry to

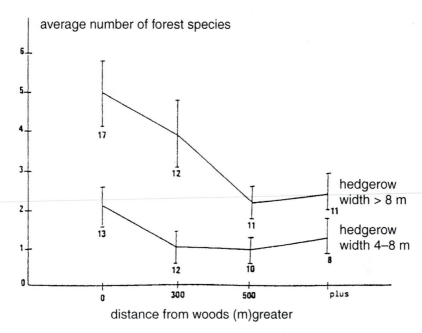

average number of forest species

distance from woods (m)greater

Fig. 20. Effect of distance from woods and width of hedgerows on the richness of forest plants (data collected in New Jersey by Baudry and Forman). The vertical bars represent the standard error and the numbers are the numbers of classes.

study the movements of *Peromyscus leucopus* released in hedgerows of three different types of structures: simple hedgerows with less than 10% linear cover of trees or shrubs; intermediate hedgerows with a continuous shrub and tree cover but trees less than 10 m high and less than 2 m wide at the base; and complex hedgerows more than 2 m wide with tall trees and a dense shrub cover. The complex hedgerows are preferred to a significant extent, notably for rapid movements and long distances.

Individuals of *Abax parallelepipedus* were tracked by radio in various elements of the hedgerow network (Charrier et al., 1997): a grove, a deep farm lane bordered by two parallel hedgerows, a hedgerow with dense vegetation (linear tree cover greater than 80%, dense herbaceous cover), and a hedgerow with less dense vegetation (holes in the tree cover, developed herbaceous stratum). Studies on the spatial distribution of this species showed that it was absent from hedgerows without trees and hedgerows in which the herbaceous stratum had disappeared (Burel and Baudry, 1989; Burel, 1991). The average distances travelled in 48 h differed significantly depending on the habitats considered (Table 3). They increased progressively with the intensity and area of the wooded cover, from the less dense hedgerow to the grove. The amplitude of the trajectories (distance between initial point and final point at the end of 2 months of tracking) and the area

Table 3. Average distance travelled in 48 h and average amplitude of trajectories in the four habitats

Habitat	Avg distance (m) travelled in 48 h	Avg amplitude of trajectories (m)	Area explored (m^2)	No. of individuals tracked
Grove	1.25	9.83	200	8
Deep lane	1.05	7.4	80	8
Dense hedgerow	0.45	3.57	20	6
Medium hedgerow	0.77	4.06	14	8

explored (convex polygon encompassing all the displacements of each individual) were generally greater in the grove and in the lane than in the hedgerows (Table 3). The shape of the trail followed, as we have seen above, is also linked to the structure of the landscape elements, the most favourable habitats (woods, deep lane) resulting in trails of more complex shapes than the others.

The key elements underlined in the three preceding examples are the width and structure (vegetation, physical structure) of corridors. It will be seen in Chapter 9 that these are the focal points of the reflection of many developments.

3.4.2. Connectivity of the network

The functioning of corridors also depends on their place in the network of linear elements. The networks are characterized by their length, the number and quality of their connections, and the quality of their elements (see Chapter 3). All of these factors define the possible routes from one point to another and thus the probabilities of individual movements.

Petit and Burel (1998) showed the importance of all these characters in explaining the spatial organization of local populations of *Abax parallelepipedus* in a hedgerow network comparing three measurements of connectivity (Fig. 21). The distribution of this species is strongly correlated with the Euclidean distance (as the crow flies) from the nearest occupied habitat, which is generally the case for subdivided populations. This Euclidean distance is a measure of the probability of arrival of disperser individuals. The correlation between distribution and isolation improves when the connectivity is measured by following the network of wooded elements (woods and hedgerows), which corresponds to the ecological needs of a species. A third measurement of the connectivity takes into account the quality of wooded habitats, which, as we have seen above, influences the intensity and quality of individual movements. For this, weighted indexes indicating the viscosity of habitats have been derived from telemetric studies. They integrate the intensity of movements and the risk of mortality for each habitat (Table 4).

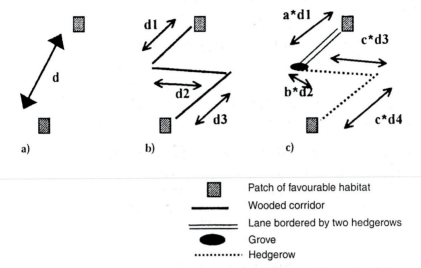

Patch of favourable habitat
Wooded corridor
Lane bordered by two hedgerows
Grove
Hedgerow

Fig. 21. Methods of measurement of connectivity. Connectivity is measured, for each sampling point, by the distance of the nearest site occupied. The three methods used are: (a) the Euclidean distance ED = d; (b) the distance along the hedgerow network ND = d1 + d2 = d3; and (c) the distance weighted by the viscosity of habitats crossed WD = a*d1 + b*d2 + c*(d3 + d4).

Table 4. Calculation of indexes of viscosity in elements of the hedgerow network

	Movement		Mortality		Viscosity
	Measurement	Index	Measurement	Index	Index
Grove	200	1	8	1	1
Deep lane	80	2.5	15	1.875	4.69
Dense hedgerow	20	10	13	1.625	16.25
Medium hedgerow	14	14.3	17	1.125	30.36

Movement: The index is the ratio between the average area explored by an individual in the grove and the area explored in the habitat.
Mortality: The measurement represents the number of individuals released in a habitat over the two months of the study, in order to ensure that there constantly remain eight survivors. The index is the ratio between the mortality measured per habitat and the mortality measured for the grove.
The index of viscosity is the product of the two preceding indexes, for each habitat.

The last measure of connectivity, which integrates not only the connectedness of the network, but also the quality of the elements that compose it, is significantly more correlated to the distribution of *Abax parallelepipedus* than the measurement using the Euclidean distance. Spatial organization and quality of elements influence the movement of individuals, and the effects are found on the population level.

Intersections are key points of the network organization. They ensure the continuity of flows between linear elements and often have particular environmental characteristics. The species richness of plants, invertebrates, or birds can be greater in the intersections than along the hedgerows (Constant et al., 1976; Baudry, 1984; Lack, 1988). The intersection effect is attributed to particular microclimatic conditions and to greater exchanges with the neighbouring elements than in other parts of the network (Forman, 1995).

3.4.3. Corridors and the functioning of metapopulations

When corridors increase the connectivity between local populations, they act on the dynamics of these populations by reducing the probabilities of extinction and favouring recolonization.

The corridor network, when it increases the rate of dispersal between patches, increases the temporal stability of interconnected populations. Szacki (1987) observes an amplitude of fluctuation three times as great for an isolated population of russet voles than for a population of the same species connected by a wooded corridor. In their model, Fahrig and Merriam (1985) underline the importance of connectivity, measured by the number of corridors linked to each patch, for the survival of a metapopulation of *Peromyscus leucopus*. This is what Brown and Kodric-Brown (1977) called the rescue effect, which expresses the fact that a significant migration towards sites that are already occupied tends to reduce the extinction rate. The data of Verboom and van Apeldoorn (1990) on the red squirrel in the Netherlands illustrate this type of functioning. The persistence of a population in a grove significantly increases with the length of hedgerows over a radius of 600 m.

Vacant patches are recolonized more rapidly in the presence of corridors, as has been demonstrated for the chipmunk (*Tamia striatus*) in Canada (Henderson et al., 1985) or for the red squirrel (van Apeldoorn et al., 1992).

The overall dynamics of metapopulations may be affected by the presence of corridors, by means of the two processes described above. Henein and Merriam (1990) showed that the spatial arrangement and quality of corridors linking the groves influence the demography of a metapopulation of *Peromyscus leucopus* at the landscape level. Corridors of poor quality, in which predation is high, may have the effect of depleting the existing populations instead of reinforcing them. Metapopulations in which the local population are linked solely by corridors of good quality are larger than those that have at least one corridor of poor quality. An increase in the number of good corridors has a positive effect on the population size, while an increase of poor corridors has a negative effect (Fig. 22). There is no relationship between the number of corridors and the size of the metapopulation when the number of interconnected patches is constant.

72   54   23

   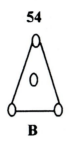   

A   B   C

Fig. 22. Size of a metapopulation for three spatial arrangements showing the effect of corridors of poor quality at the end of 10 years of simulation. The circles are patches, the solid lines are good connections, and the broken lines are poor connections. The numbers represent the size of the metapopulation in the tenth year (Henein and Merriam, 1990).

4. LANDSCAPE DYNAMICS AND THE FUNCTIONING OF POPULATIONS

In all the studies presented in this chapter, the functioning of populations has been linked to the spatial structure of landscapes. The distribution of populations changes constantly in relation with the dynamic processes of colonization and extinction, but the landscape structure is considered fixed, especially in long-term simulations. Yet, natural or anthropized landscapes are dynamic, and one may put forth the hypothesis that their evolution influences the populations, which are also variable over time. We must reconsider the idea that there is a synchronous relationship between the spatial distribution of species and the landscape patterns. The distribution of species depends on the present landscape as well as its preceding states. For example, the composition of bird communities of the Mediterranean zone is explained directly by the history of the vegetation (Blondel, 1986).

Two major processes may explain an inadequacy between the state of the landscape and the distribution of a species: delay in extinction and delay in colonization. Few field studies have validated these hypotheses, because sequences of temporal data on population distribution are rare, as are studies conducted on the landscape level. Indeed, to indicate this kind of relationship, we must sample heterogeneous spaces continuously, which is a cumbersome task of data collection.

4.1. Delay in extinction

In habitat patches isolated by a transformation in the landscape (uprooting of hedgerows in the case of hedged groves, fires, afforestation, highway construction), local populations are not extinguished immediately. Their probability of extinction increases with their isolation, because they cannot be reinforced by the arrival of disperser individuals, but they can maintain themselves over a longer or shorter time. Den Boer (1985) showed that isolated populations of carabid Coleoptera survive periods of several decades. Species of stable habitats (with a low dispersal capacity) survive around 40 years, while species of unstable environments (with a high dispersal capacity) disappear at the end of 10 years. This delayed extinction of populations can be seen in forest species that use hedgerow networks or other wooded alignments as a dispersal corridor. They can persist in landscapes when the overall structure is transformed by the uprooting of hedgerows or changes in the quality of vegetation following changes in agricultural practices.

Petit and Burel (1998) showed that the spatial distribution of *Abax parallelepipedus* is more strongly correlated to the connectivity measured in the hedgerow network of 1952 than to that measured in the 1993 network. Between 1952 and 1993, the hedgerow length diminished by 22% and the connectedness by 35%. This was expressed in an increase in the distance that an animal has to travel along the wooded network to rejoin the occupied sites, an average increase of 6%. The quality of habitats also evolved, and the proportion of intermediate hedgerows increased considerably. In 1952, 80% of displacements occurred in the woods, lanes, and dense hedgerows, as opposed to 60% in 1993. There was thus an increase in the travel time and increase in the average viscosity of lanes used by individuals. On the other hand, some populations connected by hedgerows in 1952 were isolated in 1993 (Fig. 23). All these processes explain why, in a landscape that evolves towards an opening up of the network and a reduction in hedgerow quality, the present distribution of populations is more strongly linked to the past condition of the landscape than to its present condition.

At Lalleu (see Chapter 4), Burel (1992) demonstrated that the distribution of three species of carabid Coleoptera—*Abax parallelepipedus, Chaetocarabus intricatus,* and *Steropus madidus*—depends on the recent history of the landscape. Their spatial distribution is not linked to any state of the landscape, present or past, but to the evolution of the landscape network since 1952. They are present only in the quadrats in which they found optimal conditions in 1952 (high density of hedgerows with dense vegetation, deep lanes, and groves). The opening of the hedgerow network led to an increase in parcel size, reduced connectivity, and increased heterogeneity. These three species are found in some quadrats that evolved from a similar initial state (but not in all), but they may have had different evolutionary trajectories. In 1985, their structures may have been very different (Fig. 24).

A similar procedure was used by Clergeau and Burel (1997) in wooded dykes of the Mont Saint Michel bay (Fig. 25). The tree creeper is a small

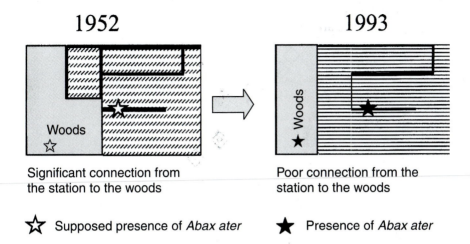

Fig. 23. Evolution of landscape between 1952 and 1993, which led to the isolation of a local population of *Abax parallelepipedus*

passerine that is not very mobile. Its movements are limited to short flights between neighbouring trees, and it remains in the same place all year round. In the polder, the favourable habitats for the tree creeper are lines of mature poplars with cracked trunks in which the birds can feed. The dykes were classified into poor, medium, and good habitats on the basis of the diameter of poplar trunks and the distance between the trees. The presence of the tree creeper was correlated to the size of favourable elements (length of planted dyke), which suggests that the birds need a certain number of old trees, planted close together, in order to establish themselves. The colonization of the polder by the tree creepers could only follow the network of favourable habitats. This species is numerous in the hedgerows of the adjacent bocage landscape and it has been hypothesized that there is a source-sink relation between the two landscapes. No diffusion effect was proved in the present configuration of polder dykes, but it exists, if one considers the spatial structure of the polder before 1992, the year in which the poplars of the Duchess Anne dyke, close to the bocage, were cut down. The authors suggest that the tree creeper colonized the polders following the wooded elements in which the lines of trees were continuous. The network of dykes served as a corridor along which the birds dispersed from the source to the recently created habitats. After elements of the network of favourable habitats were destroyed, the dykes were isolated from the source and the populations survived only in large patches that offered sufficient resources.

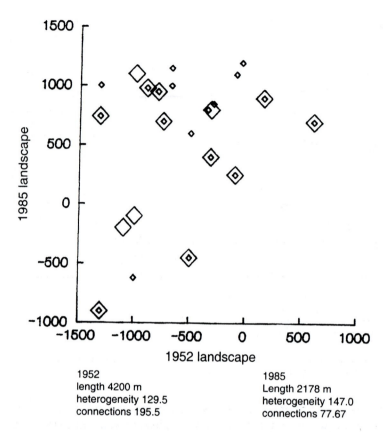

1952
length 4200 m
heterogeneity 129.5
connections 195.5

1985
Length 2178 m
heterogeneity 147.0
connections 77.67

Fig. 24. Spatial distribution of *Abax parallelepipedus* as a function of the overall landscape structure in 1952 and 1985. The landscape structures are represented by their coordinates on a multivariate analysis axis (see Chapters 3 and 4).   presence of *A. parallelepipedus*.   absence of *A. parallelepipedus*.

4.2. Delay in colonization

Another trend of recent evolution of landscapes is the abandonment of cultivated fields or pastures (Comolet, 1989; Meeus et al., 1988; Baudry and Bunce, 1991). The abandonment or reduction of agricultural activities brings about a change in the plant cover. The open fields are gradually replaced by more dense plant formations such as ferns, bushes, and thickets (Baudry and Acx, 1993). New habitats become available for species that find themselves confined to uncultivated elements of the agricultural land.

In the bocage landscape of the Pays d'Auge, in Normandy, the permanent grasslands represent 80% of the land use. In 1986, economic projections estimated the abandonment of 20% of land before the year 2000. In fact, the phenomenon was much smaller in terms of total abandonment (Laurent,

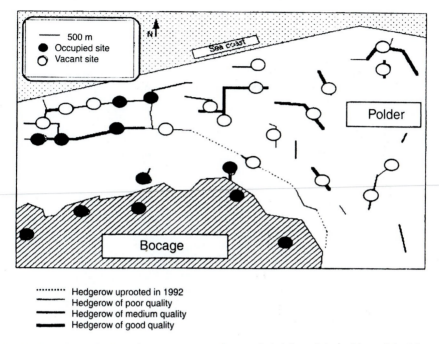

Fig. 25. Spatial distribution of tree creepers in the wooded dykes of the polders of the Mont Saint Michel bay (Clergeau and Burel, 1997)

1992), but it was expressed by a reduction of pressure on the grasslands, which led to the development of more or less permanent bramble patches (Fig. 26). These patches are unstable, and their survival depends essentially on farmers' decisions. They may be cut even after several years in case of a shortage of grass due to summer drought, for example. Between the patches, the grassy cover is maintained as a pasture. In the patches, a plant succession takes place. Herbaceous plants colonize it randomly on the landscape level but colonization on the parcel level depends on the vegetation of adjacent elements (Baudry, 1989). The forest carabid Coleoptera present in the hedgerows of the bocage find favourable conditions (shade) in these bramble patches and establish themselves there once the disperser individuals reach them (Burel, 1991).

Burel and Baudry (1994) studied the colonization of bramble patches by carabid Coleoptera of the forest. The hedgerows and dirt lanes bordered by two hedgerows are the sources. In view of the average parcel size, each new habitat is less than 200 m from a potential source. In sites that were abandoned recently, only the patches located at least 200 m from a hedgerow are colonized. The connectedness of the landscape must be maintained if colonization is to take place. Here, the connectivity is measured by the

Fig. 26. Abandoned patches in Castillon, Calvados

distance between the patch and the nearest available potential source. On land abandoned long ago, the patches are larger and they all contain carabid Coleoptera of the forest: their abundance is significantly correlated to the size of patches (Fig. 27). The presence of Coleoptera is independent of the evolution of the vegetation along the successional gradient.

Most of the time, studies on elements in the course of succession consider essentially the dynamics of the vegetation, assuming that there is a strong relationship between plants and animals (Brown, 1984). However, depending on their food needs, animals rely more or less on the quality of the vegetation. For the predator Coleoptera, the structure of the vegetation may be more important than its composition (Waliczky, 1991), while the reverse is often true for the strict herbivores (Brown and Kalff, 1986). This study confirms that the presence and abundance of polyphagous insects cannot be predicted by the gradient of vegetation. Colonization of this mosaic on the refined scale, highly unstable, depends on the landscape structure, the age of patches, and species mobility. Studies on spiders, which have a very high dispersal capacity, have shown that in these same landscapes in the Pays d'Auge, they react immediately to the growth of bramble patches (Asselin and Baudry, 1989). In less than a month, the species of the hedgerows and woods colonize the new patches.

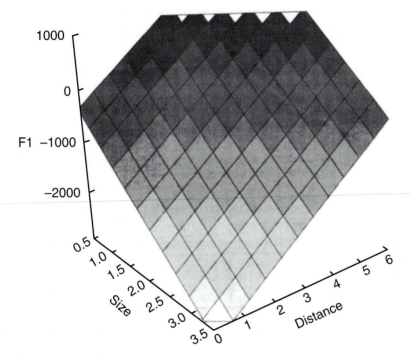

Fig. 27. Effect of patch size and distance from potential source on abundance and richness of forest carabids (Burel and Baudry, 1994)

The interval between the appearance of new habitats and their colonization is found in all landscapes, whether they have natural successions after disturbances (Pickett et al., 1987) or habitats created by man (polders, plantations). To estimate the interval, we must take into account the life history traits of species, the connectivity of the landscape, and the relevant scale on which the process must be indicated.

5.   POPULATION MODELS USED IN LANDSCAPE ECOLOGY

The impossibility of repeating a study in a controlled environment at the landscape level makes it essential that the processes of observation, modelling, and local experimentation complement one another (Fig. 28). On the basis of an empirical study of the spatial distribution and the movements of individuals, hypotheses can be put forth on the functioning of populations and the demographic models they are associated with. The validity of those hypothesis can be tested through elaboration of spatially explicit models. Models can be used to formalize the functioning and organization of the

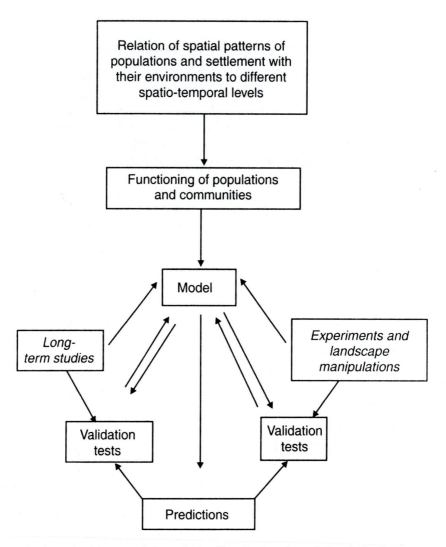

Fig. 28. Conceptual framework for analysis of functioning of populations in landscapes

system studied in a simple way (Fahrig, 1991). They can also be used to replicate the experiments on the selected landscapes (Turner et al., 1995), the complexity of which is reduced by the researcher's maps (see Chapter 3). Field experiments and manipulations of the landscape structure complement the conceptual protocol and allow us to measure, in real scale, the effects of the spatial arrangement of landscape elements on dispersal processes of species. Experimental studies (experimental model system, Wiens et al., 1993) are rare in the literature, while modelling has given rise to a large number of studies.

5.1. Objectives of models

Models linking populations and landscapes can be grouped in three major types (Martin, 2000).

1) Spatially explicit modelling of population dynamics: the models start from demographic data (Vernier and Fahrig, 1996; Henein et al., 1998).

2) Modelling of the spatial distribution of populations: models start from the relative density of individuals or data of presence-absence of local populations in the landscape elements (Vermeulen and Opdam, 1995; With and Crist, 1995, 1996; Bascompte and Sole, 1996; Boone and Hunter, 1996; Rushton et al., 1997; Tischendorf and Wissel, 1997).

3) Modelling of landscape properties that can control the distribution or dynamics of populations or individuals of a given species (estimate of maps of connectivity or permeability) (Schippers et al., 1996; Gustafson and Gardner, 1996; Schumaker, 1996).

5.1.1. Spatially explicit models of population dynamics

Spatially explicit models of population dynamics are used in rare cases in which data are available on the life history traits of species and the relations between the different activities of individuals and the spatio-temporal organization of the landscape. Henein et al. (1998) compiled data from studies that their team had conducted over more than 15 years in the agrarian landscapes of the Ottawa region, as well as demographic data collected on two small forest mammals, to model the behaviour of populations at the landscape level. One, *Peromyscus leucopus*, is more opportunistic than the other, *Tamia striatus*, and it sometimes uses elements of the mosaic to move from one grove to another. The team used 39 parameters, some of them estimated, to make the model work. Parameters relating to mortality and reproduction were adjusted as a function of the state of individuals and were obtained from the abundant literature on the biology of these two species and field studies in different habitats of the crop mosaic. Parameters linked to movement were essentially derived from field data and took into account the status of individuals and the land cover.

The simulations were done on landscapes of different structures in which the authors varied the rate of afforestation, the connectedness between hedgerows and groves, and the quality of hedgerows as corridors. The opportunist species performed better in demographic terms in any landscape structure. Connectedness was the most relevant parameter in predicting the survival of the specialist species, followed by the quality and spatial arrangement of the hedgerows.

5.1.2. Models of spatial distribution of populations or individuals

For some years, models of spatial distribution of populations or individuals have been the basis of many studies. Such models are often based on movement data obtained at the individual level. Their objective is either to obtain emerging properties at the population level by successive simulations, or to identify the spatial configurations that favour or inhibit the movement considered as a key process in the population dynamics (Schumaker, 1996). These models are mostly based on individuals (De Angelis and Gross, 1992). The level of the organism is preferred for the mechanistic study of behaviour between environment and species. Individual mechanisms are responsible for the ecological processes linked to the spatial variability of landscapes (Wiens et al., 1993).

Vermeulen and Opdam (1995) simulated the displacement of carabid Coleoptera with a low dispersal capacity along the green shoulders of highways. Data of individual movement were derived from field studies that allowed them to quantify the probabilities of movement from one habitat to another, as well as the intensity and quality of movements (Vermeulen, 1993). Mortality and fertility data were drawn from published sources. The simulations were used to test the effect of the spatial configuration (width, linearity) of highway shoulders on the efficiency of their role as corridor. Long-distance movements were more frequent in wide shoulders than in narrow shoulders. The loss of individuals towards adjacent habitats is lower and the number of individuals whose movement is pursued in this habitat higher. The use of highway shoulders as a dispersal corridor depends on the behaviour of each species.

5.1.3. Models based on functional properties of the landscape

Models based on functional properties of the landscape are very similar to the preceding type. The difference lies only in the process of simulation. In one case, the aim is to identify the spatial patterns that determine the movement (width of corridor, distance between patches), while in the other the aim is to understand how a given spatial organization influences a population. The models are used not to indicate a mechanism but to evaluate the effects of a landscape structure on a population. They are developed in the framework of risk evaluation (Akcakaya and McCarthy, 1995).

Schippers et al. (1996) modelled the displacement of badgers in Europe (*Meles meles*) in real landscapes by using a model of movement based on the individual. The objective of their simulations was to estimate the probabilities of passage between the potential habitats, that is, to define the connectivity of the landscape (Fig. 29).

5.2. Taking space into account

Models can be spatially explicit or implicit. In the former, any object or entity in the space is noted, while in the latter only some spatial parameters are taken into account, such as the number of patches, their isolation, and their size. The examples cited in the preceding sections are spatially explicit

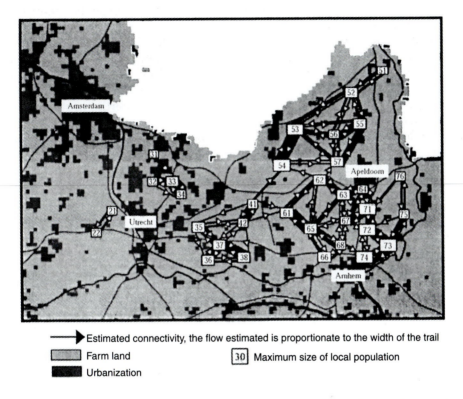

Estimated connectivity, the flow estimated is proportionate to the width of the trail

Farm land

Urbanization

30 Maximum size of local population

Fig. 29. Simulation maps of survival of badger populations in the Netherlands (Schippers et al., 1996)

models and are developed in close relation with the landscape ecology, while the metapopulation models are mostly spatially implicit (Hanski, 1997). However, the extreme simplification of the Levins model (1970) has given way to models that are complicated by the integration of an increasing number of spatial parameters. In their review of the study of metapopulations of butterflies, Thomas and Hanski (1997) show that the integration of data on size and isolation of patches, as well as the size of source populations, made possible an improvement in the simulations. From these models, they could predict where metapopulations would survive or disappear, what set of empty patches would be colonized, and what set of occupied patches would be deserted.

Spatially implicit models consider spatial measurements but work from matrices of numbers and not from a map (Hanski and Simberloff, 1997). The spatial arrangement of landscape elements, the effects of proximity, and changes in scale such as those presented in Chapter 3 are not taken into account.

Spatially explicit models are particularly developed in landscape ecology. They account for the heterogeneity of the mosaic as the researcher describes it, as a function of the processes studied. One advantage of such models is that they allow the study of processes from the global scale to the local scale and an estimate of the possible responses to such mechanisms (Dunning et al., 1995). The space being studied can be divided by means of a grid of square or hexagonal units and the models are of the cellular automat type (Fogelman-Soulie, 1983). Each cell of the grid has properties in relation to the processes studied (movement, flow); probabilities of movement to neighbouring cells are calculated by assessing the nature of the neighbouring cell and the behaviour of individuals at the margin between cells. This is the most common kind of model used (Wiens et al., 1997; With et al., 1997; Henein et al., 1998). The resolution of the map, or the size of the square, represents the lower level of the analysis, but it is possible to work at larger scales by progressively aggregating the cells.

The landscapes are generated either from real landscapes or at random with a certain number of organizational constraints. The artificial landscapes are used as neutral landscapes. When real landscapes are used, the multiple character (environment and society) of the landscape can be taken into account and the research techniques can be applied to specific cases.

7

Interspecific Relationships and Biodiversity in Landscapes

Landscape ecology has given rise to many developments relating to interactions between a population and the spatio-temporal heterogeneity of the landscape. The preceding chapters have shown the difficulties faced in collecting data in time and space that can validate the theoretical models developed in the framework of the study of fragmented populations and patch dynamics. The questions raised at higher levels—on interactions between species, study of communities—are still more complex and have generated less interest in the scientific community. However, this research is very important: to conserve landscapes, we must take into account these multi-species interactions, which are a compelling aspect of the functioning of ecological systems.

1. INTERSPECIFIC RELATIONSHIPS

1.1. Competition between species

Hanski (1995) wrote a summary of models that could be applied to the understanding of relations of competition at the landscape level. Most of the reflections are linked to patch theory and simplify the landscape mosaic for a representation of patches of favourable habitats and an unfavourable matrix.

In cases where all the patches considered are of the same kind, the asymmetry between the species may allow the construction of models. It can be supposed that there is an asymmetry in competitiveness, negatively correlated to an asymmetry in the colonization capacity. Taking as a postulate that there are random local extinctions, we can show theoretically that two species of this kind can coexist in a landscape, the less competitive species finding refuge in the patches that the other species, which is more capable of colonization, has not yet reached. Long-term studies of the distribution of invertebrates (*Daphnia*) living in puddles of sea water on the rocky coasts of Sweden and Finland (Ranta, 1979; Hanski and Ranta, 1983; Bengtsson, 1986) validated some elements of this model. There are local extinctions in the puddles. There is no difference in the colonization of occupied or vacant puddles, but the presence of another species in a puddle increases the risk of extinction. Long-term survival of these species on most of the islands studied depends on recurrent colonization, and thus on relations of competition that influence the risk of extinction.

Another hypothesis about the coexistence of competitive species in a set of patches of the same kind is based on their spatial distribution. Hanski (1981) and Atkinson and Shorrocks (1981) demonstrated that if two competing species cannot survive in a single patch, they may still survive in an archipelago of patches in which the frequency of occurrence of two species varies from one patch to another independent of the distribution of resources. This model of aggregation describes a general model of coexistence in the mosaics and has been tested only on very small mosaics for Drosophila on fungi (Shorrocks, 1990) or flies on animal carcasses (Hanski, 1987). A hypothesis may be put forth that the model can be generalized to the landscape level, but there are presently no data to validate it.

The most common pattern in landscapes is the coexistence of patches of different kinds. Some have a larger quantity of resources (patches of good quality) and others contain fewer resources (patches of poor quality). The models predict in this case that the larger and more competitive species are dominant in the most productive ("good") patches. An analysis of the distribution of several species of shrew in 93 sites of northern Europe validates this hypothesis. The largest species are abundant in the most productive patches (Hanski, 1992 in Hanski, 1995), even when the compensation between competitiveness and use of resources has not been demonstrated.

In real landscapes, in a heterogeneous mosaic, the process that occurs most commonly to allow the coexistence of competing species is the selection of the habitat. In general, when species compete in a heterogeneous landscape, their final distribution depends on the performance of each species and the quality of elements of the mosaic (Danielson, 1991). For example, if a highly competitive population is a specialist of a rare habitat or one with low productivity, in the absence of exchanges with the surrounding environment, it excludes the less competitive species from the habitat. On the other hand, if a population of a less competitive species regularly receives immigrant individuals from connected habitats, it may exclude the more competitive species.

1.2. Predation

An important part of the predation in the landscapes occurs in the ecotones. There is often a greater density of herbivores, including some game species, in the ecotones than in the interior parts of the patches. Their density attracts predators, and the ecotones are commonly called "ecological traps" because of the high rates of predation that prevail there (Gates and Harman, 1978; Angelstam, 1986; Andren and Angelstam, 1988). This effect of predation can also be linked to the use of ecotones as preferential corridors of displacement or as routes of penetration into a less favourable habitat. The impact of the predation depends on the specificity of predators. The specialists can be confined to a single type of habitat; they perceive the landscape as fragmented (Addicott et al., 1987) and the ecotones as barriers (Stamps et al., 1987). The generalist species, which penetrate into several types of habitat, perceive the landscape in its continuity while taking its heterogeneity into account; they easily cross the ecotones.

A strong hypothesis of the relationship between predation and landscape structure is that predation depends on the length of the ecotones and the size of patches, the two parameters being correlated (the larger the patch, the longer its perimeter). Andren (1995) collected the results of 40 studies that tested this hypothesis on real and dummy nests of birds. Of these, 18 were from North America, 19 from Europe, and 3 from Central America. Wilcove (1985) demonstrated a correlation between the rates of predation on dummy nests and the size of forest fragments. Among the 22 studies that considered predation in the forests or groves bordering cultivated or other open zones, only four did not validate the hypothesis of the edge effect. For two of these, the principal predators were forest species (squirrel and marten) (Nour et al., 1993), and the other two involved landscapes largely dominated by the forest (Angelstam, 1986). The increase in predation at the edges of groves is a common phenomenon in agrarian landscapes, while it is rare in forest mosaics.

The differences found in the response of predation to ecotones or to patch size can be explained by the behaviour of predators in relation to the landscape. Andren (1995) distinguished three types of predators (Fig. 1):

1) **Specialist predators**, restricted to one type of landscape element. For example, the magpie *Pica pica* or the crow *Corvus monedula* (Andren, 1992) feed essentially on nests in farm areas; the jay, the martin, and the red squirrel feed essentially on nests within groves. They may induce an increase in the ecotone predation, but this effect is limited to their preferred habitat.

2) **Generalist predators** having a preference for one habitat use several types of landscape elements but one of them more frequently. The badger, for example, prefers farm land but also uses wooded areas. This type of predator may have a significant activity on either side of the ecotone, as well as within the different habitats.

3) **Entirely generalist predators** use various landscape elements indiscriminately. The raven, for example, is the chief predator of nests in the forests of northern Europe, in the parcels of mature forest as well as in the clearings (Andren, 1992). Within the farm lands, the raven is a generalist

Selection of habitat	Habitat A	Habitat B	Increase in predation at the ecotone

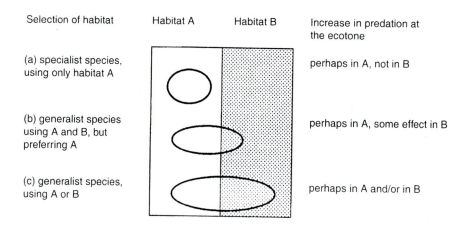

(a) specialist species, using only habitat A			perhaps in A, not in B
(b) generalist species using A and B, but preferring A			perhaps in A, some effect in B
(c) generalist species, using A or B			perhaps in A and/or in B

Fig. 1. Various types of predators as a function of their activity in different habitats and their impact on the rates of predation in the ecotone. (a) The specialist species, e.g., magpie in farm land or martin in the forest. (b) Generalists that prefer one habitat: the badger and the blue jay prefer farm land but can use the forests. (c) Generalist species such as the fox in farm land, which use grasslands and ploughed fields.

predator that can use all the elements of the mosaic. Even if these predators use the entire mosaic, they may induce an edge effect because they are often attracted by the limits between habitats.

The presence of these different types of predators is linked to the structure and composition of the landscape. Their role on the dynamics of populations of prey has been demonstrated by Delattre et al. (1996) in different landscapes of Jura, predominantly permanent grasslands. The dynamics of populations of common vole (*Microtus arvalis*) was compared in three types of landscape: open field, bocage, and village. In the villages, the numbers were low and the periods of extinction were long. In the bocages, the local extinctions were limited and the numbers varied little. In the open field zones, the density of the populations varied greatly from one year to another. The authors attribute these differences to modifications in the predation. Around the villages, the domestic cats exerted a continuous and strong predation on the small mammals (Giraudoux, 1991; Erlinge et al., 1983). The bocage zones have a great diversity of habitats and shelter a large number of predators (foxes, wild cats, raptors), particularly generalists (Andersson and Erlinge, 1977). The open field zones are less frequented by generalists (Loman, 1991) and are more favourable to specialists, which have a destabilizing effect on the populations (Heikkila et al., 1994). The landscape determines the composition of the guild of predators and indirectly controls the dynamics of the prey populations.

1.3. Pollination: long-term interactions

Many populations of pollinating insects depend on the distribution of flowering plants in patches and are thus controlled by the landscape structure. The relations between landscape and pollinators depend on the life history traits of plants and animals: the phenology of different species defines the distance the pollinators must travel between the patches of a given plant to cover their food needs. The searching ability of pollinators determines whether or not they can undertake their displacements. These life history traits do not vary independently. This allows us to link the characteristics of landscapes to the types of organisms found in them (Bronstein, 1995). The most common case is that in which the generalist pollinators are in phase or out of phase with the flowering of plant species. The pollinators are thus very strongly influenced by the phenological variability of resource plants. They can easily change their food source in a certain radius of activity. These are for the most part highly mobile species. Plants that flower for a long time, such as lavender (*Lavandula latifolia*), which flowers from mid-July to October in Spain, play a key role in the landscape. This species may be visited by 70 species of bees, butterflies, and Diptera (Herrera, 1987) (Fig. 2).

Fig. 2. Bumblebee

Few field studies have been able to confirm these conceptual models. However, the experiments of conservation or restoration of butterfly populations (Pullin, 1995) take them into account in an experimental fashion.

The control of long-term interactions (studies of interactions between parasite, host, and environment; Combes, 1998) by the spatio-temporal organization of landscapes has been studied very little so far. However, in epidemiology as well in conservation biology, the knowledge of factors that explain the dynamics of these interactions can bring about solutions to sanitary or conservation problems. Taylor and Merriam (1996) have demonstrated that the rates of parasitism in the Odonata in Nova Scotia varied with the forest fragmentation. Roland and Taylor (1997) have shown that the rates of forestation in a landscape influence the rates of parasitism by four species of caterpillar parasitoids (*Malacosoma disstria*). Giraudoux (1997) showed that the prevalence of echinococcus, for which voles are intermediary hosts, varies with the spatial organization of the landscape, and notably the proportion of permanent grasslands or fallow lands.

2. BIODIVERSITY

The maintenance of biodiversity has become a social issue that is recognized and taken into account by governments and a large number of international organizations. This social question has led to a mobilization of the scientific community; research in this domain is active and involves many disciplines (Gaston, 1996). The term *biodiversity* has many definitions derived from a simple expression, "variability of living things". The most widely cited definition is that of the US Congress Office of Technology Assessment (OTA, 1987): "Biological diversity refers to the variety and variability among living organisms and the ecological complexes in which they occur. Diversity can be defined as the number of different items and their relative frequency. For biological diversity, these items are organized at many levels, ranging from complete ecosystems to the chemical structures that are the molecular basis of heredity. Thus the term encompasses different ecosystems, species, genes, and their relative abundance."

Noss (1990) proposed a description of biodiversity according to an interlocked hierarchy from the gene to the landscape, breaking it down into three criteria: the structure, the composition, and the function (Fig. 3).

The recent evolution of landscapes under the pressure of human activities, exploitation of the tropical forest, the intensification of cultivation, and increasing urbanization deeply transform the ecological systems on the planetary level (Turner and Meyer, 1994). These changes are among the major causes of erosion of biodiversity (Solbrig, 1991) and have led to the proliferation of studies on consequences of landscape transformations. The hypotheses that are the basis of these observations are linked to general

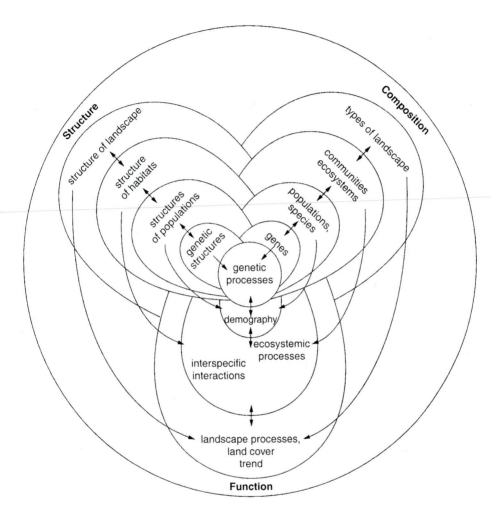

Fig. 3. Aspects of composition, structure, and functioning of biodiversity, present across interconnected spheres, each containing multiple levels of organization (Noss, 1990)

studies on biodiversity in ecosystems. The species diversity is strongly correlated to habitat diversity and is greatest at the intermediate levels of disturbances, when we consider them on the fine scales (Rosenzweig, 1995). The spatio-temporal heterogeneity of the landscape is essential to maintain the species diversity (Huston, 1995) and depends on the regime of disturbances (Turner, 1987) (Chapter 4). The successions generated by disturbances are the key to the biodiversity at all the spatial or temporal levels.

In this chapter, we present studies on the relationships between biological diversity, measured by species richness or diversity, and the spatio-temporal organization of landscapes. The most abundant data are linked to the evolution of agrarian landscapes and to successions in the "natural" or protected landscapes.

Measurement of biodiversity as it was originally defined by Wilson (1988) signifies that all the animal and plant species present in a given place are counted. Such studies, taking into account the totality of the diversity of living things, from cellular organisms to vertebrates, cannot be realized at the landscape level. Most of the time, only some taxonomic groups are studied, chosen for their role as an indicator of structure, as with birds (Balent and Courtiade, 1992), their importance in the diversity of species, as with invertebrates (Wilson, 1988), or for their patrimonial role, as with butterflies (Dennis, 1992). The simultaneous use of several groups is rare but indispensable if one wishes to evaluate the consequences of landscape transformations. The space-species relationships depend on the ecological needs of species, on their scale of perception and scale of analysis of landscapes. A single transformation will have beneficial effects for some species and negative effects for others. For example, the abandonment of land, leading to an increase in wooded area, is a threat to species of the steppes or grasslands but allows the development of species of shade and forest or thickets (Suarez Seoane, 1998). On the other hand, following the spatial extension of changes, some species will be threatened while others conserve the same functioning or adapt. For example, spiders, which have a high dispersal capacity, are less sensitive to frequent disturbances of small extent, such as the dynamics of brambles in the Pays d'Auge, because they can recolonize new habitats rapidly (Asselin and Baudry, 1989). In the same landscapes, the orchids established in the grass of chalky hills disappear when the land becomes overgrown.

To evaluate the biodiversity in a landscape, the essential dynamics, is one of the objectives of present research. Chapter 9 presents several examples of the development and management for which the social demand has outpaced the state of present scientific knowledge.

2.1. Biodiversity in the agrarian landscapes

In Western Europe, landscapes evolved according to the regions either towards intensification leading generally to a reduction of uncultivated zones, or towards abandonment of farm lands leading to the increase of uncultivated zones. These two processes induce ecological changes at the landscape level that are often considered a threat to biodiversity.

2.1.1. *Ecological consequences of the intensification of agriculture*

Most ecologists (McLaughlin and Mineau, 1995) and agronomists (Gras et al., 1989) who have analysed the effects of the intensification of agriculture concentrate on the field level. The hypothesis that underlies such research is that land cover and agricultural practices at the field level are the chief determinants of biodiversity. In agro-ecology, most of the data demonstrating the negative effects of intensification on species richness are restricted to the cultivated field or to experimental fields (Goldberg and Miller, 1990; Paoletti and Pimentel, 1992). It is in fact difficult to separate the effects due to the landscape structure from those due to farming practices in the cultivated fields or to management in the uncultivated zones. These factors are, as we have seen in Chapter 4, strongly correlated to each other. However, given the movements of individuals in the landscapes (Chapter 6), the complementarity between the uncultivated zone and the cultivated zone for a large number of species, and the variability of scales at which different species perceive the land, landscapes cannot be compared when their biodiversity is analysed at the field level.

To study a landscape with a view to evaluating its biodiversity, we must first reflect on the following: (1) the limits of the landscape, or at the very least a definition of its identity; (2) the scale of analysis of this landscape, which must be relevant to the taxa studied—in this framework, methods of change in scale must be widely applied, because the scale of ecological processes is rarely known beforehand; and (3) the protocols of sampling of living organisms in a heterogeneous landscape. Few research studies have followed this process, which requires a significant investment of time in the landscape analysis as well as in the collection of biological data. The results below are derived from a multidisciplinary study carried out on the working site of the Mont Saint Michel bay from 1991 onward.

2.1.1.1. Intensification of agriculture and species diversity

Four zones of bocage and polder located on a gradient of intensification of agriculture and opening up of the landscape (Fig. 4) were sampled (Burel et al., 1998). The various animal and plant groups were chosen as a function of their scale of perception of the landscape and their mobility and dispersal capacity. Even when there are differences that could be important between species of each group, vertebrates, small mammals, and birds have a greater mobility and larger vital domain than the invertebrates (Fig. 5). Among the invertebrates, species that fly (Diptera) use larger spaces than walking species (carabid Coleoptera of the forest). Plants have, in general, a very low displacement capacity (Fig. 6).

The species richness (number of species), species diversity (measure of the relative abundance of various species), and equitability (measure of the more or less equal distribution of species) (Maguran, 1988) were measured for all the groups studied (Table 1). The results were not all consistent with the expectations of the ecological theory, which predicted a reduction in the number of species as a function of the increase in the frequency of disturbances in zones with the most intensive agriculture, in the least

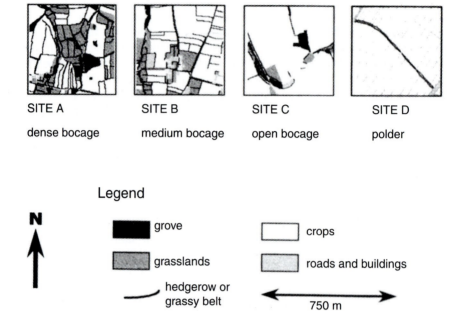

SITE A   SITE B   SITE C   SITE D

dense bocage   medium bocage   open bocage   polder

Legend

N

grove   crops

grasslands   roads and buildings

hedgerow or   750 m
grassy belt

Fig. 4. The four landscapes studied

Fig. 5. The owl provides a sample of the community of small mammals

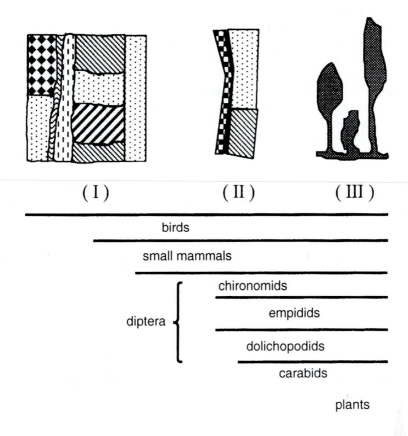

Fig. 6. Perception of space by various taxonomic groups. (I) landscape mosaic; (II) all the adjacent elements: fields and borders; (III) landscape element

Table 1. Richness (S), diversity (H'), and equitability (E) of animal and plant communities in the four landscapes studied

	Site A			Site B			Site C			Site D		
	S	H'	E	S	H'	E	S	H'	E	S	H'	E
Chironomid Diptera	8	0.56	0.53	9	0.68	0.35	5	0.85	0.47			
Empidid Dipteran	4	0.69	0.58	2	0.77	0.59	6	0.27	0.56			
Dolichopodid Diptera	7	0.41	0.25	1	0.72	0.53	5	0.40	0.45			
Carabid Coleoptera	5	0.52	0.43	1	0.50	0.44	0	0.54	0.45	8	0.67	0.45
Hibernating birds							3	0.19	0.59	2	0.20	0.42
Nesting passerines	0	0.33	0.81	5	0.32	0.84	2	0.32	0.86			
Small mammals	1	0.93	0.84	1	0.7	0.80	1	0.96	0.85	1	0.64	0.76
Herbaceous plants	89	0.56	0.87	32	0.06	0.86	71	0.42	0.87			
Woody plants	0	0.99	0.75	1	0.14	0.77	8	0.17	0.79			

heterogeneous or diversified landscapes, and in young ecological systems. Only the empidid Diptera and nesting passerines saw their richness diminish gradually along the landscape gradient. For the other groups, the richness remained stable (small mammals), changed very little (hibernating birds, trees, and shrubs), or evolved without relation with the gradient (carabid Coleoptera, herbaceous plants). The variations of indexes of diversity and of equitability were not associated with the gradient and did not give useful information about the reactions of species to the structures and dynamics of these landscapes.

These results are explained by the complexity of the landscape and the need to break down the processes and structures into hierarchic levels. The organization of communities is the result of processes operating at different spatial and temporal scales and directed by anthropic or environmental factors (Fig. 7). Some species, such as the tree creeper presented in Chapter 6, require a constant or one-time flow of immigrant individuals to colonize the recent landscape of the polder. The hedged landscape serves as a source and the polder as a habitat or sink. In the bocage landscapes, the structure of the hedgerow network allows or prevents the maintenance of metapopulations of carabid Coleoptera of the forest. Petit (personal communication) demonstrated that the abundance of populations of these

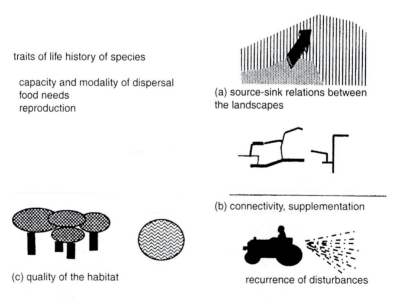

traits of life history of species

   capacity and modality of dispersal
   food needs
   reproduction

(a) source-sink relations between the landscapes

(b) connectivity, supplementation

(c) quality of the habitat

recurrence of disturbances

Fig. 7. Hierarchical representation of factors that control populations in the landscapes. (a) Landscape units may be a source or sink for species in relation with others. (b) The landscape structure, its connectedness, is necessary to the maintenance of some species. (c) The frequency and nature of disturbances define the quality of habitats at the field level and necessitate an adaptation of species to unstable environments.

species declines with the opening up of the landscape. In the same way, populations of small mammals more or less strongly associated with the wooded network (the russet vole and the wood mouse) decline with a reduction in the length of hedgerows. At the field level, the intensification of farming, and more particularly the significant and recurrent use of pesticides in the cropped areas, is the cause of the disappearance of amphibians, whose larvae cannot develop in polluted water. Frequent disturbances also cause a reduction in the average size of species of carabid Coleoptera between the bocage landscapes and the polder. The large, less mobile species characteristic of less disturbed habitats such as groves disappear, while small, more mobile species characteristic of farm land or pioneer stages of successions adapt to the disturbances and become more numerous (Fig. 8).

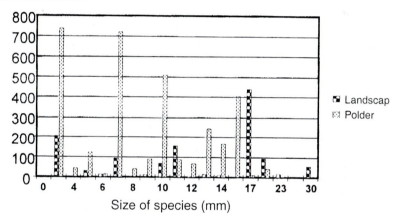

Fig. 8. Average size of carabid Coleoptera in hedged landscapes and in the polders

2.1.1.2. Plant communities in landscapes of intensive agriculture
Studies on the floristic diversity in agricultural landscapes are often concentrated on the uncultivated areas, or zones of least intensification. The flora of wooded islands and forests (Lemee, 1978), moor, grasslands (Collectif, 1996), and field boundaries (Marshall, 1989) has been analysed and has served as the basis for theoretical studies in ecology (Tilman and Downing, 1994). As in research covering several taxonomic groups, the level of approach has been most often restricted to the farm field or the landscape element. The extent of management of these lands—fertilization, grazing pressure, fallow, pastoral burning, stagnation—has been recognized as providing an explanation of the richness and diversity of plant communities. The new questions that landscape ecology poses are linked to the potential

impact of the spatial organization of these elements on their flora. Are the plant communities dependent on the connectivity, fragmentation, and spatio-temporal heterogeneity of the landscape?

From the analysis of the herbaceous flora of field boundaries of three continuous networks located in the landscape units of the bocage described above, the explanatory factors of the diversity of communities can be hierarchized (Le Coeur, 1996; Le Coeur et al., 1997). The factors retained in the study are as follows: at the local level, the structure of the field boundary vegetation, maintenance practices, and disturbances; at the field level, the land cover and the quality of management; and at the landscape level, the grain, connectivity, and heterogeneity (Fig. 9).

The structure of the tree and shrub strata influences the richness and composition of the herbaceous stratum. This result is consistent with earlier research on hedgerows and other field boundaries (Baudry, 1988; Hegarthy et al., 1994; Marshall and Arnold, 1995). The nature of the land use adjacent to the field boundary is also a factor explaining floral composition and richness. Boundaries adjacent to crops are affected by ploughing of the soil and application of pesticides, herbicides, and fertilizers. Boundaries adjacent to groves or smaller roads with little traffic are less disturbed. Those that border highways or cultivated fields are most frequently mown and crushed;

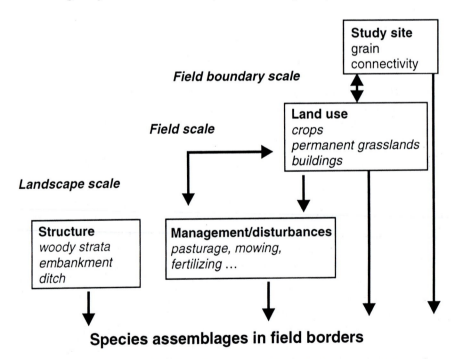

Fig. 9. Explanatory factors of the diversity of communities in the herbaceous stratum of non-cultivated linear elements of the agricultural landscape

the cut vegetation that is not removed then contributes to the enrichment of the soil by mineralization. The species richness along the borders of grasslands is thus significantly different from that along the borders of crop land. Management of the herbaceous stratum has a relatively minor effect on the landscapes studied in comparison with the results obtained on experimental quadrats (Marshall and Birnie, 1985; Parr and Way, 1988). In fact, in a bocage landscape studied in its totality, a large number of boundaries, adjacent to crops, are not accessible during the summer and thus they are not mown or treated with herbicide during the crop cycle.

This study showed that when a boundary belongs to a given landscape unit, there is a highly significant effect on the composition of the herbaceous flora of that boundary. In identical biogeographic, climatic, and geomorphological conditions, the fact of belonging to a more or less dense bocage site determines the composition of the flora. The sites differ in the size of the landscape grain, the heterogeneity of the agricultural matrix, and their recent history. All these parameters together define the identity of the landscape, as we have seen in the introduction to this work. These parameters are highly correlated among themselves because they are all dependent on the dynamics of the environment and the society (culture and technology). It is thus very difficult to identify the mechanisms responsible for these differences. It is, however, interesting to note this effect linked to the very essence of the landscape, which limits the applications of results obtained from random landscapes that do not integrate the environment or the history.

2.1.1.3. Biodiversity and forest fragmentation

As we saw in the introduction, questions about the consequences of forest fragmentation on native species, as well as on biodiversity, have given rise to numerous research studies in landscape ecology. In agricultural landscapes, the groves and forest are key elements for a large number of species; their size, form, nature, and spatial arrangement influence the biodiversity on the landscape level. The avian settlement has been thoroughly studied in forest fragments that constitute what is called the countryside.

• The morphology of a wood influences its biodiversity. The species richness of ecotones is very high (Harris, 1988), the species present in the ecotones are mostly ubiquitous and opportunistic (Hansson and Angelstam, 1991). The interior environment, on the contrary, is poor, but it shelters specialist species that are not present in the other landscape elements. The reduction of the area of forest islands leads to a loss of species by a decline of forest species (Galli et al., 1976; O'Connor, 1986; McIntyre, 1995). The large and compact woods are those that have the greatest species richness.

• The isolation of groves influences their biodiversity. The structural isolation of a grove depends on its distance from other wooded formations, the presence of corridors and barriers, and the nature of the adjacent parcels: on whether their vegetation is similar to that of the woods, and whether they are permeable for the species that live in the woods (Harris, 1984). The species richness is correlated to the density of hedgerows connected to the groves and the density of woods in the neighbourhood (Opdam and Schotman, 1987; van Dorp and Opdam, 1987). The isolation of groves has a strong impact on species that disperse over long distances and are reluctant to fly across large open spaces (Whitcomb et al., 1981).

2.1.2. Ecological consequences of land abandonment

The abandonment of farm lands, which can be defined as a decline in the agricultural pressure on a given territory, occurs in many stages. It may lead to the total abandonment of activities and gradual predominance of fallow land, or partial abandonment creating a mosaic of abandoned and cultivated fields, or a less intensive use, notably as grasslands, by a reduction of load. The consequences on the flora and fauna vary according to the initial state of the landscape and its evolutionary trend (Acx and Baudry, 1993).

2.1.2.1. Abandonment and fires

In densely wooded areas, cultivated fields represent "buffer" zones between the habitations and the forests, preventing the spread of fires. When abandoned land is left fallow, many strata of vegetation develop, the environment closes up, woody species establish themselves, and dead organic matter accumulates. Abandoned lands thus become more vulnerable to fire, so there is a greater risk of fire in the region considered.

Repeated occurrence of intense fires in the same field leads to a degradation of the soil quality, which has a significant effect on the evolution of the vegetation. In the central regions of France, for example, in old cultivated terraces subjected to repeated fire, the vegetation evolved in two different ways as a function of the landscape structure, and more particularly as a function of the stability of the terrace walls (Fig. 10).

—When poorly maintained terrace walls can no longer retain soils during runoff from rain, on land that is no longer planted, a scrubland develops.

—If there is no soil erosion, the land is quickly overrun with gorse and becomes a dense moor, with shrubs that grow up to 2 m tall (Baudry and Tatoni, 1993).

In these changing landscapes, the composition and dynamics of the vegetation depends on the history of the agriculture and natural disturbances. The spatial structure is highly organized by the relief and the cultivation history. It plays a key role in the evolution of the vegetation (Tatoni and Roche, 1994). An increase is observed in the species richness over the first 20 years after the land is abandoned. Subsequently, the number of species declines, then stabilizes.

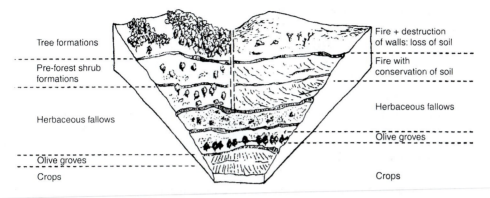

Tree formations

Pre-forest shrub
formations

Herbaceous fallows

Olive groves

Crops

Fire + destruction
of walls: loss of soil

Fire with
conservation of soil

Herbaceous fallows

Olive groves

Crops

Fig. 10. Possible evolution of a watershed when the terraces are abandoned, in the Auriol region, Bouches-du-Rhone (Vaudour et al., 1991). The abandonment of terraces often occurs from the top of the watershed to the base. Fires disturb the post-cultivation succession. After a fire, only a moor or scrubland is established.

2.1.2.2. The effects of abandonment vary according to the taxa considered

Abandonment of land, no matter what its intensity, is expressed in the establishment of new habitats in the landscape. These elements are gradually colonized as a function of the permeability of the surrounding landscape and the pool of available propagules. These two factors depend, as seen in Chapter 6, on the trends of the life history of species. Different communities thus respond differently to the evolution of the agriculture and the landscapes, and there may be gains or losses of biodiversity depending on the groups considered. The evolution also depends on the scale considered (Grubb, 1988; Wiens, 1989). The changes in community at the level of a landscape element, or at the level of a landscape as a whole, may present inverse tendencies. The increase in spatial heterogeneity of the landscape may lead to an increase in the number of species, while there is a loss in the elements considered in isolation because of effects of size or edge effects.

The consequences of abandonment in the Pays d'Auge (see section 4.2 in Chapter 6) varies according to the groups studied (Fig. 11).

Spiders

For spiders, there is no limit to dispersal on the landscape scale, and there is no effect of the spatial structure of the landscape. Among the 96 species present, 7 are found only in the prairies, 18 only in abandoned patches, and 13 only in the woods. Eleven species are common to all the habitats (Asselin and Baudry, 1989; Baudry and Asselin, 1991). The presence of bramble patches can double the number of species present in a grassland, from 25 to 50. The species added at the field level are already present at the landscape level in the woods or fallow lands. Communities of spiders react

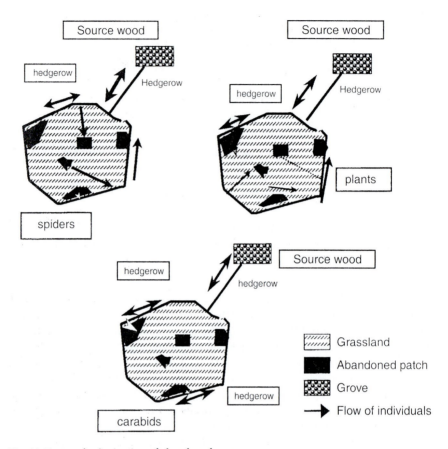

Fig. 11. Types of colonization of abandoned zone

very quickly (within a month) to fine modifications in the plant cover (such as mowing and regrowth of bramble patches).

Plants

The composition of plant communities is closely linked to the high geomorphological heterogeneity: acid soils on the plateaux, chalky soils on the slopes, and peat zones in the marl. Only some species present in the hedgerows surrounding the fields colonize the abandoned patches (Baudry, 1989). Apparently, they disperse from the hedgerows, but some are present in the seed banks. When the patches of bramble increase, the species of the hedgerow present in the boundary either establish themselves elsewhere, by vegetative or sexual reproduction, or disappear. A close correlation between the flora of hedgerows and the flora of bramble patches appears only when the patches are aggregated at the field level (Burel and Baudry, 1990). This clearly indicates that colonization by plants is a stochastic process operating at the individual patch level and the species level and not a deterministic process at the level of communities and set of patches.

Coleopteran carabids

The study of Coleopteran carabids has confirmed the presence of forest species in the agricultural space, despite a significant reduction in the number of species in comparison with the forests (10). Only two forest species are present in the groves and two in the hedgerows, their dispersal corridor, and in the bramble patches close to the hedgerows (Burel and Baudry, 1994). The grassland species such as *Amara communis*, *A. lumicollis*, and *Poecilius cupreus* are not present in the abandoned patches or in the abandoned grasslands, in which the number of species is particularly low in comparison with that found in land that is still grazed. The maintenance of hedgerows is very important, not only to provide a corridor for some species, but also to serve as a reservoir of species that can colonize abandoned land.

2.2. Biodiversity in "natural" landscapes

The term "natural" is used here for landscapes in which plant successions occur with little human intervention. These may include reserves or nature parks or regions with considerable environmental constraints that are sparsely populated: e.g., rain forests, northern steppes, mountain peaks. The dynamics of such landscapes depend essentially on biotic disturbances (epidemics, overgrazing) or abiotic disturbances (falling of trees, fires, landslides). These disturbances lead to modifications in the successions and the landscape may be defined as a changing mosaic of various successional stages. The process of succession is a key factor in understanding most of the aspects of biological diversity on a biological time scale. The factors that trigger successions lead to changes in the survival, size, and genetic variability of populations. The occurrence of successions is accompanied by changes in populations, communities, and ecological functions at the landscape level.

The spatial heterogeneity of species distribution on the landscape level is the result of the capacity of these species to survive and remain competitive over a more or less wide range of environmental conditions. For each species, optimal conditions can be defined along the gradients of resources, as well as a response in terms of abundance and fitness to variations of resources along the gradient (Austin, 1985). Species are found in environmental conditions that are very different from their optimal conditions. In the presence of competition, each species places itself in conditions that it tolerates and that are less congenial for the competing species. In a community, the ecological optimum of a species is often limited in case of abundant resources by the competition, and in case of low resources by physiological constraints (Mueller-Dombois and Ellenberg, 1974; Austin and Smith, 1989). There are many environmental zonations and gradients in landscapes (relief, soil, climate). There are gradients of resources, factors essential to the development of species, and gradients of regulation that act on the capacity of species to use or transform the resources (Austin and Smith, 1989).

The biodiversity at the landscape level is thus the result of processes of disturbance, succession, and spatial organization of environmental gradients that ensue.

Biodiversity and fire

In many ecosystems, the frequency and intensity of fires influence biodiversity at the landscape level (Forman and Boerner, 1981; Romme, 1982). Each new fire modifies the structure of the landscape mosaic and initiates new successions. Its propagation depends essentially on the topography and wind (Dupuy, 1991), and the accumulated biomass influences the speed with which it advances (Green, 1989). Many studies have associated the spatial diversity of the plant cover, the heterogeneity of the landscape mosaic, and the spread of fires from their point of origin (Romme and Knight, 1982; Turner et al., 1989; Romme and Despain, 1989). The heterogeneity of the landscape may accelerate or slow down the spread of a fire (Denslow, 1985). Some elements of the mosaic may constitute barriers to fire, while others, which are highly inflammable, may increase the speed with which it spreads. O'Neill et al. (1992) showed that the spread of fire is very closely correlated to the extension of contiguous inflammable zones. The landscape geometry, the possibility of contagion, is very often critical to the area that is burned.

To link fire with landscape heterogeneity, we must first examine the question of the scale at which heterogeneity is measured. For example, in the forest massif of Paimpont, Morvan et al. (1995) tested several spatial scales to evaluate the role of the diversity of the plant cover on the extent and recurrence of fires. During the fire in 1990, the speed at which the fire spread increased with the increase in spatial heterogeneity measured over units of 25 m (Fig. 12). Heterogeneity and rate of propagation are significantly correlated at this scale (Spearman coefficient, $r = 0.89$, $P < 0.02$). At a coarser scale, over 50 m, the relation was not significant. The heterogeneity, measured on the 25 m scale, was determined by the number of fires. A fire increases the diversity of elements of the mosaic, while a second fire reduces it. A primary disturbance fragments the landscape, but the landscape dynamics over the long term buffers this effect by integration of successions after fires in the spatial heterogeneity of plant communities.

The heterogeneity of vegetation induced by fires affects the diversity of the fauna, since the resources are modified in quality as well as their spatial

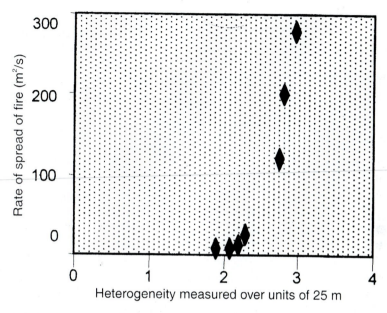

Fig. 12. Relationship between the rate of spread of fires and the heterogeneity of the landscape mosaic at a scale of resolution 25 × 25 m

arrangement. The major fire in 1988 in Yellowstone National Park created a mosaic of successional patches varying with the intensity of the fire at different places (Fig. 13). This diversity of biotic resources (pioneer species) and abiotic factors (increased insolation, wind) increased the biodiversity for the first few years after the fire. The species were faced with new environmental conditions. The effects of those conditions on the winter survival of the large ungulates, the elk (*Cervus elaphus*) and bison (*Bison bison*), were measured. The following winter, immediately after the fire, there was no food in the burned areas, while a few years later the biomass was abundant and the availability of food resources at its highest (Singer et al., 1989; Hobbs and Spowart, 1984). The severity of the winter was the most important factor in the survival of ungulates. When the winter was relatively mild, even if 60% of the park area was burned, the fire had no impact on the survival of the ungulates during the first winter. The effects of the fire were measurable only when the winter was normal or severe, whether in the first year after the fire or a few years later. The spatial arrangement of the burned or unaffected areas had a perceptible effect on the survival of these species only if the fire affected a small part of the landscape. The survival is therefore greater when there is a distribution of

Fig. 13. Yellowstone Park after the 1988 fire. In the more or less intensely burned zones, a plant mosaic offers diversified resources to ungulates.

large masses than when the mosaic is highly fragmented, which signifies that the effect of a major fire is not the same as that of several small fires (Turner et al., 1994).

8

Geochemical Flows in Landscapes

Degradation of water quality is a major environmental problem (Naiman, 1996). It is due to the presence of toxic elements (pesticides, heavy metals) as well as an excess of nutrients (nitrogen, phosphorus) (Carpenter et al., 1998). The displacement of such elements between the source zones, most often situated in the terrestrial environment, and water courses or water bodies is controlled by the landscape structure. The slope and nature of the soil directly influence the circulation of water, which is the medium that transports these elements. The presence of biogeochemical barriers that retain or transform potential pollutants brings us directly to the effects of landscape heterogeneity on their flows. This is the subject of the present chapter. Biological corridors (hedgerows, river corridors) are often biogeochemical barriers. Such barriers act in two ways: (1) physically by stopping the particulate elements, as with anti-erosion barriers, and (2) biogeochemically by transforming the elements. An especially well-studied case is that of denitrification, which is the transformation of nitrates into gaseous nitrogen. The term *buffer zone* refers particularly to this last type of phenomenon. The concept of buffer zone originated in the pioneering studies of Karr and Schlosser (1978) on land-water interfaces, even though they did not use this particular term. One of their hypotheses was that the more "natural" the vegetation along water courses, the more it contributes to maintaining water quality.

An important characteristic of these geochemical flows is that their apparent intensity depends on the scale of observation. On the scale of a large watershed, the phenomena of erosion and sediment accumulation balance each other out (Fournier and Cheverry, 1992). The fertilization of farm fields often contributes nitrogen in excess of what plants use. The excess nitrogen is leached by percolating water and reaches the water table, and subsequently the water courses. Still, it is often observed, even in areas of intensive agriculture, that there is a loss of nitrogen between the total loss in the fields and the flow in the water courses (Kesner and Meentemeyer, 1989). For this reason, the spatial dimension, the landscape structure, must be at the centre of the research. Although the watershed, a hydrological unit, is the spatial unit of reference for evaluations and development, the measurements are often taken at two scales, occasional measurements and measurements at the elementary watershed level corresponding to water courses of drainage order 1, 2, or 3.

1.   BUFFER ZONES

The existence of effects of scale and spatial heterogeneity in geochemical flows within a watershed leads to apparent loss of matter during its transfer. Such losses are due to absorption by plants in unfertilized zones such as hedgerows (Ryszkowski and Kedzoria, 1987) or forests (Lowrance et al., 1984), storage in soil organic matter, and denitrification (Correll, 1996). These phenomena occur in the buffer zones. Research in buffer zones is prolific (Haycock et al., 1996) and pertains to hydrological functioning, microbial ecology, the structure of buffer zones, their use, and their maintenance. Indeed, beyond the general principles presented below, there are many questions on the potential efficiency of such zones, the real efficiency, and the induced effects. Since water quality is an important socio-economic stake, decision-makers and managers should avoid modes of pollution management that are not only ineffective, but may even have negative ecological consequences, especially in terms of biodiversity or atmospheric pollution by N_2O (Addiscott, 1996).

1.1. Principle of buffer zones

1.1.1. *Retention of nitrogen*

Peterjohn and Correll (1984) were among the first to study the transformation of nitrates in forests at valley bottoms. Often, these were also zones without direct nitrogen input, in which nitrates were stored in plants. They were called buffer zones because they constitute a barrier against the direct input of nitrates into water courses.

Buffer zones, which are still not clearly understood, may retain 70 to 90% of nitrate inputs, or quantities that may reach 74 kg N/ha/year.

The mechanism of denitrification is as follows: In anaerobic conditions, bacteria (*Pseudomonas fluorescens*, *Alacaligenes*, and *Nitrosomonas*) transform nitrates and nitrites into gaseous nitrogen (N_2). They use oxygen from nitrates to consume carbonate matter. In aerobic conditions, these bacteria use oxygen from the air. The conditions of denitrification are found in periodically saturated zones, such as low-lying areas.

The first condition for denitrification to occur is that the water from the watershed charged with nitrates must flow into the buffer zone and remain there. This is not always the case. Sub-surface flows may pass under the buffer zone before they reach the river or nitrates may directly percolate into the water table (Haycock and Muscutt, 1995; Correll, 1996; Gold and Kellog, 1996). The presence of a drainage ditch may also allow the water from the slopes to reach the stream directly without remaining in the wet zone. Ongoing research programmes aim to determine the real efficiency of buffer zones in various conditions of climate, landscape, and agriculture.[1]

1.1.2. *Other buffer effects*

Apart from nitrogen retention, buffer zones are also barriers for sediments, pesticides, and phosphorus (Lowrance et al., 1984; Correll, 1996). In their review of work on sediment traps in low-lying zones, Dillaha and Inamdar (1996) reported that a herbaceous buffer 5 m wide along a slope of 7% can reduce the transport of clay to 83%. The functioning of a buffer is affected by fine particles that seal up the soil pores and increase runoff. The authors emphasize the need to distribute nutrient inputs as uniformly as possible. The creation of rivulets, by agricultural equipment or by the movement of cattle, nullifies the efficiency of a buffer. This efficiency does not vary linearly with the width of the zones. For example, to improve the retention of sediment from 90% to 95% along a slope of 2%, the width of a forest buffer must be increased from 30.5 m to 61 m.

With respect to phosphorus, the studies compiled by Uusi-Kamppa et al. (1996) are based on retention rates varying from 20% to 90%. The reduction of water runoff speed by the buffer is an essential factor of efficiency.

Generally, the authors (cf. Haycock et al., 1996) acknowledge that a buffer zone must comprise several zones in order to be effective. The zone in direct contact with the slope must trap the sediments. Denitrification must occur in a zone further downstream that is really hydromorphic, and the zone close to the water course must have a permanent vegetation that absorbs nutrients (Lowrance, 1996). Scientists also agree that buffer zones are essential along the water courses of lesser drainage order (1 to 4) because it is here that they can be most effective. Indeed, in these places in the landscape, the ratio of runoff area to buffer area is minimal. As we have mentioned in Chapter 4, further downstream the water courses themselves deposit sediments in the alluvial zones. In the Armorican landscapes, the corridors of wet grasslands bordered with embankments are landscape elements inherently favourable to denitrification. Figure 1 shows the set-up for a study on denitrification in such a corridor. The arrangement comprises piezometers in which the nitrogen levels of water are measured. The limnigraph is used to follow variations in the depth of the water table. The automatic data recording station records data continuously.

---

[1]The list of programmes can be found on the internet site of the Buffer Zones Information Centre, http://www.qest.demon.co.uk/bzchp.htm.

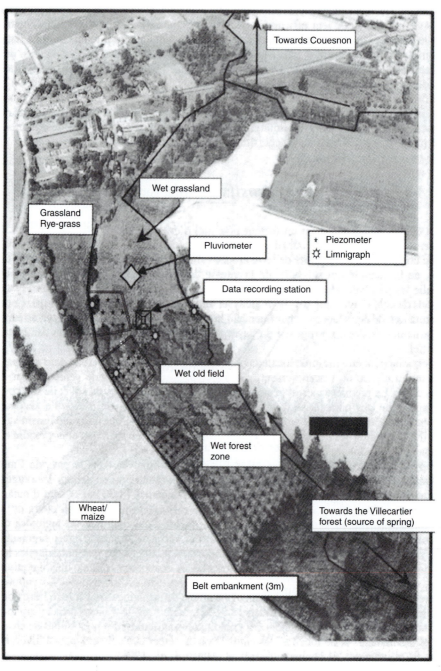

Fig. 1. Field study on denitrification along a water course of order 2. Set-up: G. Pinay and J.C. Clement. Photo: Air Papillon. This study is located in site B described in section 4.2.2 of Chapter 4.

2. EROSION PHENOMENA AND LANDSCAPE STRUCTURE

Water circulation is a potential factor of erosion, from the terrestrial environment to the valley, from watersheds to streams, or from upstream to downstream of the bed and of the flood plain of a river. The quantity of soil eroded is a function of the kinetic energy of the water, the quantity that runs off, the speed of runoff, and the facility with which the soil particles are washed down, and thus the soil texture and structure. In this discussion, we will set aside the study of soils to analyse the effects of landscape structure. Landscape structure acts in two ways: influencing the quantity and speed of the flowing water and holding back the eroded particles. The rugosity of the landscape to water movement is an essential parameter.

Erosion can be measured locally, as the quantity of soil lost in a field, or on the watershed level, as the quantity of solid flow transported by the water course. The first measure results from local land characteristics and cover as well as the characteristics upstream that control the incoming water. The second measure integrates all the effects of the watershed structure. For a large watershed, the solid flows must be measured at different points on the water course to take into account the phenomena internal to the water course and its corridor.

Papy and Douyer (1991) analysed the catastrophic erosions in the Caux landscape (Normandy, France) over a period of 30 years. They showed that apart from storms, these catastrophes occurred on large impluvia (> 1000 ha) and were severe to the extent that the area susceptible to runoff was large. During the period from 1955 to 1989, the evolution of land cover was such that the susceptibility of the land to gullying decreased (with an expansion in grassland areas) and then increased. Occasions of catastrophic erosion corresponded to periods in which the susceptibility to gullying was the greatest. With an equivalent overall land cover, one way to increase the rugosity of a landscape is to plant different crops, alternating spring crops (sensitive to erosion) with winter crops on the slopes (Papy and Souchere, 1993). The feasibility of such a project involves modelling of the functioning of farms, possibilities for changing the rotation, and location of the crops or plant cover for the winter on the basis especially of the work load (Martin et al., 1998).

In bocage zones, hedgerows, embankments, and ditches help increase the rugosity of the landscape. The levelling of one side or the other of an embankment perpendicular to the slope indicates that that side halts the eroded particles in the upstream part (Merot and Ruellan, 1980; Pihan, 1976). It is an important function of soil protection of the bocage. From an analysis of various cases of erosion in the Armorican Massif, Baudry (1988) considered the entire bocage network. Some hedgerows or embankments showing no sign of soil accumulation may play an important role in preventing water from reaching the fields. The hedgerows parallel to the slope, channelling the runoff water, also play a role. Overall, the structure of the hedgerow-ditch-embankment network is more important than the length of these elements. This was demonstrated by Merot et al. (1999) using a hydrological model. In a dense bocage, there may be fewer hydrologically useful hedgerows than in a moderately dense bocage. Moreover, Merot and Bruneau (1993) showed the importance of boundary embankments of the wet zone in flood control and relationships between the valley bottom and the slope. A boundary embankment creates a barrier between the slope and the wet zone, preventing the rise of the water layer on the slope and slowing the runoff, thus mitigating floods. However, it has no effect during severe floods.

Thus, a bocage watershed can be represented as a function of the water circulation and flood control (Fig. 2).

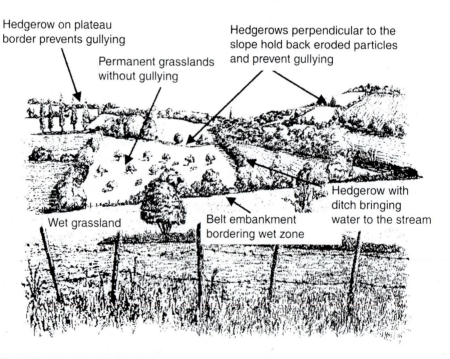

Fig. 2. The diversity of hedgerows and land cover plays a role in the control of gullying and erosion in a hedged slope.

3. TRANSFERS IN WATERSHEDS

3.1. Calculation of mineral balances

Risser (1989) retraced the approaches of nutrient flows in landscapes from the ecosystems approach of watersheds developed by Likens and Bormann (1975) and Lowrance et al. (1985), the former on forest watersheds and the latter on agricultural watersheds. These studies demonstrated that the sum of outputs of all nutrients was less than the sum of inputs. In consequence, some landscape elements and structures retained nutrients or allowed transfers to the gaseous state in the case of nitrogen.

The studies of Likens and Bormann (1995) were of great significance in ecology because they were among the first long-term studies on a watershed, dating from 1967. These authors were able to accumulate a long series of data, essential in examining the variety of phenomena over time, and they were also able to experiment on the landscape scale. They demonstrated thus the increases in nutrient loss after deforestation. From their studies, we retain some conclusions on the landscape scale. First of all, they observe an accumulation of nitrogen by microorganisms during the development of a forest towards mature stages. The abrasion of the relief is due mainly to the dissolving of elements and, to a lesser extent, to the loss of particles.

Lowrance et al. (1985) studied four watersheds in Georgia (USA) of an area of 1500 to 2000 ha, with 36 to 54% of their land occupied by agriculture, the remainder being essentially forest. The annual precipitation was around 1200 mm, of which less than a third was runoff. The authors established input and output balances for nitrogen, phosphorus, potassium, calcium, magnesium, and chlorine (taking into account rain and fertilization for inputs, and harvest and runoff into streams for outputs). In the watersheds, water and ions circulated essentially by a sub-surface flow. With the exception of chlorine, all the elements had a negative balance; the outputs were less than the inputs. With respect to nitrogen, the outputs of organic N were perceptibly the same as in all watersheds, and the outputs of N-NO$_3$ were 1.5 to 4.4 times greater in the two basins having the highest proportion of agricultural land. In 1981, which was a dry year, the differences reached a factor of 50. The annual crops contributed most to nitrate loss. The permanent grasslands contributed to losses of ammoniac. The authors noted that chlorine, a biologically inactive element, was retained in small quantities, while the other elements were susceptible of being stored or returning to the atmosphere in the form of gas in the case of nitrogen. There were thus mechanisms that made the watershed a functional unit, not only hydrologically, but also biogeochemically.

In an arid Mediterranean context, Bellot and Golley (1989) studied the inputs and outputs of an irrigated watershed of 5000 ha. The inputs of water were around 1500 mm/year, two thirds of which was from irrigation. Irrigation water was the primary source of the input of chemical elements, except for N and P, which come from fertilizers. It brought in a large amount of calcium and sulphates, respectively 816 and 166 kg/ha/year. The total annual output of ions was greater than 10 t/ha. The calcium losses from the irrigated basin were greater than 4 t. In contrast, the nitrogen and phosphorus balance was negative, which signified that these elements may accumulate.

The use of geographic information systems has allowed spatially explicit analysis of phenomena and progress beyond balances. It is possible to locate the flows, detect the barriers, and take into account distances between the sources of elements and the sinks.

3.2. Structural approach

Baudry and Berthelot (unpublished) proposed an approach of structural relationships between sources and sinks of nitrogen and barrier effects of buffer zones in a watershed on the following principle: it is hypothesized that the pollutant impact of a field is a function of the risks of loss in a given place and the distance of this place from the stream. The phenomenon of denitrification may thus be considered an increase in the distance from the field to the stream: the greater the denitrification, the more a significant "rugosity" appears in the landscape, and thus the less intense the impact of a possible pollution source.

We place ourselves here in a perspective of structural analysis, not in an analysis of processes. What we look for is the correlation between landscape structures and states of the water of a stream. The hypothesis of an effect of the distance between the pollution source and the stream may be hydrologically false, but that is not important. Similarly, an increase in "distance" has no functional meaning. It is the possible denitrification in the zones of high rugosity that is the process in question.

The relationship of phenomena that are easy to understand (water quality, landscape structure) has at least two advantages: (1) it can be used to test the relevance of the search for purifying mechanisms and (2) it has a heuristic value in operational terms. If the "landscape structure" effect exists, there is a need, in principle, to call for caution and preservation of this structure.

The principles of the structural model are presented in Fig. 3. The hedgerows perpendicular to the slope and the hydromorphic zones are barriers. Each parcel is assigned a risk of being a pollution source. The objective of the model is not to make a quantitative prediction of nitrate losses in the watershed, but to classify the losses according to the structures (land cover + buffer zone) presenting more or less risk. In this context, the risks of loss in each field correspond only to a calculation of risks that may occur in each crop in a given pedo-climatic context. The model thus requires

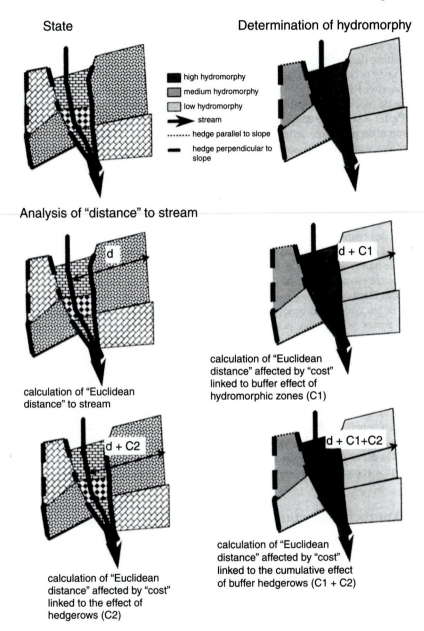

Fig. 3. Principles of structural analysis

only minimal information: land cover, presence of hedgerows, and delimitation of hydromorphic zones.

To implement this model, we have also used the *varcost* function of Idrisi©, the distances being calculated from the stream and the rugosity map being made up of hedgerows and hydromorphic zones. Subsequently, we multiply the map of distances by the map of potential losses to obtain an impact map from which the average value of a pixel can be calculated, which can be used to classify the watersheds.

As an example, the maps used earlier for landscape fragmentation and permeability are presented. Stages 4 and 5 are considered with different crops and hedgerows plus the water courses, and the potential impacts in the different situations (crop rotation, presence of hedgerows) are analysed.

Landscapes of stage 5, the most deforested, are found to be potentially more "polluting", no matter what the situation. There is also a difference in potential source between the two years of crop succession. The buffer effect of hydromorphic zones is higher than that of hedgerows.

Figure 4 represents the potentials of "pollution" in different situations.

3.3. Functional approaches

Functional approaches attempt to begin from a realistic representation of flows and to quantify the pollution sources. Nevertheless, they remain highly qualitative and are mostly used in the comparison of one watershed with another.

Poiani et al. (1996) developed an approach of nutrient flows from agricultural land towards wetland zones. The model determined a potential diffuse pollution index from soil characteristics, potentially leachable nitrogen, and the position of these nitrogen sources in the watershed. This model was tested in nine watersheds in the central part of the state of New York (USA), occupied mainly by agriculture and forest. The authors were then able to order the watersheds according to the potential levels of nitrogen inputs into wetland zones. This model was not validated, but the authors considered it realistic, considering the available data. The process is similar to that proposed by Baudry and Berthelot and presented above, but Poiani et al. (1996) integrated internal mechanisms of soil and estimates of nitrogen inputs. However, the Baudry and Berthelot model achieved a preliminary validation on seven elementary watersheds of northern Ille-et-Vilaine (site C, described in section 4.2.2 of Chapter 4). A rank correlation showed that the index of potential impact of crops could be used to classify these watersheds according to the annual nitrogen flow measured in 1996.

Comeleo et al. (1996) used an analogous process to link the contamination of sediments of 25 sub-estuaries, of area less than 26,000 ha, from the Chesapeake Bay (USA) and the land cover of slopes by heavy metals and organic pollutants. The authors defined five types of land cover: buildings and roads, grassy cover, forest cover, bare soil, and water. The model also comprised variables expressing stresses on the system such as the annual water flow and the flows of heavy metals. The results indicate that these

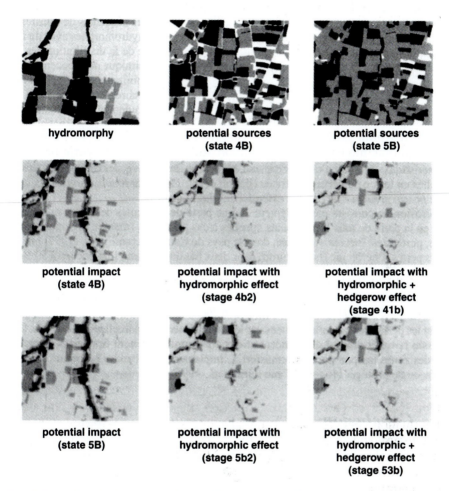

Fig. 4. Example of buffer effect and potential sources of "pollution" of water courses on a territory

flows and urban development have an effect on the pollutant level in sediments. The effects were more marked when the sources of stress were located less than 10 km from the estuary. In this study also, the rank correlation coefficients were calculated. The distributions of variables, as well as the dependence of observations, necessitated the transformation of observed values into ordinal values.

The study of pollutant flow in a watershed with intensive cattle farming in Brittany (Cheverry, 1998) showed the variety of spatial approaches possible: from the study of land cover, to the heterogeneity of the contribution of parcels in terms of nitrogen loss or denitrification, to the study of soils. It can be seen how spatial structures and agricultural activities interact to produce pollutant flows. The annual average concentration of nitrate increased from 10 mg/l in 1975 to values around 50 mg/l in 1990 (Can, 1998). It was related with the increase of area cropped with maize, which doubled during this period, and the number of pigs, which increased from 20,000 to 90,000. Merot (1998), in concluding studies, highlighted the importance of certain spatial parameters such as the role of topography in the organization of soils and the division of the watershed according to modalities of transfer of water: well-drained zones with vertical transfers and hydromorphic zones with runoff. This can be used to propose a topographic model of soil distribution and the genesis of hydric flows, as well as the geochemical control of concentrations. These studies have refined the problem of the minimal area of a representative watershed. In the present case, an area of 200 ha was needed to obtain stable measurements. In this watershed on schist, the lateral sub-surface flows were considerable. The flows are thus sensitive to spatial heterogeneity and mechanisms such as denitrification may act. On the other hand, in sedimentary basins, such as those of Lorraine, the vertical flows are predominant and the characteristics of the water from a source are in direct relation with the characteristics of the parcels in terms of nitrate loss (Benoit et al., 1997). *The location of crops in the watershed is an essential factor in the nitrogen flows.*

Erosion is not only a loss of soil particles but also a loss of all the elements attached to them (e.g., organic matter, phosphorus, pesticides, metals). In the Caux countryside, Lecomte et al. (1997) studied the losses of herbicide (Isoproturon) in two watersheds, one occupied 50% by woods and permanent grasslands, and the other almost entirely ploughed. They took measurements on parcels and on the water released. Geographic information systems were used to integrate the data on herbicide treatment and hydrological functioning. In the watershed under cultivation, the inputs into the released water seemed to be losses from fields diluted by runoff from untreated fields. In the wooded and grassy watershed, the concentrations in the water reflected the losses of product in the buffer zones during the transfer. The planting of grassy belts can be one way of protecting water courses (Gril et al., 1996).

4. CONCLUSION

The study of physicochemical flows is still poorly developed in the context of landscape ecology. It remains the prerogative of hydrology and hydrobiology. However, there is an important potential to develop concepts such as those of source, sink, and barrier (or connectivity). Moreover, it is one way of linking nutrient or pollutant flows to the spatial distribution of species.

Part IV

Applications to Landscape Management

9

Application of Landscape Ecology Concepts to Landscape Management and Design

From its origin, landscape ecology has been linked to problems of land development and management, as we have seen throughout this work. The questions posed to scientists involve the development of natural areas (reserves, national parks), agricultural lands (consolidation of parcels, river contracts), or urban spaces (green corridors). In France, for example, the 1976 law on conservation of nature, in requiring impact studies for major development works, favoured the elaboration of methods to evaluate the consequences of transformations of landscape structures on ecological processes.

The central place of land in landscape ecology has made it a special field of research for the definition of principles of land management. The concepts, methods, and results presented in this work highlight the importance of spatial structures, but they also emphasize the need to study these structures in relation to combinations of processes or particular phenomena, such as the displacement of animals or the circulation of water. The visible elements and structures in a landscape are also closely dependent on organizational and evolutionary processes that are not necessarily spatial. In the context of development, social groups that are antagonistic or cooperative are necessarily active, and economic and cultural dimensions must be taken into account. Landscape ecology offers a point of view and tools with which to test alternative scenarios of results to negotiate.

This chapter presents general principles as well as applications in order to answer the following question: *In terms of particular objectives, what are the elements that should be maintained, introduced, or modified in a landscape? What form and area should they have, and how should they be organized with respect to each other? What might be the consequences on the other functions of the landscape?*

In this chapter, we present some examples in which concepts of landscape ecology are broadly applied in development or management policies.

Before applying the concepts of landscape ecology, it is essential to review some of the major and recurrent ideas developed in the preceding chapters:

—The spatio-temporal description of the environment depends on the processes studied.

—The scale of analysis must be relevant to the scale of the process studied.

—The connectivity is dependent on the spatio-temporal structure and traits of life history of the species considered.

—Landscapes are dynamic and they are not necessarily in conformity with the dynamics of ecological processes.

—The dynamics of anthropized landscapes is difficult to predict at fine scales.

1. THE CORRIDOR CONCEPT APPLIED TO DEVELOPMENT

The concept of a corridor, a route of circulation for disperser individuals, was very quickly used by developers who found it a space-saving means to mitigate the negative effects of fragmentation. More generally, the recognition of the instability of animal and plant populations gave rise to concepts relating to conservation of nature and more recently to maintenance of biodiversity. Developers frequently referred to central zones, buffer zones, and corridors; the reservation, even a managed reservation, of well-defined territories was not considered sufficient for the maintenance of populations.

Some questions arise in the establishment of corridors to maintain biodiversity: What is a corridor for fauna or for flora? What species will be favoured by the establishment of a corridor, and what species will suffer on account of it? Is a narrow corridor with vegetation dominated by annual species better than no corridor at all? How should we define the width, vegetation to be planted, and the modalities of management, for the corridor to be as effective as possible?

Questions of scale are very important in the evaluation of the function a corridor serves with respect to biodiversity. Although a hedgerow and the Isthmus of Panama are both corridors, they have different functions.

As we have seen in Chapter 6, corridors serve many functions at the landscape level. It is only for populations or individuals that we have precise knowledge of the relationships between corridor and organisms. There is no empirical study evaluating the effects of a corridor on biodiversity as a whole. However, we must keep in mind the functioning of landscapes, even if corridors are planted most of the time to conserve an endangered species (Diamond, 1976; Soule and Gilpin, 1991).

Hobbs and Wilson (1998) underline the recent progress in research on the role of corridors as a conduit for fauna. Such research is based mostly on modelling approaches or empirical data, and very rarely on experimentation (Lindenmayer and Nix, 1993). It has resulted in the identification of a series of factors that may play a role in the use of corridors by fauna (Fig. 1). However, the lack of experimental data is still considered by some a reason to question the role of these elements.

The establishment of corridors is integrated into an innovative approach to conservation. The concept of reserves in the broad sense is a fundamental tool for the protection, maintenance, or reinforcement of ecological functions, the heterogeneity, the fragmentation, and ultimately the biodiversity. The species, communities, and habitats that traditionally constitute the unique objectives of protected zones are now considered components of open ecological systems, dynamic and heterogeneous. In the contemporary perspective of conservation, the reserves concept takes into account ecological processes, turnover of organisms, and the structure of communities. It also considers the landscape context, the relationships with neighbouring systems that may or may not be anthropized.

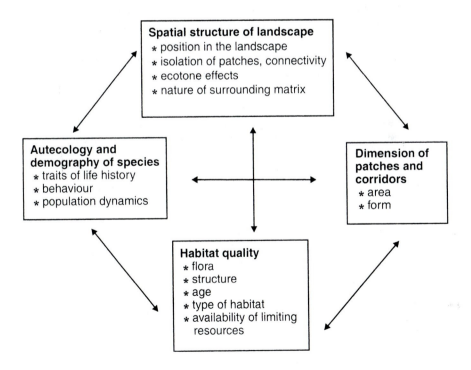

Fig. 1. Major factors influencing the use of corridors by fauna, identified from a bibliographic review (Lindemayer and Nix, 1993)

Reserves must make it possible to protect and manage natural or semi-natural landscapes that are representative of regional landscapes. For this, we must reason on the regional level in terms of networks. These networks, called bioreserves in the United States, are made up of several central zones in which the protective measures are strictest, and a series of concentric zones that serve as buffer zones, where the constraints of management are less strict, and corridors linking these zones (Barrett and Barrett, 1996). The level of protection in each zone depends on its relationship with the central zone and conflicts of potential uses. These concepts dominate the major present projects of conservation (e.g., Natura 2000 network, ECONET).

1.1. Form and nature of corridors

The quality of habitats within corridors is a fundamental element in any type of landscape, whatever the species considered. Research studies have shown it for carabid Coleoptera of the forest (Charrier et al., 1997), the white-footed mouse (Merriam and Lanoue, 1990), and forest grasses (Forman and Godron, 1986). For these forest species, the complexity of the structure and the vegetation determine the survival capacities at the local level.

Corridor width is considered fundamental even though few results are available to support this hypothesis; narrow corridors such as hedgerows do serve as a conduit for many species. However, we can reasonably hypothesize that wider corridors are favourable to the displacement of species in the interior environment. To an administrator's question on how wide a corridor should be, most scientists would respond, the wider it is, the better it will function (Noss, 1987).

On the basis of present understanding of populations in fragmented environments, Forman (1995) proposed an evaluation of the efficiency of corridors as a function of their quality and their width (Fig. 2). A corridor linking two patches and bordered with habitats of medium to good quality is supposed to favour the highest rate of movement between the patches (Fig. 2a). A reduction in the connectivity of the corridor and its quality as well as that of nearby habitats must reduce the intensity of the flow of individuals. For example, the environmental gradient along the corridor will reduce the movement, as a function of different conditions encountered by the species. The presence of barriers, more or less hostile environments, between patches that do ensure connectivity also limits flows.

Gaps in the corridors are avoided by many species, and they thus interrupt the flows (Yahner, 1983; Opdam et al., 1985; van Dorp and Opdam, 1987). In fact, many mammals living in trees need a continuous line of trees to move from one wood to another (Dmowski and Koziakiewicz, 1990; Henein and Merriam, 1990; Merriam and Lanoue, 1990).

The form of a corridor, straight, slightly curved, or highly curved, also affects the flow of organisms. Soule and Gilpin (1991) hypothesize that

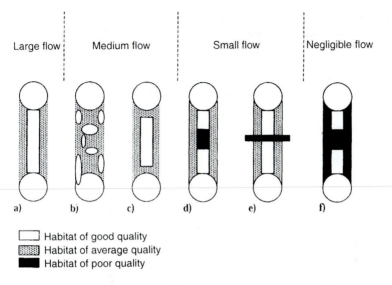

Fig. 2. Supposed intensity of flows of individuals between two patches as a function of the connectivity and quality of elements (Forman, 1995): (a) connected corridor, (b) set of small patches, (c) unconnected corridor, (d) corridor with a gap, (e) corridor with a barrier, (f) corridor interrupted by a barrier in an environment of poor quality.

straight corridors are more effective than curved ones because an animal need not search for them or modify its direction to follow them. This is particularly relevant for juveniles, which do not know the trails.

1.2. The European network of corridors and the pan-European strategy

European governments have observed an overall decline in biodiversity associated with recent changes in the landscape. The reduction of natural or semi-natural habitats, as well as the fragmentation of many large areas into increasingly isolated islands, have been responsible for this condition. Wetland area has fallen to less than half in some countries over the past 40 years. The Mediterranean forest covers only 10% of its original area. The coastal and alpine zones are threatened. Intensive agriculture has reduced areas of grassland, moor, and woods throughout Europe.

1.2.1. ECONET

A strategy has been developed to define an "ecological network" on the continental level, drawing on the concepts and theories explained in the earlier chapters of this work. Based on the island biogeography theory, the functioning of populations, and the spatio-temporal organization of landscapes by human activities, it aims to provide populations the structures needed for their survival in landscapes that have been intensely exploited by man for a long time.

The credibility of the ecological network concept was sustained by many nature conservation organizations across Europe. Several concrete initiatives have already been taken by some countries (Nowicki et al., 1996). At the European level, the proposition to create a European ecological network, ECONET, was made in 1991 and ratified by 46 countries in 1995. These countries engaged themselves to establish the network within 10 years.

In 1991, Bischoff and Jongman (1991) proposed a plan for this network under the title Tentative Ecological Main Structure (Fig. 3). The principles that dominated the plan were the maintenance of central zones and the establishment of connecting zones. The central zones are the principal zones protected in each state: national parks, landscape conservation areas, zones of international interest for the conservation of nature. The connections are defined according to the ecological characteristics of the natural reserves they link together. At the European level, the connections lie on the major migration routes of birds or large mammals. Regional or local corridors may be included in a later phase of the project.

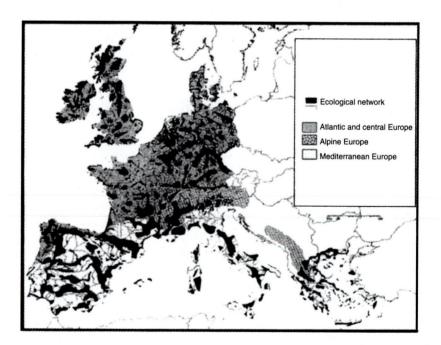

Fig. 3. Map of Tentative Ecological Main Structure in the European Community (Bischoff and Jongman, 1991)

In 1995, the European countries decided to bring this project within a pan-European strategy for the maintenance of biological and landscape diversity. The strategy draws on European as well as international projects and organizations: the European network Natura 2000, the conventions of Bern, Bonn, and Ramsar, the biogenetic reserves of the European Commission, the reserves of the International Union for Conservation of Nature, and the ECONET model.

1.2.2. *The pan-European strategy for maintenance of biological and landscape diversity*

The strategy for maintenance of biological and landscape diversity has four objectives (Nowicki et al., 1996):

—to reduce threats to biological and landscape diversity at the European level;

—to increase the resilience of biological and landscape diversity in Europe;

—to reinforce ecological coherence; and

—to ensure the awareness of citizens about the maintenance of biological and landscape diversity and their participation in it.

To meet these objectives, the strategy involves six major actions that must be realized in the next 20 years:

— conservation and restoration of key ecosystems, habitats, species, and landscape elements by means of the operation of ECONET;

—reasoned and sustainable use and management of biodiversity and landscapes;

—integration of biodiversity and landscapes in all sectors of the economy;

—improvement of information available to the public;

—improvement of investment and understanding at the state level; and

—assurance of adequate finance for the implementation of this strategy.

The specific actions to be undertaken in the first four years were:

—establishment of an ecological network;

—integration of objectives in the other sectors;

—promotion of the project among the public and authorities;

—conservation of landscapes, specifically coastal and marine ecosystems, rivers and riverine zones, continental wetlands, grassland ecosystems, forest ecosystems, and mountain ecosystems; and

—action for the conservation of endangered species.

1.2.3. *An example of a national network: the Netherlands*

The Netherlands Parliament approved a nature policy plan in 1990. Its key objective was the development of a network of ecological corridors in the next 20 to 30 years.

The Netherlands lost a significant part of its natural wealth during the 20th century. Efforts on the part of the government and private organizations were not able to prevent this deterioration. From 1900 onward, the area of natural zones diminished by 75%. The remaining area was small and fragmented: more than 75% of the natural or semi-natural zones on sandy soil had a width less than 200 m. During the past 50 years, 30% of the higher plant species and around 50% of butterfly species declined in abundance, or even disappeared (Lammers and van Zadeldhoff, 1996).

As in the ECONET network, this plan consisted of a set of central zones, zones of natural development, and ecological corridors (Fig. 4).

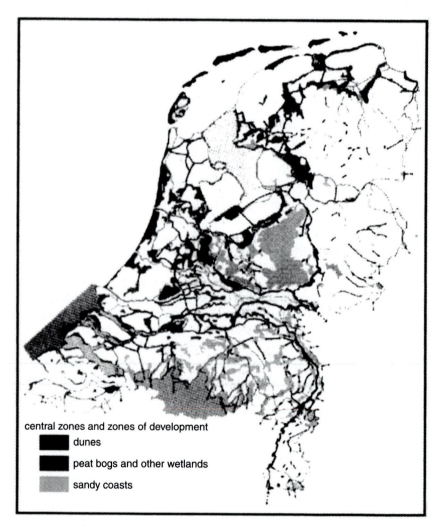

central zones and zones of development

   dunes

   peat bogs and other wetlands

   sandy coasts

Fig. 4. The Netherlands national ecological network (Lammers and van Zadeldhoff, 1996)

The central zones are larger than 500 ha and their present ecological value is of national or international importance.

Zones of natural development are areas in which new habitats are created, which are meant to complement the network. These are mainly wet permanent grasslands and marsh lands.

The corridors are either natural or constructed. They facilitate the movement of organisms between the central zones and consist of hedgerows, dykes, canals, roadsides, or the banks of streams.

The most important aspect of this project has been its adoption by Parliament, and thus its acknowledgement as a political stake. The idea of a coherent network, presented on a readable map, proved to be a unifying concept and motivation for politicians and scientists as well as the public. The elements that slowed down the establishment of the network were related to low environmental quality and the lack of precise objectives on the regional and local level. To overcome these difficulties, a list of target species, key habitats, and management strategies were defined. These were used to fine-tune the priorities for the future development of the network, quantitatively (hectares per species to be protected) as well as qualitatively (choice of habitats to be protected, restored, or created). The cost-benefit of this network for society as a whole will be evaluated (Lammers, 1994).

1.3. Corridors between national parks or reserves: a tool for conservation of species

The establishment of corridors for management of fauna dates from the beginning of the 20th century in the United States and in Great Britain. Before 1950, conservators of wildlife and managers of water fowl had advocated the efficacy of linear elements such as hedgerows or forest belts along river banks (Harris and Scheck, 1991). Corridor networks soon became the basis of conservation of other species: game birds, squirrels, ungulates (Pedevillano and Wright, 1987), the large carnivores (Maehr, 1990), and elephants in Africa and Asia (Philley et al., 1985). In Europe, together with the awareness of the importance of hedgerow networks in the development of landscapes especially for game birds (Dowdeswell, 1987), a considerable effort was made to establish corridors for the amphibians (Langton, 1989).

The function of corridors has been recognized unequivocally for about 20 years in wildlife protection policies. The national parks and other nature reserves are no longer considered in isolation, but as part of a continuous group of protected zones, in order to conserve the biodiversity (Wilcove and May, 1986; Diamond, 1988; Hunter et al., 1988). In land development plans, conservators of nature have become aware of the futility of trying to conserve species in highly anthropized zones without establishing an integrated system of parks, buffer zones, and corridors. This is true for cities (Stearns, 1995; Taylor et al., 1995; Walmsley, 1995), agricultural zones

(Vos and Opdam, 1993), and forest massifs (Maguire, 1987; Harris and Harris, 1997). These principles were generally acknowledged during the 1980s (Harris, 1985).

Corridors are established in landscapes on the basis of many different criteria and could be the source of conflicts of use. A corridor put into place for an electric powerline, the green embankments of a highway, or a hiking trail will not necessarily meet the needs of flora or fauna. The development must be carried out with a precise objective in order to be effective, but in order to be long-lasting it must take into account other cultural aspects of the landscape, in the broad sense.

1.3.1. Examples of general principles for establishment of corridors

Scientists and land managers agree that there is no universal definition of corridor, that its form, nature, and maintenance depend on the species considered and the state of the environment. However, some principles necessary for the establishment of corridors have been stated in the past few years and serve as a basis. They are derived from our understanding of the needs of species as a function of their traits of life history, and they rely heavily on a legislative framework and technological feasibility. There is no general rule, but only particular cases, detailed in a "fragmented" literature. Such information is found in scientific articles as well as in reports on transport administration, the environment, and agriculture, and in publications of symposium proceedings. This dispersal of sources makes it difficult to assess holistically the importance of corridors in the management of species, and few follow-up studies are available at present. Observatories of corridors need to be set up to evaluate their role with respect to the target species, their integration in the landscape, and their acceptance by society.

Harris and Scheck (1991) compiled some texts on the establishment, management, size, and form of corridors. Some extracts from their work are presented below.

Establishment, location

Restoration of mining sites: Replanted belts must be connected to sites that were never disturbed by quarrying activities. This is vital to allow a large number of species to recolonize the rehabilitated spaces (King et al., 1985).

Crossing of caribou: The number of animals crossing a road or a pipeline may be reduced by the combination of these two types of corridor in a single place. In fact, the frequency of crossing is lowest when the traffic on a road running along an above-ground pipeline is greater than 3 vehicles a minute (Curatolo and Mirphy, 1986).

Management, maintenance

Controlled fire: Controlled fire is predicted to burn every three years in zones located in high-altitude pine forests. To reinforce the aesthetics of the landscape, some of the corridors are located close to highways (Gehrken, 1975).

Crossing of vehicles: If the corridors must be crossed by roads providing access to forest parcels that will be exploited, such roads must be as few and as narrow as possible. Access lanes other than those required for protection against fire must be closed after exploitation (Carr et al., 1984).

Width and length

To put into place effective corridors for elephants and other large animals in Sri Lanka, the following rules were articulated in the government's policy: The corridor must be at least 5 km wide, which corresponds to the migration route and a buffer zone to protect it on either side. Each buffer zone must be at least 500 m wide in a forest plantation, or 1.5 km wide in a grazed grassland. On the border of the grazed zone, a road for access to pump vehicles is necessary. The migration route can be developed by adding watering holes and opening up the vegetation or planting vegetation if necessary. The corridor must be policed by guards every 6 km (Rudran et al., 1980). This policy led to the establishment of corridor networks on the national level (Fig. 5).

Andersson and Erlinge (1977) showed that the trench created by an electric powerline in a forest massif must be more than 60 m wide to serve as a corridor offering birds the conditions of an open environment. On the other hand, wooded corridors in an agricultural zone must be wider than 100 m for the survival of some tree species characteristic of interior environments, such as beech in the state of Wisconsin (USA) (Noss, 1983).

1.3.2. Some corridors established for fauna

Corridor for fauna of the Rio Grande

This large-scale project was set up by the US Fish and Wildlife Service and the Texas state parks and wildlife department. It also involved the cooperation of non-government organizations in the southwestern United States. The project aimed to link a dozen natural reserves by corridors, over a total length of 750 km. This system of reserves and connections served not only to protect the big cats of the region (panther, jaguar, lynx), but also to consolidate the populations of several other endangered species. The average width of connections established is 10 km, which is much greater than what is usually called a corridor for fauna in land development plans.

An international corridor between Italy and Switzerland

It is acknowledged that isolated reserves will not suffice to protect biodiversity. The idea of connection has been extended to landscape units, as for example between the Stelvo National Park in Italy and the northernmost national park in Switzerland (Fig. 6). This landscape link, created in 1977, was designed to maintain or restore migration routes of the elk (*Cervus elaphus*).

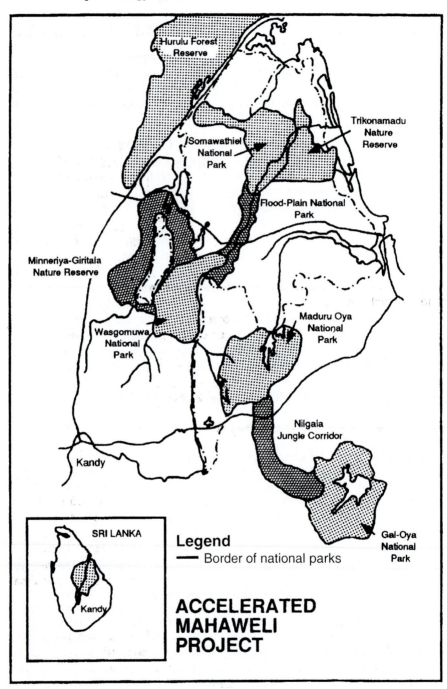

Fig. 5. One of the largest corridor networks for fauna was developed in Sri Lanka for elephants and other forest species (Harris and Scheck, 1991)

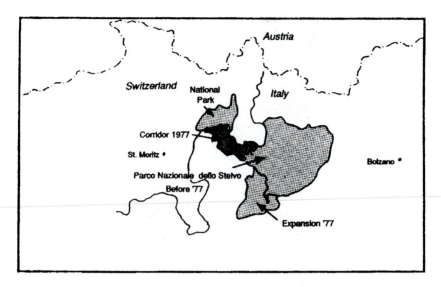

Fig. 6. Landscape corridor between Switzerland and Italy

1.4. The greenways movement

Greenway and *corridor* are very similar terms. A hedgerow is a corridor for the ecologist who studies the movements of forest species. It is considered a greenway by the administrator or manager who plants a linear element for environmental reasons. The term *greenway* emerged in the 1950s and was largely used by administrators, landscape architects, and conservationists (Little, 1990). But the establishment of such elements dates from more than a century ago, especially in the United States.

1.4.1. A brief history

As early as 1860, Frederick Law Olmsted underlined the benefits provided by open linear spaces, which allow access to parks of large cities and increase their positive effects. He carried out his ideas in plans for Berkeley (California), New York, Chicago, Buffalo, Boston, and other cities. In Boston, the objectives of administrators combined aesthetics and management of the quality and quantity of water flows. They redesigned the valley to increase the flood plain in the peat zone, thus facilitating the drainage of land from the slopes. These works were combined with the establishment of a system to purify city wastewater, which until then had simply been released into the river (Fig. 7).

In England, the concept of *garden city* developed at the end of the 19th century. The city centre, essentially residential, was surrounded by a green

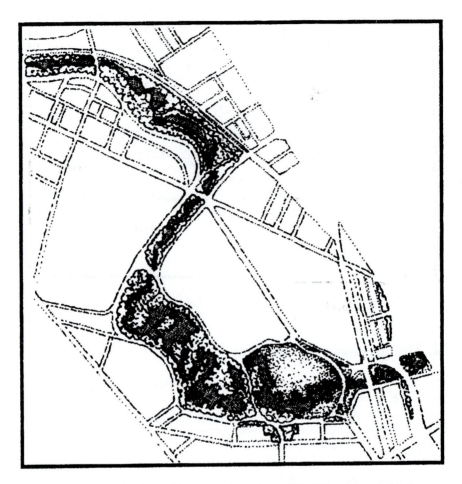

Fig. 7. Olmsted's plan for the Muddy river and peat bogs of Black Bay, integrated in the green belt (Emerald Necklace)

belt about 100 m wide. Beyond that, after the commercial and industrial zones, the garden city was surrounded by farms and forests. These concepts were applied with more or less success to the management of many cities in Great Britain (Smith, 1993). The concept of green belt was taken up in many other countries. In France, it appeared in urban development plans for large cities in the 1970s.

These first conceptions of greenways were rapidly reinforced with the development of ideas on ecological corridors, especially in the valleys and in zones with highly marked relief (Lewis, 1964, in Smith, 1993).

From the 1960s onward, problems of surface water pollution led to the establishment of many greenways in the riverine zones. The restoration of water quality and aquatic habitats became a management objective. The conservation of buffer zones around water courses is one of the means implemented to meet this demand. In France, the establishment of permanent green belts along rivers could be integrated in the framework of river contracts, permanent fallows, and development plans on the level of departments or watershed agencies.

1.4.2. Definition and diversity of greenways

The term *greenways*, essentially used in the United States, was defined by Little (1990) as a linear space established along a natural corridor, such as the bank of a large river, a river valley, a ridge, or a space running along a railway, canal, or scenic road, or even a link between urban parks, nature reserves, and heritage or historic sites (Plate 12). This definition applies to many types of greenways. On the one hand, planners propose mostly greenways for recreational and aesthetic value to respond to the needs of the public. On the other hand, conservation biologists and landscape ecologists propose the management of greenways to conserve or restore the ecological integrity of landscapes. These objectives are different but not necessarily mutually exclusive. Even though some of them serve only one function, either conservation or recreation, most serve both.

The ecological functions of greenways are those of corridors, which have been described in detail in Part III of this book. The social functions are mostly linked to the development of open-air recreation near large cities. Greenways also improve landscape aesthetics and reinforce the identity of communities by linking parks and cultural and historic sites.

1.4.3. Examples

The Florida greenways

Because of the rapid extension of urbanization, the fragmentation of natural habitats in Florida has increased at a disturbingly rapid rate. The state, the federal forest service, nature conservation organizations, and other public and private bodies combined their resources to maintain the connectivity between natural habitats at the state level. Among the most important achievements are the following (Fig. 8):

—*The faunistic corridor of Pinhook Swamp*, which linked two nature reserves: Georgia's Okefenokee National Wildlife Refuge and Florida's Osceola National Forest. These areas together represent a total of 300,000 ha and constitute a favourable site for the reintroduction of the wolf and Florida panther.

—*The Suwannee river*, which flows from the swamps of Okefenokee and Pinhook towards the Gulf of Mexico. Large areas were bought by an organization similar to the French watershed agencies, to protect the natural zones and the water quality. A corridor for fauna was considered, linking the Okefenokee complex to the coastal zones.

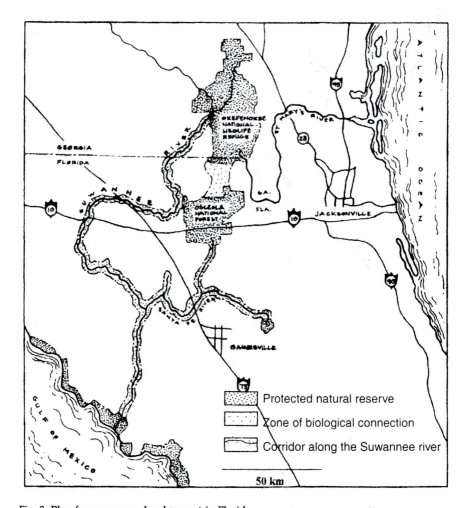

Fig. 8. Plan for greenway development in Florida

—*The greenway of the Wekiva river*, which connects several nature reserves of northern Florida, in the region of Orlando. In this zone, wildlife protection efforts are focused on the black bear.

—*A set of underground passes* for fauna was constructed on route 75, which cuts through the present habitat of the Florida panther.

Development of a new town—Woodlands, Texas

In the early 1970s, Woodlands, a new town, was built north of Houston. The objective of urban planners was to establish a community improving the daily life of inhabitants and respecting the aesthetics and ecological sustainability of the environment. An ecological plan was proposed, and its major axes were riverine corridors:

—The streams and their valleys were protected in order to maintain a natural drainage system that would be cheaper than a traditional constructed network and congenial to faunistic corridors and conservation of open areas for recreation.

—The highways and residential areas were concentrated on poorly drained soils; the runoff waters flowed directly toward permeable soils, which could absorb the excess water.

—The connectivity of natural habitats was favoured to ensure the movement of animals.

—A set of regulations defined locally the rules to be followed in protecting the habitats and the fauna and ensuring the infiltration of waters.

2. CONSIDERING LANDSCAPE ECOLOGY CONCEPTS IN ESTABLISHING TRANSPORTATION INFRASTRUCTURES

Roads, highways, canals, and metalled roads are linear elements that may serve as barriers, conduits, habitats, sinks, or sources in the landscapes they run through. The impacts of these infrastructures vary as a function of the ecological processes considered. We present in this chapter a combination of results on the impacts on fauna and flora, and propositions for a multi-scale process of risk evaluation and setting up of compensatory measures.

2.1. Impacts of a linear infrastructure

2.1.1. Modification of habitats

The construction of a transport route contributes to the reduction and fragmentation of habitats. The negative effects are particularly serious for specialist species with a large vital domain, such as the elk, which is greatly affected by a reduction of less than 5000 ha in a wooded area in which it has found refuge (Grege, 1996).

Simultaneously, there is a creation of new habitats: the green embankments on either side of the highway or metalled road. These environments may have an ecological function that is not negligible. They are most often left in an early stage of succession and serve as a habitat for many species, some of which are locally rare. Sarthou (1996) inventoried adults of the Diptera Syrphidae on the embankments of two highways of southwestern France. The survey of numbers, the calculation of species richness and diversity, overall and per functional group, and the presence of remarkable species show that these environments are used heavily by the winged syrphids in preference to the more or less anthropized biotopes of adjacent environments. In Great Britain, roadsides constitute a reproductive site for all species of reptiles, nearly half the mammal and butterfly species, and about a quarter of the bird species (Bennett, 1991).

The relationships that species maintain with this new environment are variable. They may avoid it, integrate it within their vital domain, or live in it. The repellent effect that highways have on the black bear (*Ursus americanus*) in wooded regions of the Adirondacks (New York State) is such that the density of the highway network is a good estimator of the density of populations (Brocke et al., 1990). The bat and certain raptors (such as the common kestrel in France) have used highways as a preferred hunting territory.

2.1.2. Emissions and source effect

In general, plants and animals living close to roads have higher lead levels than other organisms: this is particularly true for earthworms and other detritivores, as well as for insectivorous small mammals. Nevertheless, no symptom or trace of lead toxicity or poisoning has been demonstrated (Bennett, 1991).

Highway traffic is also a source of noise disturbance. In this case the functional width of the corridor is much more important than that of the road and depends on the sensitivity of various species to noise disturbance. Some, such as the badger or the beaver, are not very sensitive, while others such as the lapwing (*Vanellus vanellus*) or the black-tailed godwit (*Limosa limosa*) perceive a disturbance more than 2000 m on either side of a highway with average daily traffic of 54,000 cars (van der Zande et al., 1980).

Roadsides may also be a source of individuals or propagules of auxiliary, adventitious, or pest species.

All these effects have variable ranges in the adjacent landscape elements (Fig. 9).

2.1.3. Sink effects

Seasonal migrations and movements from one landscape element to another for reproduction are essential to the maintenance of populations. An increase in mortality during these displacements has an impact on the population dynamics. For example, the mortality associated with crossing a road leads to a reduction in the population density of amphibians (Fahrig et al., 1994).

Collisions are frequent for all the organisms that come near roads or their embankments: insects, small mammals, granivorous birds, predators, and necrophages. For common species of small mammals or birds, only a low percentage of the population is killed annually (Hodson and Snow, 1965; Bennett, 1991). This loss is easily compensated for by reproduction. However, road collisions may have a significant impact on populations of large mammals or some rare species. In Florida, for example, vehicles are the main cause of mortality of several species with a low reproductive rate,

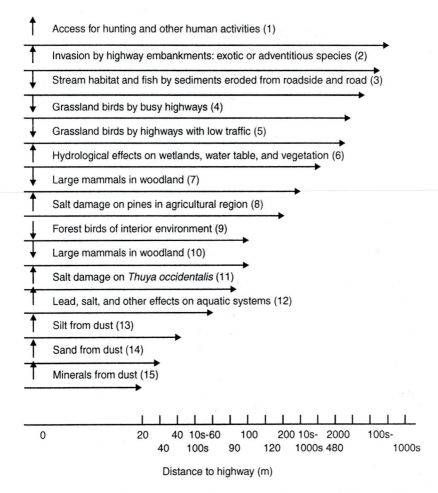

Fig. 9. Effects of a highway corridor on the adjacent matrix (Forman, 1995). The small arrows at left indicate an increase or decrease. Note that the horizontal axis is not linear. (1) Mech et al., 1998; Brocke et al., 1990; (4) van der Zande et al., 1980; Rost and Bailey, 1979; (8) Hofstra and Hall, 1971; (9) Ferris, 1979; (10) Rost and Bailey, 1979; (11) Hofstra and Hall, 1971; (13–15) Tamm and Troedsson, 1955; (2), (3), (6), and (12) estimated by Forman, 1995.

such as the black bear or the Florida panther (an endangered species) (Harris and Gallagher, 1989).

2.1.4. Isolation

Isolation caused by the construction of a road is linked to the association of sink effects and the barrier effect, both of which modify the landscape permeability. The barrier effect may be due to the exposed area of the road, the alteration of bordering environments, or vehicular emissions. The response of animals to this barrier effect depends on their mobility and their tolerance of such constraints.

Isolation has an unfavourable effect on the density of populations and their chances of survival. It may have a significant impact on the survival of metapopulations because it interrupts the flows between sites favourable to local populations and reduces the rates of colonization of vacant sites. The felling caused by the construction of a road is also significant for the functioning of populations when it interrupts the flows between the elements of a functional unit. For example, it may prevent access to reproductive sites, essential resources such as areas of remission or renewal for populations of ungulates.

2.1.5. Connection

Like any linear element, roads may also serve as conduits for the dispersal of organisms. Vehicles displace a large number of seeds. In a car wash in Australia, 259 plant species were identified, some of which were transported over 100 km (Wace, 1977). Some animals, such as snakes and frogs, may also be dispersed in this way (Bennett, 1991). The secondary road axes may be followed in the night by predators (foxes, wolves) or ungulates, if the road is narrow and the traffic sparse (Forman, 1995). The open space above the road may serve as a preferred axis of displacement for bats, especially in forest massifs (Crome and Richards, 1988).

Roadsides constitute dispersal corridors for species such as some small mammals that are not highly sensitive to the artificiality of environments. Highways that form continuous corridors over several hundreds of kilometres may be the cause of dispersal beyond the regional scale. In Illinois, the geographic distribution of the russet vole was extended in this manner over more than 100 km in six years, following the grassy vegetation on highway embankments (Getz et al., 1978).

2.2. Measures of reduction and compensation

2.2.1. Choice of layout

A careful choice of layout for an infrastructure may allow us to limit its harmful effects. At present, spatially explicit models of population dynamics can be used to evaluate the risks linked to establishment of a barrier, a filter, or a conduit in a landscape (Akcakaya and McCarthy, 1995).

The RAMAS/GIS model is an example of the type of tool available at present. From maps that indicate habitats favourable to the species analysed, demographic and spatial data on populations and simulation models evaluate the risks of extinction and the estimated survival time of metapopulations (Fig. 10). The scale of the map depends on life history traits of the species considered, mobility, migration, and other factors, and

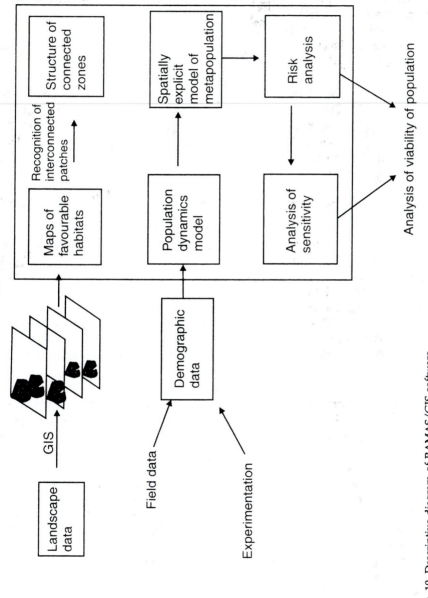

Fig. 10. Descriptive diagram of RAMAS/GIS software

the procedure of development. For example, in France, the highway layouts are studied at various scales as a function of different phases of their development. First, preliminary studies are done on some zones 1 km wide. At this stage it is possible to evaluate the effect of layouts on the fragmentation of large forest massifs, the barrier effect between large complementary environments such as a marsh and its watershed. The study considers the effect of felling on a large scale. Second, the pre-project summary is used to study more precisely the 1 km zone that was chosen. At this stage, the choice of a layout of 300 m that will be subjected to public hearing must take into account the impacts on a certain number of key species characteristic of the landscapes involved as well as on the flows in general. The scale of analysis must go beyond the framework of kilometres and is fixed by the landscape structure and the scale of perception of species. Finally, in the framework of the pre-project study, the layout of the road is fixed in the range of 300 m, and the location of compensatory works is determined. The scale of simulations for evaluation of the impact of development must be the same as for the preceding step.

2.2.2. Compensatory measures: reduction of barrier effect and mortality

To mitigate the barrier effect and mortality, two types of infrastructure are used: passages for fauna reducing the barrier effect and mortality, and closures increasing the barrier effect and reducing the mortality. The passages designed for small fauna (reptiles, frogs, small mammals) are ducts or small tunnels dug in the embankments, often coupled with hydraulic works near the water courses. To balance out the negative impact of an infrastructure on a population of amphibians whose reproductive territory is often isolated, at least 50% of the animals must follow the compensatory route during their migration (Verboom, 1995).

Construction methods also have an impact. The option often taken in France, of constructing roads with clearings and embankments, is the most destructive. The biological or physical flows are systematically interrupted by the work, which is practically impermeable. On the contrary, the construction of bridges over valleys or tunnels for mining disturbs ecological flows much less.

Compensatory mechanisms suitable for large fauna are large overpasses or underpasses. To increase their efficiency, Clergeau (1993) proposed to integrate them in the existing landscape to restore some of its connectivity. To be effective for many species, they must be wide and planted, and the structure of their vegetation must be diverse.

More generally, these passages may be included in zones of biological connectivity (Clergeau and Desire, 1999), the objective of which is to take into account a maximum number of species to conserve the ecological functioning of a landscape.

3. THE DEVELOPMENT OF RURAL LANDSCAPES

Development works generally concern large, heterogeneous spaces and a long time period. By definition, they modify the spatial structures. They are distinguished from management operations, which concern mostly the internal structure of elements, their maintenance in the existing conditions, or their individual evolution and which are conducted in cyclical fashion over short time periods. For example, grassy belts and hedgerows meant to protect a water course must remain for many years and have particular positions, and their planting is a development operation. Each year or every two years, the vegetation must be pruned and trees or shrubs must be cut or replanted according to the mode of management.

Bocage landscapes, as we have seen throughout this work, are characterized by the presence of networks of tree hedgerows. They are constructed by people on land reserved for agricultural production. They are generally landscapes that people wish to conserve for aesthetic, cultural, agronomic, and ecological reasons. In this section, we present ways in which one can maintain or even restore a certain number of processes in the framework of management operations, especially those concerning biodiversity and flows of water and nutrients.

In France, impact studies, following the law on conservation of nature, have been mandatory since January 1978 for land management operations of any scale or cost.

3.1. Principles of ecological engineering

Two major categories of mechanisms are particularly involved: those linked to biodiversity and those linked to water quality. Before their general principles are presented, two points must be emphasized:

(1) A bocage landscape is not limited to a set of hedgerows forming a network. It also comprises a mosaic of farm fields associated with the hedgerow network. For a long time, the effect of hedgerows on fields was emphasized, especially in bioclimate studies. Recent studies also indicate an effect of the field on its boundary and on the structure of the hedgerow. This relationship is established through the mode of management, which differs perceptibly according to the use of the parcel. There is also a relationship with the type of farmer as a function of the farmer's personal attitude toward hedgerows, system of production, and organization of work.

(2) This brings us to the second point: the management of hedgerows, embankments, and ditches before and after management operations. That is, principles of ecological engineering must be developed at two scales: that of the landscape, most often collective (uprooting or planting of hedgerows), and that of the individual manager (annual maintenance,

harvesting of wood). The individual manager is often a farmer but may be a collective, as along the highways.

3.2.   Structures and mechanisms relating to biodiversity

The principal results of studies in landscape ecology that are applicable to rural management operations are as follows:

—The species composition of a hedgerow depends not only on its structure (number of strata of vegetation, tree cover, presence of an embankment) but also on its place in the landscape, particularly the species composition of the hedgerows and groves with which it is connected. That is, there are ecological functions (species flows) within the network.

—This species composition also depends on overall structures (heterogeneity) and local structures (use of fields) of the landscape mosaic, as well as their history. Landscapes having different structures will have different species compositions.

—The corridor effect, possible in hedgerows with dense vegetation for species preferring the forest environment (birds, Coleoptera), is a barrier for species of open spaces (Lepidoptera).

From the point of view of management principles, there are three points that emerge:

—The species to be favoured in a given landscape must be decided on.

—The management must be planned not in terms of hedgerow quantities, but in terms of structure of the network and the mosaic.

—A diversity of landscapes or landscape units of a few hundred hectares may be useful, but that remains to be tested.

3.3.   Structures and mechanisms relating to water quality

The water resource is of major concern. The recent evolution of agricultural activities and land management, for agricultural and other purposes, has led to a general degradation of the quality of this resource. Despite this observation, our understanding of mechanisms linking the use and management of land and the water quality is fragmentary. We would like to highlight a few points.

Many bocage landscape are characterized by the presence of ditches that have an essential role in the control of water circulation by their barrier effects (protection of parcels downstream by ditches perpendicular to the slope) or draining effects (ditches parallel to the slope). The first, associated with a hedgerow or even an embankment, has an anti-erosive value. This has two types of positive consequences—the protection of soils and the protection of water, since the eroded particles and the associated products (phosphorus, pesticides) do not reach the water course.

In regions in which the sub-surface flows are significant (ancient massifs), the hedgerows located on the border of the low-lying wet zones play an essential hydrological role. In particular, they prevent (or limit) the rise of the water table towards the watershed in episodes of severe floods.

Generally, hedgerows perpendicular to the slope have a hydrological function, but they are effective in limiting floods only when the soil is not completely saturated with water.

Despite many affirmations, the role of hedgerows in the protection of water against nitrates has not been demonstrated. There are a few studies showing the low level of nitrates in water under hedgerows, but the reason for it is not well understood. The comparison of water quality in bocage watersheds and non-bocage watersheds is not yet thorough enough for us to evaluate the respective roles of bocage and agricultural practices.

The role of hydromorphic zones of low-lying areas in the denitrification of surface and sub-surface water is being increasingly better understood. The maintenance of the hydromorphic character of these zones, their use in grasslands, and the maintenance of hedgerow belts are advocated. From the perspective of the manager, it is essential to:

—maintain or restore hedgerows and ditches on the boundaries of wet zones and

—maintain or complete the hedgerow networks that control water circulation and prevent erosion.

Whether it is to regulate mechanisms for preservation of biodiversity or maintenance of water quality, management operations must be planned on the landscape scale, in terms of the structure of the network and the mosaic and in relation with the geomorphology.

3.4. Implementation of new modes of management

The shift from individual evaluation of hedgerows, which was common originally in the establishment of impact studies, to the recognition of their functioning within a landscape constitutes a significant evolution in principles of ecological engineering. This evolution was made possible by the development of new scientific processes and concepts as well as new technologies such as geographic information systems.

These technologies must be implemented by study centres; the organization of information in a spatially explicit manner is essential.

On the research site, the development of spatially explicit models of hydrology and landscape ecology (landscape dynamics, population dynamics) must bring about simulations that can be used to test management alternatives: *What would happen if...?*

Figure 11 presents maps of hedgerow quality in terms of biological corridor and relationship to hydromorphic zones. This is done in a diagnostic phase. In the management proposal phase, it is necessary to consider the

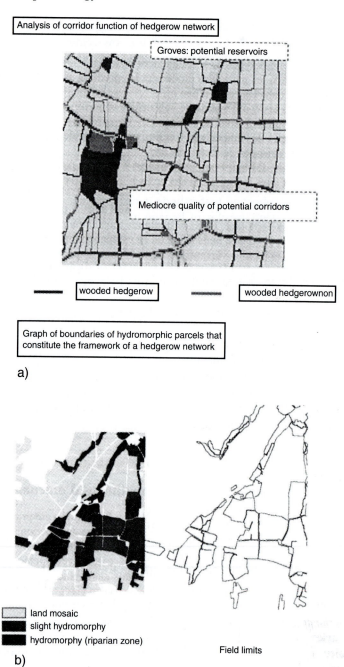

Analysis of corridor function of hedgerow network

Groves: potential reservoirs

Mediocre quality of potential corridors

—— wooded hedgerow   —— wooded hedgerownon

Graph of boundaries of hydromorphic parcels that constitute the framework of a hedgerow network

a)

land mosaic
slight hydromorphy
hydromorphy (riparian zone)

Field limits

b)

Fig. 11. Analysis of a hedgerow network (a) from the perspective of efficiency of biological corridor network and (b) from the perspective of relationships between hedgerows and hydromorphic zones (INRA, SAD-Armorique, CNRS UMR 6365, Ministry of Environment).

Plate 12. Two views of the Garden State Parkway, New Jersey (USA), one of the first greenways constructed in the New York region

(a)

(b)

(c)

Plate 13. Diversity of landscapes of Ercee in Lame, Ille-et-Vilaine, France, a test community for the consideration of aesthetic and ecological aspects in a land development plan (Baudry and Burel, 1984).

different functions of hedgerows and embankments along with social constraints. The hedgerows and embankments bordering hydromorphic zones, the essential role of which in water circulation and the biogeochemical dynamics has been presented in Chapter 8, are a priority. They constitute the framework of a network. Deep lanes, because of their biological importance, constitute the other priority (Baudry and Jouin, in press).

It is also necessary to establish follow-up procedures for management operations, to test whether, in relation to the clearly defined objectives, the management options followed proved to be relevant.

Finally, we must rapidly integrate the evolution of agricultural activities in the evaluation of management projects and take into account the modalities of hedgerow management. There is an engineering technology to be developed in this domain, with one difficulty: maintaining the diversity of practices. We cannot, for ecological and agricultural reasons, imagine a uniform technological itinerary for the management of hedgerows.

3.5. Landscape law: aesthetics and ecological functioning

The French law 93–24 of 8 January 1993 on the protection and evaluation of landscapes offered many regulatory mechanisms to protect and restore uncultivated zones, and more particularly hedgerows and groves in agricultural landscapes. This law aimed to protect landscapes in their many dimensions: aesthetic, cultural, and functional. The methods of landscape evaluation and management to meet the ambitions of the legislator are still in the experimental stage.

During impact studies on land consolidation or other types of land reorganization, the criteria commonly taken into account to conserve or eradicate wooded elements were based, as we have seen in the preceding section, on agronomic properties of the land, the ecological quality of elements, and the spatio-temporal structure of the landscape. The aesthetic aspects and uses other than agricultural were rarely taken into account. This simplification of uses, concentrating on the principal activity and ecological function, constituted an interesting preliminary approach (Baudry and Burel, 1984) but was insufficient at the time the rural landscapes were transformed. The principal function of these landscapes is not only agricultural production but also other varied uses, whether for conservation of nature, protection of the environment, recreation, or production.

Burel and Baudry (1995) compared on a test site the effects of landscapes on ecological functioning and on the perception that different agencies have of a land management plan at the municipal level (Plate 13). This study was based on investigations into the local actors and ecological field studies. The perception of the wooded condition of the landscape was different in farmers and non-farmers. The former perceived the landscape primarily as a tool of production and looked for homogeneous parcels of simple shape

Fig. 12. Simulation of hedgerow suppression. (A) Present state of landscape. (B) In the simulation the hedgerow on the ridge has been uprooted and some trees have been replanted in the remaining hedgerows. Farmers and non-farmers agree in preferring landscape A and consider the uprooting of the hedgerow along the ridge unacceptable in the context of a bocage landscape.

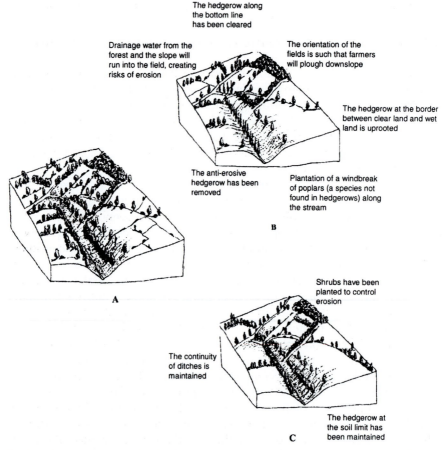

Fig. 13. Alternative options for the development of a bocage landscape. (A) The initial state. (B) Mainly visual aspects were taken into account. (C) Ecological processes are maintained.

and size compatible with farm machinery. The latter preferred the wooded ambience of hedgerow landscapes and articulated a demand for nature. However, the two groups agreed on the need to conserve a certain number of hedgerows, which represent the identity of their landscapes. The most sensitive places were the ridges (Fig. 12), the roadsides, and the outskirts of villages.

Most respondents were ignorant of the functional role of the hedgerow network and were sensitive only to its visual effect.

From the results of this investigation and ecological studies, several scenarios were proposed on aesthetic and ecological bases (Fig. 13), and management principles were elaborated. The integration of various points of view by managers is one of the objectives of a holistic approach to landscapes (Naveh and Lieberman, 1984; Zonneveld, 1995).

Bibliography

Acot, P. (1988). *Histoire de l'écologie.* PUF, Paris.

Acx, A.S. (1991). Hétérogénéité spatiale des pratiques agricoles dans les polders du Mont-Saint Michel. Rennes, ENSAR.

Acx, A.S. and Baudry, J. (1993). *Écologie et friches dans les paysages agricoles.* Ministry of the Environment.

Addicott, J.F., Aho, J.M. and Antolin, M.F. (1987). Ecological neighbourhoods: scaling environmental patterns. *Oikos,* 49: 340–346.

Addiscott, T.M. (1996). A critical review of the vlaue of buffer zone environments as pollution control tool. *In:* N.E. Haycock, T.P. Burt, K.W.T. Goulding and G. Pinay, *Buffer Zones: Their Processes and Potential in Water Protection.* Quest Environmental, Harpenden, pp. 236–243.

Agger, P. and Brandt, J. (1988). Dynamics of small biotopes in Danish agricultural landscapes. *Landscape Ecology,* 1: 227–240.

Akçakaya, H.R. and McCarthy, M.A. (1995). Linking landscape data with population viability analysis: management options for the helmeted oneyeater *Lichenostomus melanops cassidix. Biological Conservation,* 73: 169–176.

Alet, B. (1986). L'oiseau dans le géosystème. Essai de cartographie de l'avifaune dans le massif de Grésigne (Tarn). *Rev. Géogr. des Pyrénées et du Sud-Ouest,* 57: 343–362.

Allen, T.F.H. and Hoekstra, T.W. (1994). *Nested and non-nested hierarchies: a significant distinction for ecological systems.* Systems methodologies and isomorphies. Proc. Soc. General Systems Research Int. Conf., May 1994, Intersystems Publications, pp. 1–7.

Allen, T.F.H. and Starr, T.B. (1982). *Hierarchy. Perspectives for Ecological Complexity.* The University of Chicago Press.

Allen T.H. and Hoekstra, T.W. (1992). *Toward a Unified Ecology.* Columbia University Press, New York.

Amoros, C. and Bravard, J.P. (1985). L'intégration du temps dans les recherches méthodologiques appliquées à la gestion écologique des vallées fluviales. *Rev. Française des Sciences de l'eau,* 4: 349–364.

Andersson, M. and Erlinge, S. (1977). Influence of predation on rodent populations. *Oikos,* 29: 591–597.

Andren, H. (1992). Corvid density and nest predation in relation to forest fragmentation. A landscape perspective. *Ecology,* 73: 794–804.

Andren, H. (1994). Effects of habitat fragmentation on birds and mammals in landscape with different proportions of suitable habitat: a review. *Oikos,* 71: 355–366.

Andren, H. (1995). Effects of landscape composition on predation rates at habitat edges. *In:* L. Hansson, L. Fahrig and G. Merriam, *Mosaic Landscapes and Ecological Processes.* Chapman Hall, London, pp. 225–255.

Andren, H. and Angelstam, P. (1988). Elevated predation rates as an effect in habitat islands: experimental evidence. *Ecology,* 69: 544–547.

Angelstam, P. (1986). Predation on ground nesting birds' nest in relation to predator densities and habitat edge. *Oikos,* 47: 365–373.

Anon. (1974). Carte d'occupation du sol au 1/100 000. Photointerprétation 1974. Parid, IGN—Ministère de la Protection de la Nature et de l'Environnement. Mission de l'environnement rural et urbain.

Anon. (1990). Étude de l'évolution des modes d'occupations agricoles du sol du Marais Poitevin et des Marais de Charente-Maritime. Occupation des sols. État 90. Données SPOT 2 Sept. 1990.

Asselin, A. and Baudry, J. (1989). Les aranéides dans un escape agricole en mutation. *Acta Œcologica Œcologia Applicata*, 10: 143–156.

Atkinson, W.D. and Shorrocks, B. (1981). Competition on a divided and ephemeral resource: a simulation model. *Journal of Animal Ecology*, 54: 507–518.

Atlan, H. (1992). *L'organisation biologique et la théorie de l'information*. Hermann, Paris.

Auger, P., Baudry J. and Fournier F., eds. (1992). *Hiérarchies et échelles en écologie*. Naturalia Publications.

Auricoste, C., Deffontaines, J.P., Fiorelli, J.L., Langlet, A. and Osty, P.L. (1993). *Friches, parcours et activités d'élevage: le cas des Vosges et des Causses*, Institut National de la Recherche Agronomique, Paris.

Austad, I. (1990). Tree pollarding in western Norway. *In:* I. Birks and H. Hilary, *The Cultural Landscape: Past, Present and Future*. Cambridge University Press, Cambridge, pp. 11–30.

Austin, M.P. (1985). Continuum concept, ordination methods, and niche theory. *Annual Review of Ecology and Systematics*, 16: 39–61.

Austin, M.P. and Smith, T.M. (1989). A new model for the continuum concept. *Vegetatio*, 83: 35–47.

Baker, W.L. (1989). A review of models of landscape change. *Landscape Ecology*, 2: 111–135.

Baldock, D., Beaufoy, G., Brouwer, F. and Godeschalk, F. (1996). *Farming at the margins: abandonment or redeployment of agricultural land in Europe*. Institute for European Environmental Policy, Agricultural Economics Research Institute, London, The Hague, 202 pp.

Balent, G. (1987). *Structure, fonctionnement, et évolution d'un système pastoral. Le pâturage vu comme un facteur écologique piloté, dans les Pyrénées Centrales*. Rennes, Thesis, Université de Rennes, p. 150.

Balent, G. and Barrué-Pastor, M. (1986). "Pratiques pastorales et stratégies foncières dans le processus de déprise de l'élevage montagnard en vallée d'Oô (Pyrénées Centrales)." *Revue géographique des Pyrénées et du Sud-Ouest*, 57(3): 403–447.

Balent, G. and Courtiade, B. (1992). Modelling bird communities/landscape patterns realationships in a rural area of South-Western France. *Landscape Ecology*, 6(3): 195–121.

Barbault, R. (1981). *Biologie des Populations et des Peuplements*. Masson, Paris.

Barbault, R. (1992). *Écologie des Peuplements*. Masson, Paris.

Barbero, M., Bonin, G., Loisel, R. and Quézel, P. (1990). Changes and disturbances of forest ecosystems caused by human activities in the western part of the mediterranean basin. *Vegetatio*, 87: 151–173.

Bariou, R. (1978). *Manuel de télédétection—Photographies aériennes—Images radar—Thermographies—Satellites*. SODIPE, Paris.

Barnaud, G. and Lefeuvre, J.C. (1992). L'écologie avec ou sans l'homme? *In:* Collectif, *Entre nature et société. Les passeurs de frontières*, CNRS, Paris, pp. 69–112.

Barrett, N.E. and Barrett, J.P. (1996). Reserve design and the new conservation theory. *In:* S.T.A. Pickett, *The Ecological Basis of Conservation: Heterogeneity, Ecosystems, and Biodiversity*. Chapman Hall, New York, pp. 236–251.

Bascompte, J. and Solé, R.V. (1996). Habitat fragmentation and extinction thresholds in spatially explicit models. *Journal of Animal Ecology*, 65: 465–473.

Bauchau, V. and Le Boulangé, E. (1991). Biologie de populations de rongeurs forestiers dans un espace hétérogène. *In:* M. Le Berre and L. Le Guelte, *Le Rongeur et l'Espace*, Chabaud, Paris, pp. 275–283.

Baudry, J. (1984). *Effects of landscape structure on biological communities: the case of hedgerow network landscapes*. First international seminar on methodology in landscape ecological research and planning, Roskilde, Denmark, Roskilde Universitetsforlag GeoRuc, pp. 55–65.

Baudry, J. (1985). *Utilisation des concepts de landscape ecology pour l'analyse de l'espace rural. Utilisation du sol et des bocages.* Thesis, Universite de Rennes.

Baudry, J. (1988). Structure et fonctionnement écologique des paysages: cas des bocages. *Bulletin d'Écologie,* 19: 523–530.

Baudry, J. (1989). *Colonization of grassland under extensification by hedgerow species.* Brighton Crop Protection Conference, Brighton, Thornton Heath, Surrey, pp. 765–774.

Baudry, J. (1989). Interactions between agricultural and ecological systems at the landscape level. *Agricultural Systems and Environment,* 27: 119–130.

Baudry, J. (1992). Approche spatiale des phénoménes écologiques: détection des effets d'échelles. *In:* P. Auger, J. Baudry and F. Fournier, *Hiérarchies et Échelles en Écologie,* Naturalia Publications, pp. 157–171.

Baudry, J. (1992). Dépendance d'échelle d'espace et de temps dans la perception des changements d'utilisation des terres. *In:* P. Auger, J. Baudry and F. Fournier, *Hiérarchies et Échelles en Écologie,* Naturalia Publications, pp. 101–113.

Baudry, J. (1992). Introduction Générale. *In:* P. Auger, J. Baudry and F. Fournier, *Hiérarchies et Échelles en Écologie,* Naturalia Publications, pp. 9–18.

Baudry, J. (1993). Landscape dynamics and farming systems: problems of relating patterns and predicting ecological changes. *In:* R.G.H. Bunce, L. Rykszkowski and M.G. Paoletti, *Landscape Ecology and Agroecosystems,* Lewis, London, pp. 21–40.

Baudry, J. (1997). Quelle place pour les activités humaines dans la recherche en écologie du paysage? *Œcologia Mediterranea,* 23(1–2): 69–70.

Baudry, J. and Acx, A-S., eds. (1993). Ecologie et friches dans les paysages agricoles. Ministry of the Environment, La Documentation Française.

Baudry, J., Alard, D., Thenail, C., Poudevigne, I., Leconte, D., Bourcier, J.-F. and Girard, C.M. (1997). Gestion de la biodiversité dans les prairies dans une région d'élevage bovin: le Pays d'Auge, France. *Acta Botanica Gallica,* 143: 367–381.

Baudry, J. and Asselin, A. (1991). Effects of low grazing pressure on some ecological patterns in Normandy. *Options Méditerranéennes,* A15: 103–109.

Baudry, J. and Baudry-Burel, F. (1982). La mesure de la diversité spatiale: Utilisation dans les evaluations d'impact. Acta Oecol. *Œcologia Applicata,* 3: 177–190.

Baudry, J. and Bunce, R.G.H., eds. (1991). Land abandonment and its role in conservation. *Options Méditerranéennes.* A15, p. 148.

Baudry, J. and Bunce, R.G.H. (2001). An overview of the landscape ecology of hedgerows. *In:* C. Barr and P. Petit, eds., *Hedgerows of the World: Their Ecological Functions in Different Landscapes,* IALE, UK, pp. 3–16.

Baudry, J., Bunce, R.G.H. and Burel, F. (2000). Hedgerow diversity: an international perspective on their origin, function, and management. *Journal of Environmental Management,* 60: 7–22.

Baudry, J. and Burel, F. (1984). Landscape project: Remembrement: Landscape consolidation in France. *Landscape Planning,* 11: 235–241.

Baudry, J. and Burel, F. (1985). Système écologique, espace et théorie de l'information. *In:* V. Berdoulay and M. Phipps, *Paysage et Système de l'Organisation Écologique à l'Organisation Visuelle.* Université d'Ottawa, pp. 87–102.

Baudry, J. and Burel, F. (1998). Dispersal, movement, connectivity and land use processes. *In:* J. Dover and R.G.H. Bunce, eds., *Key Concepts in Landscape Ecology,* UK IALE, pp. 323–340.

Baudry, J. and Burel, F. (in press). Trophic flows and heterogeneity in agricultural landscapes. *In:* G.A. Polis, M.E. Power and G.R. Huxel, eds., *Food Webs at the Landscape Level,* Chicago University Press.

Baudry, J., Burel, F., Aviron, S., Martin, M., Ouin, A., Pain, G. and Thenail, C. (in press). Temporal variability of connectivity in agricultural landscapes: how farming activities help? *Landscape Ecology.*

Baudry, J., Burel, F., Thenail, C. and Le Coeur, D. (2000). A holistic landscape ecological study of the interactions between farming activities and ecological patterns in Brittany, France. *Landscape and Urban Planning,* 50: 119–128.

Baudry, J. and Denis, D. (1995). *Chloé: Utilitaire d'analyse de l'hétérogénéité d'une image (fichiers image IDRISi)*. Rennes, INRA, SAD-Armorique.

Baudry, J. and Jouin, A. (en préparation). *L'arbre en Réseau: Connaissance et Gestion des Paysages Bocagers*, INRA, Paris.

Baudry, J. and Jouin, A., eds. (in press). De la haie au bocage: structure, fonctionnement et gestion. Ministry of Ecology and Sustainable Development, INRA, Paris.

Baudry, J., Jouin, A. and Thenail, C. (1998). La diversité des bordures de champ dans les exploitations agricoles de pays de bocage. *Études et Recherches sur les Systèmes Agraires*. 31: 117–134.

Baudry, J., Laurent, C. and Denis, D. (1991). *A hierarchical framework for studying land cover patterns changes from an ecological and an economical stand point concepts and results in the Normandy research*. Comparisons of landscape pattern dynamics in European rural areas. EUROMAB Research Program., Ukraine/Normandy, pp. 104–114.

Baudry, J., Laurent, C., Thenail, C., Denis, D. and Burel, F. (1999). Driving factors of land-use diversity and landscape patterns at multiple scales—A case study in Normandy, France. *In:* R. Krönert, J. Baudry, I.R. Bowler and A. Reenberg, eds., *Land-Use Changes and Their Environmental Impact in Rural Areas in Europe*, Parthenon Publishing, Lancs, pp. 103–119.

Baudry, J. and Merriam, H.G. (1988). Connectivity and connectedness: functional versus structural patterns in landscapes. *In:* K.F. Schreiber, *Connectivity in Landscape Ecology*. Proc. 2nd IALE seminar. *Münstersche Geographische Arbeiten*, 29: 23–28.

Baudry, J. and Papy, F. (2001). The role of landscape heterogeneity in the sustainability of cropping systems. *In:* J. Nösberger, H.H. Geiger and P.C. Struik, eds., *Crop Science—Progress and Prospects*. Cabi Publishing, Oxon, pp. 243–259.

Baudry, J. and Tatoni, T. (1993). Changes in landscape patterns and vegetation dynamics in Provence, France. *Landscape and Urban Planning*, 24: 153–159.

Baudry, J. and Thenail, C. (1999). Ecologie et agronomie des bocages: construction des objets d'observation. *In:* S. Wycherek, ed., *Paysages Agraires et Environnement: Principes Écologiques de Gestion en Europe et au Canada*, CNRS, Paris, pp. 129–138.

Baudry, J. and Zhenrong, Y. (1999). Landscape pattern changes in two subtropical Chinese villages as related to farming policies. *Critical Review in Plant Sciences*, 18(3): 373–380.

Bazin, G., Larrère, G.R., De Montard, F.X., Lafarge, M. and Loiseau, P. (1983). *Système Agraire et Pratiques Paysannes dans les Monts Dômes*, INRA, Paris.

Bellot, J. and Golley, F.B. (1989). Nutrient input and output of an irrigated agroecosystem in an arid mediterranean landscape. *Agriculture, Ecosystem, Environment*, 25: 175–186.

Bengtsson, J. (1986). Life histories and interspecific competition between three Daphnia species in rockpools. *Journal of Animal Ecology*, 55: 641–655.

Bennett, A.F. (1991). Roads, roadsides and wildlife conservation: a review. *In:* D.A. Saunders and R.J. Hobbs, *The Role of Corridors*, Surrey Beatty Sons, Chiing Norton, Australia, pp. 99–117.

Bennett, F. (1990). Habitat corridors, their role in wildlife management and conservation. *Conservation and environment*.

Bennett, G., ed. (1996). *Cultural Landscapes: the Conservation Challenge in a Changing Europe*. Institute for European Environmental Policy, London.

Benoît, M., Deffontaines, J.P., Gras, F., Bienaimé, E. and Riela-Cosserat, R. (1997). Agriculture et qualité de l'éau—Une approache interdisciplinaire de la pollution par les nitrates d'un bassin d'alimentation. *Cahiers Agriculture*, 6: 97–105.

Benzécri, J.-P. and Benzécri, F. (1984). *Pratique de l'Analyse des Données: l Analyse des Correspondances et Classification, exposé élémentaire*. Dunod, Paris.

Berdoulay, V. and Phipps, M., eds. (1985). *Paysage et Système*. Université d'Ottawa, Ottawa.

Berger, J.F., Magnin, F., Thiébault, S. and Argant, J. (1997). Essai de paléoécologie d'un paysage: le bassin valdainais (Drôme, France) à l'holocène. *Ecologia Mediterranea*, 23: 145–167.

Berglund, B.E., ed. (1991). *The Cultural Landscape during 6000 Years in Southern Sweden—the Ystad Project.* Munskgaard International Booksellers and Publishers Ecological bulletin, Copenhagen.

Berkes, F. and Folke, C., eds. (1998). *Linking Social and Ecological Systems. Management Practices and Social Mechanisms for Building Resilience.* Cambridge University Press, Cambridge.

Beroutchachvili, N. and Radvanyi, J. (1978). La structure verticale des géosystèmes. *RGPSO,* 49: 181–198.

Berque, A. (1995). *Les Raisons du Paysage.* Hazan.

Bertrand, G. (1975). Pour une histoire écologique de la France rurale. *In: Histoire de la France Rurale.* Le Seuil, Paris, vol. 1, pp. 34–113.

Bertrand, G. (1978). Le paysage entre la nature et la société. *RGPSO,* 49: 239–258.

Birks, H.H., Birks, H.J.B., Kaland, P.E. and Moe, D., eds. (1988). *The Cultural Landscape. Past, Present and Future.* Cambridge University Press, Cambridge.

Bischoff, N.T. and Jongman, R.H.G. (1991). *Development of Rural Areas in Europe: the Claim for Nature.* Netherlands Scientific Council for Government Policy, The Hague.

Blanc-Pamart, C. (1986). Dialoguer avec le paysage ou comment l'espace écologique est vu et pratiqué par les communautés rurales des hautes terres malgaches. *In:* Chatelin and G. Riou, *Milieux et Paysages,* Masson, coll. recherche en géographie, Paris, p. 154.

Blondel, J. (1986). *Biogéographie Évolutive.* Masson, Paris.

Blondel, J. (1995). *Biogéographie Approche Écologique et Évolutive.* Masson, Paris.

Blondel, J., Perret, P., Maister, M. and Dias, P. (1992). Do harlequin Mediterranean environments function as source-sink for blue tits (*Parus caeruleus* L.). *Landscape Ecology,* 6: 213–219.

Bonn, F. and Rochon, G. (1992). *Précis de Télédétection—volume 1: Principes et Méthodes.* Presses universitaires, du Québec/AUPELF, Québec.

Boone, R.B. and Hunter, M.L. (1996). Using diffusion models to simulate the effects of land use on grizzly bear dispersal in the rocky mountains. *Landscape Ecology,* 11: 51–64.

Boorman, S.A. and Levitt, P.R. (1973). Group selection on the boundary of a stable population. *Theoretical Population Biology,* 4: 85–128.

Bormann, F.H. and Likens, G.E. (1979). Catastrophic disturbance and the steady state of northern hardwood forests, *American Scientist,* 67: 660–669.

Boujot, C. (1995). Les Marais de Dol-Châteauneuf: ou la "nature" d'un quiproquo. *In:* C. Voisenat, *Paysage au Pluriel: pour une Approche Ethnologique des Paysages.* Ed. de la Maison des sciences de l'homme, Paris, XVI: 3–18.

Bowles, M.L. and Whelan, C.J., eds. (1994). *Restoration of Endangered Species: Conceptual Issues, Planning and Implementation.* Cambridge University Press, Cambridge.

Breuning-Madsen, H., Reenberg, A. and Holst, K. (1990). Mapping potentially marginal land in Denmark. *Soil Use and Management,* 6: 114–119.

Brocke, R.H., O'Pezio, J.P. and Gustafson, K.A. (1990). A forest management scheme mitigating impact of road networks on sensitive wildlife species. *In: Is Forest Fragmentation a Management Issue in the Northeast?* US Forest Service, Radnor, Pennsylvania.

Bronstein, J.L. (1995). The plant-pollinator landscape. *In:* L. Hansson, L. Fahrig and G. Merriam, *Mosaic Landscapes and Ecological Processes.* Chapman Hall, London, pp. 256–288.

Brown, J. and Kodric-Brown, A. (1977). Turnover rates in insular biogeography: effect of immigration on extinction. *Ecology,* 58: 445–449.

Brown, J.H. (1971). Mammals on mountaintops: Nonequilibrium insular biogeography, *American Naturalist,* 105: 467–478.

Brown, V.K. (1984). Secondary successions: Insect-Plant Relationships. *Bioscience,* 34: 711–717.

Brown, V.K. and Kalff, J. (1986). Successional communities of plants and phytophagous Coleoptera. *Journal of Ecology,* 74: 963–976.

Brunel, E. and Cancela Da Fonseca, J.R. (1979). Concept de la diversité dans les écosystèmes complexes. *Bulletin d'écologie,* 10: 147–163.

Brunet, P. (1992). *L'Atlas des Paysages Ruraux de France.* De Monza, Paris.

Burel, F. (1989). Landscape structure effects on carabid beetles spatial patterns in Western France. *Landscape Ecology*, 2: 215–226.

Burel, F. (1991). *Dynamique d'un Paysage: Réseaux et Flux Biologiques*. Univesité de Rennes, vol. 1, p. 235.

Burel, F. (1991). Ecological consequences of land abandonment on carabid beetles distribution in two contrasted grassland areas. *Options Méditeranéennes*, A15: 111–119.

Burel, F. (1992). Effect of landscape structure and dynamics on carabids biodiversity in Brittany France. *Landscape Ecology*, 6: 161–194.

Burel, F. (1996). Hedgerows and their role in agricutural landscapes. *Critical Review in Plant Sciences*, 15(2): 169–190.

Burel, F. and Baudry, J. (1989). Hedgerow network patterns and process in France. *In:* ZIS and FRTT. *Changing Landscapes: an Ecological Perspective*, Springer-Verlag, New York, pp. 99–120.

Burel, F. and Baudry, J. (1990). Structural dynamic of a hedgerow network landscape in Brittany France. *Landscape Ecology*, 4: 197–210.

Burel, F. and Baudry, J. (1990). Hedgerow networks as habitats for colonization of abandoned agricultural land. *In:* R.H.G. Bunce and D.C. Howard, *Species Dispersal in Agricultural Environments*. Belhaven Press, Lymington, pp. 238–255.

Burel, F. and Baudry, J. (1994). Reaction of ground beetles to vegetation changes following grassland derelictions. *Acta Œcologica*, 15: 401–415.

Burel, F. and Baudry, J. (1995). Social aesthetical and ecological aspects of hedgerow in rural landscapes as a framework for greenways. *Landscape and Urban Planning*, 33: 327–340.

Burel, F., Baudry, J. and Lefeuvre, J.C. (1993). Landscape structure and water fluxes. *In:* R.G.H. Bunce, L. Ryszkowski and M.G. Paoletti, *Landscape Ecology and Agroecosystems*, Lewis Publishers, Boca Raton, pp. 41–47.

Burel, F., Baudry, J., Delettre, Y., Petit, S. and Morvan, N. (2000). Relating insect movements to farming systems in dynamic landscapes. *In:* B. Ekbom, M.E. Irwin and Y. Robert, eds. *Interchanges of Insects between Agricultural and Surrounding Landscapes*, Kluwer Academic Publishers, Dordrecht, pp. 5–32.

Burel, F., Baudry, J., Thenail, C. and Le Coeur, D. (2000). Relationships between farming systems and ecological patterns along a gradient of bocage landscapes. *In:* U. Mander and R.H.G. Jongman, eds. *Conséquences of Land Use Changes*, WIT Press, 5: 227–246.

Burel, F., Butet, A., Canard, A., Clergeau, P., Constant, P., Eybert, M.C., Freytet, T., Le Garff, B., Paillat, G., Petit, S., Tiberghien, G., Ysnel, F., Baudry, J., Acx, A.S., Chevallier, F., Delmas, P., Rivière, J.M., Morzadec, M.T., Langouët, L., Bizien C. and Lefeuvre, J.C. (1995). *Conséquences des transformations de l'espace et des pratiques agricoles en Baie du Mont-Saint-Michel*. Paris, CNRS, PIren, Comité Systèmes Ruraux.

Burel, F., Baudry, J., Butet, A., Clergeau, P., Delettre, Y., Le Cœur, D., Dubs, F., Morvan, N., Paillat, G., Petit, S., Thenail, C., Lefeuvere, J.C. and Brunel, E. (1998). Comparative biodiversity along a gradient of agricultural landscapes. *Acta Œcologica*, 19: 47–60.

Burgess, R.L. and Sharpe, D.M., eds. (1981). *Forest Island Dynamics in Man-dominated Landscapes*. Ecological Studies, Springer-Verlag, New York, 310 pp.

Burgess, R.L. and Sharpe, D.M. (1981). Introduction. *In:* R.L. Burgess and D.M. Sharpe, *Forest Island Dynamics in Man-dominated Landscapes*, Springer-Verlag, New York, 41: 1–5.

Burrough, P.A. (1987). Spatial aspects of ecological data. *In:* R.H.G. Jongman, C.J.F. ter Braak and V. Tofr. *Data Analysis in Community and Landscape Ecology*. Center for Agricultural Publishing and Documentation (Pudoc), Wageningen, The Netherlands, pp. 213–274.

Can, C. (1998). Transfert des polluants vers l'eau. *In:* C. Cheverry, *Agriculture Intensive et Qualité des Eaux*. INRA Éditions, Paris, pp. 233–247.

Canévet, C. (1992). *Le Modèle Agricole Breton: Histoire et Géographie d'une Révolution Agroalimentaire*. Presses Universitaires de Rennes, Rennes, p. 397.

Carcaillet, C. (1997). Évolution de la végétation pendant l'holocène dans la haute vallée de la Maurienne (Alpes du Nord-Quest): un programme multidisciplinaire de paléoécologie du paysage. *Ecologia Mediterranea*, 23: 131–144.

Carpenter, S.R., Caraco, N.F., Correll, D.L., Howarth, R.W., Sharpley, A.N. and Smith, V.H. (1998). Nonpoint pollution of surface waters with phosphorus and nitrogen. *Ecological Applications*, 8: 559–568.

Carr, G.W., Horrocks, G.F.B., Opie, A.M., Triggs, B.E. and Schulz, M. (1984). *Flora and Fauna of the Coast Range Forest Block, East Gippsland, Victoria*. Melbourne, Victoria, Department of Conservation, Forests and Lands. State Forest and Land Services.

Carson, R. (1962). *Silent Spring*, Houghton Mifflin, Boston.

Cassegrain, A. and Barmoy, S. (1995). *Organisation du travail et gestion des haies dans les exploitations agricoles en zones bocagères*. Mémoire de fin d'études. INRA, SAD-Armorique. Rennes, École Nationale Supérieure d'Horticulture de Versailles, 138 pp. + annexes.

Chapelot, J. (1980). Archéologie du paysage. *In*: E. universalis. *Symposium*, 17: 278–280.

Charrier, S. (1994). *Déplacements d'Abax ater* (Coleoptera carabidae) *dans un paysage agricole dynamique, conséquences sur le fonctionnement d'une métapopulation*. Dijon, ENESAD.

Charrier, S., Petit, S. and Burel, F. (1997). Movements of *Abax parallelepipedus* (Coleoptera, Carabidae) in woody habitats of a hedgerow network landscape: a radio-tracing study. *Agriculture, Ecosystems, Environment*, 61: 133–144.

Chatelin, Y. and Riou, G. (1986). *Milieux et Paysages*. Masson, Paris.

Cheverry, C., ed. (1998). *Agriculture Intensive et Qualité des Eaux*. Science Update, INRA Editions, Paris, 297 pp.

Christensen, N.L., Agee, J.K., Brussard, P.F., Hughes, J., Knight, D.H., Minshall, G.W., Peek, J.M., Pyne, S.J., Swanson, F.J., Thomas, J.W., Wells, S., Williams, S.E. and Wrright, H.A. (1989). Interpreting the Yellowstone fires of 1988. *BioScience*, 39(10): 678–685.

Christian, C.S. (1952). Regional land survey. Journal of the Australian Institute of Agricultural Science, 18: 140–147.

Ciceri, M.F., Marchand, B. and Rimbert, S. (1977). *Introduction à l'Analyse de l'Espace*, Masson, Paris.

Clément, B. (1978). *Contribution à l'étude phytoécologique des monts d'Arrée. Organisation et cartographie des biocénoses, évolution et productivité des landes*. Thesis, Université de Rennes.

Clements, F.E. (1936). Nature and structure of the climax. *The Journal of Ecology*, 24: 252–284.

Clergeau, P. (1993). Utilisation des concepts de l'écologie du paysage pour l'élaboration d'un nouveau type de passage à faune. *Gibier Faune Sauvage*, 10: 47–57.

Clergeau, P. and Burel, F. (1997). The role of spatio-temporal connectivity on tree creeper distribution at the landscape level. *Landscape and Urban Planning*, 38: 37–43.

Clergeau, P. and Désiré, G. (1999). Biodiversité, paysage et aménagement: du corridor à la zone de connexion biologique. *Mappemonde*, 5: 18–23.

Cliff, A.D. and Ord, J.K. (1981). *Spatial Processes: Models and Applications*. Pion Limited, London.

Collectif (1996). *Écosystèmes prairiaux*. Bulletin de la société botanique de France, Condé sur Noiraud.

Collectif, ed. (1995). *Paysage au pluriel. Pour une approche ethnologique des paysages*. Collection Ethonologie de la France, Éditions de la Maison des sciences de l'homme, Paris, 250 pp.

Combes, C. (1998). Les interactions durables: parasites, hôtes, environnement. *In*: Collectif. *Recherches actuelles sur l'écologie et l'évolution: de la biologie évolutive à la biosphère*. Centre National de la Recherche Scientifique, Paris, pp. 35–42.

Comeleo, R.L., Paul, J.F., August, P.V., Copeland, J., Baker, C., Hale, S.S. and Latimer, R.W. (1996). Relationships between watershed stressors and sediment contamination in Chesapeake Bay estuaries. *Landscape Ecology*, 11: 307–319.

Comolet, A. (1989). Prospective à long terme de la déprise agricole et environnement. Ministry of Environment, Paris, SRETIE.

Constant, P., Eybert, M.C. and Maheo, R. (1976). Avifaune reproductrice du bocage de l'Ouest. *In:* CNRS INRA, Ensa et Université de Rennes. *Les Bocages: Histoire, Ecologie, Économie.* INRA, Rennes, pp. 327–332.

Conway, G.R. and Pretty, J.N. (1991). *Unwelcome Harvest: Agriculture and Pollution.* London.

Coppens, Y. (1983). *Le Singe, l'Afrique et l'Homme.* Fayard, Paris.

Corbet, G.B. (1975). *Examples of short and long term changes of dental pattern in Scottish voles* (Rodentia, Microtinae), *Journal of Animal Ecology,* 56: 17–21.

Correll, D.L. (1996). Buffer zones and water quality protection: general principles. *In:* N.E. Haycock, T.P. Burt, K.W.T. Goulding and G. Pinay. *Buffer Zones: Their Processes and Potential in Water Protection,* Quest Environmental, Harpenden, pp. 7–20.

Cowles, H.C. (1899). The ecological relations of the vegetation of the sand dunes of Lake Michigan. *Botanical Gazette,* 27: 95–117, 167–202, 281–308, 361–391.

Craighead, J.J. (1991). Yellowstone in transition. *In:* R.B. Keiter and M.S. Boyce, *The Greater Yellowstone Ecosystem: Redefining America's Wilderness Heritage,* Yale University Press, New Haven, pp. 27–39.

Crist, T.O., Guertin, D.S., Wiens, J.A. and Milne, B.T. (1992). Animal movement in heterogeneous landscapes: an experiment with Eleodes beetles in shortgrass prairie. *Functional Ecology,* 6: 536–544.

Crome, F.H.J. and Richards, G.C. (1988). Bats and gaps: microchiropteran community structure in a Queensland rainforest. *Ecology,* 69: 1960–1969.

Crumley, C.L. and Marquardt, W.H. eds., (1987). *Regional Dynamics. Burgundian Landscapes in Historical Perspective,* Academic Press, INC, San Diego, California.

Curatolo, J.A. and Mirphy, S.M. (1986). The effects of pipelines, roads, and traffic on the movements of caribou (*Rangifer tarandus*). *Canadian Field Naturalist,* 100: 218–224.

Daget, P. (1975). Le diagnostic phyto-écologiqué appliqué l'aménagement des régions de moyenne montagne: cas de La Margeride. *In:* G. Long, *Diagnostic Phyto-écologique et Aménagement du Territoire,* vol. 2, Masson, Paris, pp. 42–55.

Danielson, B.J. (1991). Communities in a landscape: the influence of habitat heterogeneity on the interactions between species. *The American Naturalist,* 138: 1105–1120.

DeAngelis, D.L. and Gross, L.J., eds. (1992). *Individual Based Models and Approaches in Ecology: Populations, Communities and Ecosystems.* Chapman Hall, New York.

De Gennes, P.G. (1990). Continu et discontinu: l'exemple de la percolation. *In: Symposium: les Enjeux.* Encyclopaedia Universalis, Paris, p. 744.

De Gennes, P.G. and Guillon, eds. (1977–1989). *L'ordre du Chaos.* Pour la Science/Belin, Paris, 203 pp.

De Golia, J. (1993). *Fire a Force of Nature.* KC Publications Inc., Las Vegas.

Décamps, H. (1984). Toward a landscape ecology of river valleys. *In:* Coeley and F. Golley. *Trends in Ecological Research for the 1980's.* Plenum Publishing Corporation, pp. 5–7.

Décamps, H., Fortuné, M., Gazelle, F. and Pautou, G. (1988). Historical influence of man on the riparian dynamics of a fluvial landscape. *Landscape Ecology,* 1(3): 163–173.

Décamps, H. and Naiman, R.J. (1989). L'écologie des fleuves. *La Recherche,* 208: 310–319.

Deffontaines, J.P. (1993). The farm field, a focus for interdisciplinary studies. From ecophysiology to the human sciences. *In:* J. Brossier, L. de Bonneval and E. Landais, *System Studies in Agriculture and Rural Development,* INRA Editions, Paris, pp. 19–30.

Deffontaines, J.P. (1996). Enjeux spatiaux en agronomie. *C.R. Académie d'Agriculture,* 82: 5–14.

Deffontaines, J.P. (1996). Du paysage comme moyen de connaissance de l'activité agricole à l'activité agricole comme moyen de production du paysage. L'agriculteur producteur de paysage. Un point de vue d'agronome. *C.R. Académie d'Agriculture,* 82: 57–69.

Deffontaines, J.P. and Lardon, S., eds. (1994). *Itinéraires cartographiques et développement.* Espaces ruraux, INRA Publications, Paris, 136 pp.

Deffontaines J.P., Thenail, C. and Baudry, J. (1995). Agricultural systems and landscape patterns: how can we build a relationship? *Landscape and Urban Planning,* 31: 3–10.

Delattre, P., Giraudoux, P., Baudry, J., Quéré, J.P. and Fichet, E. (1996). Effect of landscape structure on common vole (*Microtus arvalis*) distribution and abundance at several space scales. *Landscape Ecology*, 11: 279–288.

Delelis-Dussollier, A. (1973). *Contribution à l'étude des haies, des fourrés préforestiers, des manteaux sylvatiques de France*. Thesis, Université de Lille.

Den Boer, P.J. (1985). Fluctuations of density and survival of carabid populations. *Œcologia*, 67: 322–330.

Deniel, J. (1965). Les talus et l'aménagement de l'espace rural. *Penn ar bed*, 41: 41–54.

Dennis, R., ed. (1992). *The Ecology of Butterflies in Britain*. Oxford University Press, Oxford.

Denslow, J.S. (1985). Disturbance-mediated coexistence of species. *In:* S.T.A. Pickett and P.S. White. *The Ecology of Natural Disturbance and Patch Dynamics*. Academic Press Inc., New York, pp. 307–324.

Desender, K. (1982). Ecological and faunal studies on coleoptera in agricultural land. II hibernation of carabidae in agro-ecosystems. *Pedobiologia*, 23: 295–303.

Despain, D.G. (1990). *Yellowstone Vegetation*. Roberts Rinehart Publishers, Boulder, Co., 239 pp.

Di Castri, F. (1981). L'écologie: naissance d'une science de l'homme et de la nature. *Le Courrier de l'UNESCO*, 34: 6–11.

Di Pietro, F. (1996). *Durabilité et organisation du paysage: Application des concepts de l'écologie systémique au diagnostic de la gestion pastorale du territoire des vallées des Pyrénées centrales— France*. Université Paul Sabatier, Toulouse.

Diamond, J.M. (1976). Island biogeography and conservation: strategy and limitations. *Science*, 193: 1027–1029.

Diamond, J.M. (1988). Urban extinction of birds. *Nature*, 332: 393–394.

Dillaha, T.A. and Inamdar, S.P. (1996). Buffer zones as sediment traps or sources. *In:* N.E. Haycock, T.P. Burt, K.W.T. Goulding and G. Pinay, *Buffer Zones: Their Processes and Potential in Water Protection*, Quest Environmental, Harpenden, pp. 33–42.

Dmowski, K. and Koziakiewicz, M. (1990). Influence of a shrub corridor on movements of passerine birds to a lake littoral zone. *Landscape Ecology*, 4: 98–108.

Dowdeswell, W.H. (1987). *Hedgerows and Verges*. Allen, Unwin Ltd., London.

Dub, F. (1997). *Avifaune nicheuse des paysages de bocage: échelles d'analyse et échelles de réponse*. Sciences Biologiques, Rennes, thesis, Université de Rennes, 1: 230.

Ducruc, J.P. (1980). *Le Système Ecologique, Unité de Base de la Cartographie Écologique*. Ottawa, Environnement Canada, p. 152.

Ducruc, J.P. (1985). Le système écologique: un niveau privilégié du paysage. *In:* V. Berdoulay and M. Phipps, *Paysage et Système*, Editions de l'Université d'Ottawa. Ottawa, pp. 23–32.

Duncan, P. (1983). Determinants of the use of habitats by horses in a Mediterranean wetland. *Journal of Animal Ecology* 52: 93–109.

Dunning, J.B., Danielson, B.J. and Pulliam, H.R. (1992). Ecological processes that affect populations in complex landscapes. *Oikos*, 65: 169–175.

Dunning, J.B., and Stewart, J.D.J., Danielson, B.J., Noon, B.R., Root, T.L., Lamberson, R.H. and Stevens, E.E. (1995). Spatially explicit population models: current forms and future uses. *Ecological Applications*, 5: 3–11.

Dupuy, J.L. (1991). *Modélisation prédictive de la propagation des incendies de forêts*. INRA Avignon— Université Claude Bernard Lyon.

Duvigneaud, P. (1980). *La Synthèse Écologique*. Doin, Paris.

Erlinge, S., Goransson, G., Hogstedt, G., Liberg, O., Loman, J.N., Nillson, I.T., Von Schantz, T. and Sylven, M. (1983). Predation as a regulating factor on small rodent populations in southern Sweden. *Oikos*, 40: 36–52.

Eybert, M.C. (1985). *Dynamique évolutive des passereaux des landes armoricaines. Cas particulier d'une population de Linotte mélodieuse* (Acanthis cannabina L.). Thesis, Université de Rennes, p. 336.

Fabos, J.G. and Ahern, J., eds. (1995). *Greenways: the Beginning of an International Movement*. Elsevier, Amsterdam.

Fahrig, L. (1991). Simulation methods for developing general landscape level hypotheses of single species dynamics. *In:* M.G. Turner and R.H. Gardner, *Quantitative Methods in Landscape Ecology*, Springer-Verlag, New York, pp. 417–442.

Fahrig, L. and Merriam, H.G. (1985). Habitat patch connectivity and population survival. *Ecology,* 66: 1762–1768.

Fahrig, L., Peldar, J.H., Pope, S.E., Taylor, P.D. and Wegner, J.F. (1994). Effect of road traffic on amphibian density. *Biological Conservation,* 73: 177–182.

Falardeau, G. and Desgranges, J.L. (1991). Sélection de l'habitat et fluctuations récentes des populations d'oiscaux des milieux agricoles du Québec. *Canadian Field-Naturalist,* 105: 469–482.

Farina, A. (1998). *Principles and Methods in Landscape Ecology.* Chapman Hall, London.

Farjon, J.M.J., Bulens, J.D. and Prins, A.H. (1995). A national survey of ecological potentials for developing natural areas in the Netherlands. *In:* J.F. Schoute, P.A. Finke, F.R. Veenklaas and H.P. Wolfert, *Scenario Studies for the Rural Environment,* Kluwer Academic Publisher, Dordrecht, pp. 427–435.

Fernandez Ales, R., Martin, A. and Ortega, F. (1992). Recent changes in landscape structure and function in a mediterranean region of SW Spain (1950–1984). *Landscape Ecology,* 7(1): 3–19.

Ferris, C.R. (1979). Effects of interstate 95 on breeding birds in Northern Maine. *Journal of Wildlife Management,* 43: 421–427.

Flaherty, M. and Smit, B. (1982). An assessment of land classification techniques in planning for agricultural land use. *Journal Environmental Management,* 15: 323–332.

Fogelman-Soulié, F. (1983). Réseaux d'automates et morphologie. *In:* P. Dumouchel and J.P. Dupuy. *L'auto-organisation, de la physique au politique.* Éditions du Seuil, Paris, pp. 101–114.

Forman, R.T.T. (1995). *Land Mosaic. The Ecology of Landscapes and Regions.* Cambridge University Press, Cambridge.

Forman, R.T.T. and Baudry, J. (1994). Hedgerows and hedgerow networks in landscape ecology. *Environmental Management,* 8: 499–510.

Forman, R.T.T. and Boerner, R.E. (1981). Fire frequency in the pine barrens of New Jersey. *Bulletin Torrey Botanical Club,* 108: 34–50.

Forman, R.T.T., Galli, A.E. and Leck, C.F. (1976). Forest size and avian diversity in New Jersey woodlots with some land use implication. *Œcologia* (Berlin), 26: 1–8.

Forman, R.T.T. and Godron, M. (1981). Patches and structural components for a landscape ecology. *BioScience,* 31: 733–740.

Forman, R.T.T. and Godron, M. (1986). *Landscape Ecology.* John Wiley and Sons, New York.

Fournier, F. and Cheverry, C. (1992). Les échelles spatiales d'étude du rôle du sol dans l'environnement. *In:* P. Auger, J. Baudry and F. Fournier, *Hiérarchies et Échelles en Écologie,* Naturalia Publication, pp. 21–41.

Frampton, G.K., Cilgi, T., Fry, G.L.A. and Wratten, S.D. (1995). Effects of grassy banks on the dispersal of some carabid beetles (Coleoptera: Carabidae) on farmland. *Biological Conservation,* 71: 347–355.

Freemark, K.E. and Merriam, H.G. (1986). Importance of area and habitat heterogeneity to bird assemblages in temperate forest fragments. *Biological Conservation,* 36: 115–141.

Fresco, L.O., Stroosnijder, L., Bouma, J. and van Keulen, H., eds. (1994). *The Future of the Land: Mobilising and Integrating Knowledge for Land Use Option.* Wiley Sons, West Sussex.

Fry, G. (1995). Landscape ecology and insect movement in arable ecosystems. *In:* D.M. Glen, M.P. Greaves and H.M. Anderson, *Ecology and Integrated Farming Systems,* John Wiley and Sons, Chichester, pp. 177–202.

Fry, G.L.A. (1994). The role of field margins in the landscape. *In:* N. Boatman, *Field Margins: Integrating Agriculture and Conservation,* University of Warwick, Coventry, pp. 31–40.

Gallego, F.J. (1982). Codage flou en analyse des correspondances. *Les Cahiers d'Analyse des Données,* VII(4): 413–430.

Galli, A.E., Leck, C.F. and Forman, R.T.T. (1976). Avian distribution patterns in forest islands of different sizes in central New Jersey. *The Auk*, 93: 356–364.

Gardner, R.H., Milne, B.T., Turner, M.G. and O'Neill, R.V. (1987). Neutral models for the analysis of broad scale landscape patterns. *Landscape Ecology*, 1: 19–28.

Gardner, R.H., O'Neill, R.V., Turner, M.G. and Dale, V.H. (1989). Quantifying scale-dependent effects of animal movement with simple percolation models. *Landscape Ecology*, 3: 217–228.

Gaston, K.J., ed. (1996). *Biodiversity: a Biology of Numbers and Difference*. Blackwell Science Ltd., Oxford.

Gates, J.E. and Harman, D.M. (1978). Avian nest dispersion and fledging success in field-forest ecotones. *Ecology*, 59: 871–883.

Gebhardt, A. (1988). Évolution du paysage agraire au cours du sub-atlantique dans la région de Redon (Morbihan, France) apport de la micromorphologie. *Bulletin de l'association française pour l'Étude du Quaternaire*, pp. 197–203.

Gebhardt, A. (1990). *Évolution du paléopaysage agricole dans le nord-ouest de la France. Apport de la micromorphologie*. Doct. Thesis, Université de Rennes.

Gehrken, G.A. (1975). *Travel Corridor of Wild Turkey Management*. National wild turkey symposium, Austin, Texas chapter of the Wildlife Society.

Génard, M. and Lescourret, F. (1985). Caractères insulaires de l'avifaune des Alpes du sud et des Pyrénées orientales. *Œcologia Generalis*, 6: 209–235.

Getz, L.L., Cole, F.R. and Gates, D.L. (1978). Interstate roadsides as dispersal routes for *Microtus pennsylvanicus*. *Journal of Mammology*, 59: 208–212.

Giles, R.H. and Koeln, G.T. (1983). Land and cropland primeness concepts and methods of determination. *Environmental Management*, 7: 129–142.

Gilpin, M. and Hanski, I., eds. (1991). *Metapopulation Dynamics: Empirical and Theoretical Investigations*. Academic Press, London.

Giot, P.R., Briard, J. and Pape, L. (1995). *Protohistoire de la Bretagne*. Ed. Ouest-France, Rennes, 423 pp.

Giot, P.-R., L'Helgouac'h, J. and Monnier, J.-L. (1979). *Préhistoire de la Bretagne*. Ouest-France Université, Rennes.

Giraudoux, P. (1991). *Utilisation de l'espace par le hôte du ténia multiloculaire* (Echinococcus multilocularis): *conséquences épidémiologiques*. Thesis, Université de Dijon.

Giraudoux, P. (1997). Gestion de l'espace et transmission de l'échinococcose alvéolaire. *Revue Bilingue: Science et Techniques* (12–16).

Girel, J. and Pautou, G. (1996). The influence of sedimentation on vegetation structure. *In:* N.E. Haycock, T.P. Burt, K.W.T. Goulding and G. Pinay, *Buffer Zones: Their Processes and Potential in Water Protection*, Quest Environmental, Harpenden, pp. 93–112.

Gleason, H.A. (1926). The individualistic concept of the plant association. *Bulletin of the Torrey Botanical Club*, 53: 7–26.

Gleick, J. (1991). *La Théorie du Chaos*. Flammarion, Paris.

Gold, A.J. and Kellog, D.Q. (1996). Modelling internal processes of riparian buffer zones. In: N.E. Haycock, T.P. Burt, K.W.T. Goulding and G. Pinay, *Buffer Zones: Their Processes and Potential in Water Protection*, Quest Environmental, Harpenden, pp. 192–207.

Goldberg, D.E. and Miller, T.E. (1990). Effects of different resource additions on species diversity in an annual plant community. *Ecology*, 71: 213–225.

Golley, F.B. (1993). *A History of Ecosystem Concept in Ecology: More than the Sum of the Parts*. Vail-Ballou Press, Binghamton, New York.

Gras, R., Benoît, M., Deffontaines, J.P., Duru, M., Lafarge, M., Langlet, A. and Osty, P.L. (1989). *Le Fait Technique en Agronomie. Activité Agricole, Concepts et Méthodes d'Étude*. INRA, L'Harmattan, Paris.

Grassberger, P. (1991). La percolation ou la géométrie de la contagion. *La Recherche*, 22: 640–646.

Green, B. (1996). *Countryside Conservation: Landscape Ecology, Planning and Management*. E, FN SPON, London.

Green, D.G. (1989). Simulated effects of fire, dispersal and spatial pattern on competition within forest mosaics. *Vegetatio*, 82: 139–153.

Grege (1996). *Méthode de conduite des études préalables à la réalisation des passages pour la grande faune*. Grege pour le compte de l'Union des Sociétés d'Autoroutes à péage.

Gril, J.J., Real, B., Patty, L., Fagot, M. and Perret, I. (1996). Grassed buffer zones to limit contamination of surface waters by pesticides: research and action in France. *In:* N.E. Haycock, T.P. Burt, K.W.T. Goulding and G. Pinay, *Buffer Zones: Their Processes and Potential in Water Protection*, Quest Environmental, Harpenden, pp. 70–73.

Groupe d'histoire des forêts françaises (1986). Du pollen au cadastre. *Hommes et Terres du Nord*, pp. 2–3.

Grubb, P.J. (1988). The uncoupling of disturbance and recruitment, two kinds of seed bank, and persistence of plant populations at regional and local scales. *Ann. Zool. Fenici*, 25: 23–36.

Guinochet, M. (1973). *Phytosociologie*. Masson, Paris.

Gustafson, E.J. and Gardner, R.H. (1996). The effect of landscape heterogeneity on the probability of patch colonization. *Ecology*, 77: 94–107.

Haines-Young, R. and Chopping, M. (1996). Quantifying landscape structure: a review of landscape indices and their application to forested landscapes. *Progress in Physical Geography*, 20: 418–445.

Haines-Young, R., Green D.R. and Cousins, S.H. (eds.) (1993). Landscape ecology and GIS. London, Taylor & Francis. 288 p.

Haining, R. (1990). *Spatial Data Analysis in the Social and Environmental Sciences*. Cambridge University Press, Cambridge.

Hanski, I. (1981). Coexistence of competitors in patchy environments with and without predation. *Oikos*, 36: 306–312.

Hanski, I. (1987). Carrion fly community dynamics: patchiness, seasonality and coexistence. *Ecological Entomology*, 12: 257–266.

Hanski, I. (1989). Metapopulation dynamics: does it help to have more of the same? *TREE*, 4: 113–114.

Hanski, I. (1991). Single-species metapopulation dynamics: concepts, models and observations. *In:* M. Gilpin and I. Hanski, *Metapopulation Dynamics: Empirical and Theoretical Investigations*, Academic Press, San Diego, pp. 17–38.

Hanski, I. (1995). Effects of landscape patterns on competitive interactions. *In:* L. Hansson, L. Fahrig and G. Merriam, *Mosaic Landscapes and Ecological Processes*, Chapman Hall, London, pp. 203–224.

Hanski, I. (1997). Metapopulation dynamics: from concepts and observations to predictive models. *In:* I. Hanski and M. Gilpin, *Metapopulation Biology: Ecology, Genetics and Evolution*, Academic Press, San Diego, pp. 69–92.

Hanski, I. and Gilpin, M. (1991). Metapopulation dynamics: brief history and conceptual domain. *In:* M. Gilpin and I. Hanski, *Metapopulation Dynamics: Empirical and Theoretical Investigations*, Academic Press, London, pp. 3–16.

Hanski, I. and Gilpin, M.E. (1997). *Metapopulation Biology: Ecology, Genetics and Evolution*. Academic Press, San Diego.

Hanski, I. and Ranta, E. (1983). Coexistence in a patchy environment: three species of Daphnia in rockpools. *Journal of Animal Ecology*, 52: 263–279.

Hanski, I. and Simberloff, D. (1997). The metapopulation approach. *In:* I. Hanski and M.E. Gilpin, *Metapopulation Biology: Ecology, Genetics and Evolution*, Academic Press, London, pp. 5–26.

Hansson, L. (1967). Index line catches as a basis of population studies on small mammals. *Oikos*, 18: 261–276.

Hansson, L. and Angelstam, P. (1991). Landscape ecology as a theoretical basis for nature conservation. *Landscape Ecology*, 5: 191–201.

Harms, B.H., Knaapen, J.P., Rademakers, J.G. (1992). Landscape planning for nature restoration: comparing regional scenarios. *In:* C.C. Vos and P. Opdam, *Landscape Ecology of a Stressed Environment*. Chapman Hall, pp. 195–218.

Harms, W.B., Knaapen, J.P. and Roos-Klein-Lankhorst, J. (1995). COSMO: A decision-support system for the central open space, The Netherlands. *In:* J.F. Schoute, P.A. Finke, F.R. Veenklaas and H.P. Wolfert, *Scenario Studies for the Rural Environment*, Kluwer Academic Publisher, Dordrecht, pp. 437–444.

Harris, E. and Harris, J. (1997). *Wildlife Conservation in Managed Woodland and Forests*. Research Studies Press Ltd., John Wiley and Sons Inc., New York.

Harris, L.D. (1984). *The Fragmented Forest*. The University of Chicago Press, Chicago.

Harris, L.D. (1985). *Conservation Corridors: a Highway System for Wildlife*. Florida Conservation Foundation, Winter Park, Fl.

Harris, L.D. (1988). Edge effects and conservation of biotic diversity. *Conservation Biology* 2: 330–332.

Harris, L.D. and Gallagher, P.B. (1989). New initiatives for wildlife conservation: the need for movement corridors. *In:* G. Mackintosh, *Preserving Communities and Corridors*, Defenders of Wildlife, Washington, D.C., pp. 11–34.

Harris, L.D. and Scheck, J. (1991). From implications to applications: the dispersal corridor principle applied to the conservation of biological diversity. *In:* D.A. Saunders and R.J. Hobbs. *Nature Conservation 2: the Role of Corridors*, Surrey Beatty and Sons, Chipping Norton, pp. 189–220.

Harrison, S. (1991). Local extinction in a metapopulation context: an empirical evaluation. *In:* M. Gilpin and I. Hanski, *Metapopulation Dynamics: Empirical and Theoretical Investigations*. Academic Press, San Diego, pp. 73–88.

Hastings, A. and Harrison, S. (1994). Metapopulation dynamics and genetics. *Annual Review in Ecology and Systematics*, 25: 167–188.

Hastings, A. and Wolin, C.L. (1989). Within-patch dynamics in a metapopulation. *Ecology*, 70: 1261–1266.

Hastings, H.M. and Sugihara, G. (1993). *Fractals: a User's Guide for the Natural Sciences*. Oxford University Press, Oxford, 235 pp.

Haycock, N.E., Burt, T.P., Goulding, K.W.T. and Pinay, G., eds. (1996). *Buffer Zones: Their Processes and Potential in Water Protection*. Quest Environmental, Harpenden, 326 pp.

Haycock, N.E. and Muscutt, A.D. (1995). Landscape management strategies for the control of diffuse pollution. *Landscape and Urban Planning*, 31(1–3): 313–322.

Hegarthy, C.A., McAdam, J.H. and Cooper, A. (1994). Factors influencing the plant species composition of hedges implications for management in environmentally sensitive areas. *In:* N. Boatman, *Field Margins Integrating Agriculture and Conservation*, British Crop Protection Council, Farnham, pp. 227–234.

Heikkilä, J., Below, A. and Hanski, I. (1994). Synchronous dynamics of microtine rodent populations on islands in Lake Inari in northern Fennoscandia: evidence for regulation by mustelid predators. *Oikos*, 70: 245–252.

Heinselmann, M.L. (1973). Fire in the virgin forests of the Boundary Waters Canoe Area, Minnesota. *Quat. Res.* (NY), 3: 329–382.

Helliwell, D.R. (1976). The effects of size and isolation on the conservation value of wooded sites in Britain. *Journal of Biogeography*, 3: 409–416.

Henderson, M.T., Merriam, H.G. and Wegner, J. (1985). Patchy environments and species survival: chipmunks (*Tamia striatus*) in an agricultural mosaic. *Biological Conservation*, 31: 95–106.

Henein, K. and Merriam, G. (1990). The elements of connectivity where corridor quality is variable. *Landscape Ecology*, 4(2): 157–171.

Henein, K., Wegner, J. and Merriam, H.G. (1998). Population effects of landscape model manipulation on two behaviourally different woodland small mammals. *Oikos*, 81: 168–186.

Herkert, J.R. (1994). The effects of habitat fragmentation on midwestern grassland bird communities. *Ecological Applications*, 4(3): 461–471.

Herrera, C.M. (1987). Components of pollinator "quality": comparative analysis of a diverse insect assemblage. *Oikos*, 50: 79–90.

Hill, A.R. (1975). Ecosystem stability in relation to stresses caused by human activities. *Canadian Geographer*, 19: 206–220.

Hill, A.R. (1983). Denitrification: its importance in a river draining an intensively cropped watershed. *Agriculture, Ecosystems and Environment*, 10: 47–62.

Hobbs, N.T. and Spowart, R.A. (1984). Effects of prescribed fre on nutrition of mountain sheep and mule deer during winter and spring. *Journal of Wildlife Management*, 45: 551–560.

Hobbs, R.J. (1992). The role of corridors in conservation: solution or bandwagon? *TREE*, 7: 389–392.

Hobbs, R.J. and Wilson, A.M. (1988). Corridors: theory, practice and the achievement of conservation objectives. *In:* J.W. Dover and R.G.H. Bunce. *Key Concepts in Landscape Ecology* IALE UK, Preston, pp. 265–279.

Hodson, N.L. and Snow, D.W. (1965). The road death enquiriy, 1960–61. *Bird Study*, 12: 90–99.

Hofstra, G. and Hall, R. (1971). Injury on roadside trees: leaf injury on pine and white cedar in relation to foliar levels of sodium and chloride. *Canadian Journal of Botany*, 49: 613–622.

Hooper, M.D. (1976). Études historiques et biologiques des haies anglaises. *In:* INRA, CNRS, ENSA and Université de Rennes, *Les Bocages: Histoire, Écologie, Économie*. INRA. Rennes, pp. 225–228.

Hoskins, W.G. (1955). *The Making of the English Landscape*. Penguin Books Limited, Harmondsworth.

Hubert, B. (1991). Changing land uses in Provence (France) Multiple use as a management tool. *In:* J. Baudry and R.G.H. Bunce, *Land Abandonment and Its Role in Conservation*, CIHEAM, Zaragoza, A15: 31–52.

Hudson, W.E., ed. (1991). *Landscape Linkages and Biodiversity*. Island Press, Washington, D.C.

Huffman, E. and Dumanski, J. (1983). *Agricultural Land Use Systems of the Regional Municipality of Ottawa-Carleton*. LRRI Contribution 82-07. Research Branch, Agriculture Canada, Ottawa.

Hulshoff, R.M. (1995). Landscape indices describing a Dutch landscape. *Landscape Ecology*, 10: 101–112.

Hunter, M.L., Jacobson Jr., G.L. and Webb, T.W.I. (1988). Paleoecology and the coarse-filter approach to maintaining biological diversity. *Conservation Biology*, 2: 375–385.

Huston, M.A. (1995). *Biological Diversity: the Coexistence of Species on Changing Landscapes*. Cambridge University Press, Cambridge.

Ims, R.A. (1995). Movement patterns related to spatial structure. *In:* L. Hansson, L. Fahrig and G. Merriam, *Mosaic Landscapes and Ecological Processes*, Chapman Hall, London, pp. 85–109.

INRA, CNRS, ENSA and Université de Rennes (1976). *Les Bocages: Histoire, Écologie, Économie*. Table ronde CNRS: les écosystèmes bocagers, Rennes, INRA-Rennes.

INRA, ENSSAA (groupe de recherche) 1977). *Pays, Paysans, Paysages dans les vosges du Sud*. INRA, Enssaa. Paris, Dijon.

INRAP (1986). *Lectures du Paysage*. Foucher.

Johnson, A.R., Milne, B.T. and Wiens, J.A. (1992). Animal movements and population dynamics in heterogeneous landscapes. *Landscape Ecology*, 7: 63–75.

Johnson, W.C. (1998). River regulation and landscape change. *In:* J.W. Dover and R.G.H. Bunce. *Key Concepts in Landscape Ecology*. IALE (UK), Preston, pp. 3–18.

Johnson, W.C. and Adkisson, C.S. (1985). Dispersal of beechnuts by blue jays in fragmented landscapes. *American Midland Naturalist*, 113: 319–324.

Johnson, W.C., Sharpe, D.M., DeAngelis, D.L., Fields, D.E. and Olson, R.J. (1981). Modeling seed dispersal and forest island dynamics. *In:* R.L. Burgess and D.M. Sharpe, *Forest Island Dynamics in Man-dominated Landscapes*, Springer-Verlag, New York, pp. 215–239.

Johnston, C.A. (1988). *Geographic Information Systems in Ecology*. Blackwell Science Oxford, p. 239.

Jurdant, M., Beaubien, J., Bélair, J.L., Dionne, J.C. and Gerardin, V. (1972). *Carte écologique de la région du Sagueny-Lac-Saint-Jean. Notice explicative. vol. 1: L'environnement et ses ressources: identification, analyse et évaluation*. Quebec Environnement Canada.

Kareiva, P. (1985). Finding and losing host plants by Phyllotreta: patch size and surrounding habitats. *Ecology*, 66: 1809–1816.

Karr, J.R. and Schlosser, I.J. (1978). Water resources and the land-water interface. *Science*, 201: 229–234.

Keiter, R.B. and Boyce, M.S., eds. (1991). *The Greater Yellowstone Ecosystem: Redefining America's Wilderness Heritage*. Yale University Press, New Haven.

Kesner, B.T. and Meentemeyer, V. (1989). A regional analysis of total nitrogen in an agricultural landscape. *Landscape Ecology*, 2: 151–163.

King, T., Stout, R. and Gilbert, T. (1985). *Habitat reclamation guidelines: a series of recommendations for fish and wildlife habitat enhancement on phosphate mined land and other disturbed sites*. Bartow, Fl., Office of Environmental Services, Florida Game and Fresh Water Fish Commission.

Klir, G.J. and Folger, T.A. (1988). *Fuzzy Sets, Uncertainty, and Information*. Prentice Hall, Englewoods Cliffs, New Jersey.

Knight, D.H. (1994). *Mountains and Plains: the Ecology of Wyoming Landscapes*. Yale University Press, New Haven, Connecticut.

Knight, D.H. and Wallace, L.L. (1989). The Yellowstone fires: issues in landscape ecology *BioScience*, 39: 700–706.

Koestler, A. (1967). *The Ghost in the Machine*. Macmillan, New York.

Kolasa, J. and Pickett, S.T.A., eds. (1991). *Ecological Heterogeneity*. Ecological Studies, Springer-Verlag, New York.

Kotliar, N.B. and Wiens, J.A. (1990). Multiple scales of patchiness and patch structure: a hierarchical framework for the study of heterogeneity. *Oikos*, 59: 253–260.

Kozakiewicz, M. (1993). Habitat isolation and ecological barriers: the effect on small mammal populations and communities. *Acta Theriologica*, 38: 363–367.

Kristensen, S.P., Thenail, C. and Kristensen, L. (2001). Farmers' involvement in landscape activities: An analysis of the relationship between farm location, farm characteristics and landscape changes in two study areas in Jutland, Denmark. *Journal of Environmental Management*, 61(4): 301–318.

Krönert, R., Baudry, J., Bowler, I.R. and Reenberg, A., eds. (1999). Land-use changes and their environmental impact in rural areas in Europe. Mand and Biosphere series. UNESCO and the Parthenon Publishing Group, Pearl River, Paris.

Krummel, J.R., Gardner, R.H., Sugihara, G., O'Neil, R.V. and Colemen, P.R. (1987). Landscape patterns in a disturbed environment. *Oikos*, 48: 321–324.

Kuussaari, M., Nieminen, M. and Hanski, I. (1996). An experimental study of migration in the butterfly *Melitaea cinxia*. *Journal of Animal Ecology*.

Lack, P.C. (1988). Hedge intersections and breeding bird distribution in farmland. *Bird Study*, 35: 133–136.

Lammers, G.W. and van Zadeldhoff, F.J. (1996). The Dutch ecological network. *In:* P. Nowicki, G. Bennett and D. Middleton, *Perspectives on Ecological Networks*, ECNC Publications, Arnhem, Netherlands, pp. 101–113.

Lammers, W. (1994). A new strategy in nature policy: towards a national ecological network in the netherlands. *In:* E.A. Cook and H.N. Van Lier, *Landscape Planning and Ecological Networks*, Elsevier Sciences B.V., Amsterdam, pp. 283–307.

Lamotte, M. (1978). La savane préforestière de Lanto, Côte d'Ivoire. *In:* M. Lamotte and F. Bourlière, *Problèmes d'Écologie: Écosystèmes Terrestres*, Masson, Paris, pp. 231–311.

Landers, J.L., Hamilton, R.J., Johnson, A.S. and Marcington, R.L. (1979). Foods and habitat of black bears in southeastern North Carolina. *Journal of Wildlife Management*, 43: 143–153.

Langouët, L. (1995). La colonisation médiévale du marais de Dol entre le Guyoult et Saint Broladre. Approches géographique et archéologique. *In*: L. Langouët and M.T. Morzadec.-Kerfourn. *Baie du Mont Saint Michel et Marais de Dol. Milieux naturels et peuplements dans le passé*. Centre Régional d'Archéologie d'Alet, Saint Malo, pp. 111–118.

Langouët, L. (1994). *La colonisation médiévale du Marais de Dol entre le Guyoult et Saint Broladre. Approches géographique et archéologique. L'approche sédimentaire: l'histoire pour comprendre le présent et prévoir le futur*. Paris, Rapport CNRS, Programme environnement, Comité Systèmes Ruraux.

Langton, T.E.S., ed. (1989). *Amphibians and Roads*. ACO Polymer Products Ltd., Shefford, Bedfordshire, England.

Larsonneur, C. (1989). La Baie du Mont-Saint-Michel. *Bulletin de l'institut de géologie du bassin d'Aquitaine*, 10: 19–21.

Larsonneur, C. and L'Homer, A. (1982). *La Baie du Mont-Saint-Michel. Voyage d'étude*. Paris, Association des Sédimentologistes Français.

Lauga, J. and Joachim, J. (1992). Modelling the effects of forest fragmentation on some species of forest-breeding birds. *Landscape Ecology*: 183–194.

Laurent, C. (1992). L'agriculture et son territoire dans la crise. Analyse et démenti des prévisions sur la déprise des terres agricoles à partir d'observations réalisées dans le Pays d'Auge. *Structures productives et système mondial*. Thesis, Université de Paris, VII: 554.

Laurent, C. and Bowler, I., eds. (1997). *CAP and the Regions: Building a Multidisciplinary Framework for the Analysis of the EU Agricultural Space*. INRA Editions, Paris.

Laurent, C., Langlet, A., Chevallier, C., Jullian, P., Maigrot, J.L. and Ponchelet, D. (1994). Ménages, activités agricoles et utilisation du territoire: du local au global a travers les RGA. *Cahiers d'études et de recherches francophones. Agricultures*, 3: 93–107.

Lavorel, S., Gardner, R.H. and O'Neill, R.V. (1993). Analysis of patterns in hierarchically structured landscapes. *Oikos*, 67: 521–528.

Le Cœur, D. (1996). La végétation des éléments linéaires non cultivés des paysages agricoles: identification à plusieurs échelles spatiales, des facteurs de la richesse et de la composition floristiques des peuplements. Sciences Biologiques. Rennes, Thesis, de l'Université de Rennes.

Le Cœur, D., Baudry, J. and Burel, F. (1997). Field margins plant assemblages: variation partitionning between local and landscape factors. *Landscape and Urban Planning*, 37: 57–72.

Le Coeur, D., Baudry, J., Burel, F. and Thenail, C. (2002). Why and how we should study field boundaries biodiversity in an agrarian landscape context. *Agriculture, Ecosystems & Environment*, 89(1–2): 23–40.

Lebart, L., Morineau, A. and Tabard, N. (1977). *Techniques de la description statistique*, Dunod. Paris.

Lebeau, R. (1979). *Les Grands Types de Structures Agraires dans le Monde*. Masson, Paris.

Lebeaux, M.O. (1985). ADDAD, Association pour le Développement et la Diffusion de l'Analyse des Données.

Lecomte, V., Le Bissonnais, Y., Renaux, B., Couturier, A. and Ligneau, L. (1997). Erosion hydrique et transfert de produits phytosanitaires dans les eaux de ruissellement. *Cahiers Agriculture*, 6: 175–183.

Leduc, A., Prairie, Y.T. and Bergeron, Y. (1994). Fractal dimension estimates of a fragmented landscape: sources of variability. *Landscape Ecology*, 9: 279–286.

Lefeuvre, J.C. (1979). Les études d'impact un an après les décrets d'application de la loi sur la protection de la nature. *Combat Nature*, 75: 10–12.

Lefeuvre, J.C. (1980). Genèse et présentation du colloque (colloque franco-anglais sur l'écologie des Landes). *Bulletin d'Écologie*, 11: 135–146.

Lefeuvre, J.C. (1986). *Des arbres et des hommes: le bocage, la haie, le bois*. Ministry of Environment, Ministry of Agriculture, Basse Normandie, DRAE.

Lefeuvre, J.C. and Barnaud, G. (1988). Écologie du paysage: mythe ou réalité? *Bulletin d'Écologie*, 19: 493–522.

Lefeuvre, J.C., Missonnier, J. and Robert, Y. (1976). Caractérisation zoologique. Écologie animale (des bocages), Rapport de synthèse. *In:* CNRS INRA, ENSA and Université de Rennes, *Les Bocages: Histoire, Écologie, Économie*, INRA, Rennes, pp. 315–326.

Lefeuvre, J.C., Raffin, J.P. and de Beaufort, F. (1979). Protection, conservation de la nature et développement. *In:* J.C. Lefeuvre, G. Long and G. Ricou. *Les connaissances scientifiques écologiques, le développement et la gestion des ressources et de l'espace*. Écologie et Développement, CNRS, pp. 169–192.

Legendre, P. and Fortin, M.J. (1989). Spatial pattern and ecological analysis. *Vegetatio*, 80: 107–138.

Legrand, I. (1995). *Dynamique du paysage de polders en Baie du Mont-Saint-Michel*. Rennes, Université de Paris IV.

Le Lannou, M. (1950). *Géographie de la Bretagne, vol. 2. Economie et population*. Plihon, Rennes.

Lemée, G. (1978). La hêtraie naturelle de Fontainebleau. *In:* M. Lamotte and F. Bourlière, *Structure et Fonctionnement des Écosystèmes Terrestres*, Masson, Paris, pp. 75–125.

Lenclud, G. (1995). Ethnologie et paysage. *In:* C. Voisenat, *Paysage au Pluriel: pour une Approche Ethnologique des Paysages*. Maison des Sciences de l'Homme, Paris. XVI: 3–18.

Levin, S.A. (1989). Challenges in the development of a theory of a community and ecosystem structure and function. *In:* J. Roughgarden, R.M. May and S.A. Levin, *Perspectives in Ecological Theory*, Princeton University Press, Princeton, N.J., pp. 242–255.

Levins, R. (1970). Extinctions. *In:* M. Gertenhaber (ed.) *Some Mathematical Questions in Biology*, American Mathematics Society, Providence, Rhode Island, 2: 77–107.

Likens, G.E. and Bormann, F.H. (1975). Nutrient hydrologic interactions: an experimental approach in New-England landscapes. *In:* A.D. Hasler, *Coupling Land and Water Systems*, Springer-Verlag, New York, 10: 5–29.

Likens, G.E. and Bormann, F.H. (1995). *Biogeochemistry of a Forested Ecosystem*. Springer-Verlag, New York.

Lima, S.L. and Zollner, P.A. (1996). Towards a behavioral ecology of ecological landscapes. *Trends in Ecology and Evolution*, 11: 3–6.

Lindenmayer, D.B. and Nix, H.A. (1993). Ecological principles for the design of wildlife corridors. *Conservation Biology*, 7: 627–630.

Little, C.E. (1990). *Greenways for America*, John Hopkins University Press, Baltimore.

Lizet, B. (1991). De la campagne à la nature ordinaire. *Etudes Rurales*, pp. 121–124.

Loman, J. (1991). Small mammal and raptor density in habitat islands; area effects in a south Swedish agricultural landscape. *Landscape Ecology*, 5: 183–189.

Long, G. (1974). *Diagnostic phyto-écologique du territoire*. Masson, Paris.

Long, G. (1975). *Diagnostic phyto-écologique du territoire*. Masson, Paris.

Lowrance, R.R. (1996). The potential role of riparian forests as buffer zones. *In:* N.E. Haycock, T.P. Burt, K.W.T. Goulding and G. Pinay, *Buffer Zones: Their Processes and Potential in Water Protection*, Quest Environmental, Harpenden, pp. 128–133.

Lowrance, R.R., Leonard, R.A., Asmussen, L.E. and Todd, R.L. (1985). Nutrient budgets for agricultural watersheds in the southeastern coastal plain. *Ecology*, 66(1): 287–296.

Lowrance, R., Todd, R., Fail, J. Jr., Hendrickson, O. Jr., Leonard, R. and Asmussen, L. (1984). Riparian forests as nutrient filters in agricultural watersheds. *BioScience*, 34(6): 374–377.

Lubchenco, J., Olson, L., Brubaker, B., Carpenter, S.R., Holland, M.M., Hubell, S.P., Levin, S.A., MacMahon, J.A., Matson, P.A. Melillo, J.M., Mooney, H.A., Peterson, C.H., Pulliam, H.R., Real, L.A., Regal, P.J. and Risser, P.G. (1991). The sustainable biosphere initiative: an ecological research agenda. *Ecology*, 72(2): 371–412.

Luginbuhl, Y. (1989). Paysages, textes de représentation du paysage du siècle des lumières à nos jours, *La Manufacture*.

Lühning, A. (1984). *Koppelwirstschaft un knicks—eine neue wirtschaftweise und ihre auswirkung auf die landschaft in Schleswig-Holstein seit dem 18. Jh.* CIMA 7 association internationale des musées de l'agriculture.

MacArthur, R.H. and Wilson, E.O. (1963). An equilibrium theory of insular zoogeography. *Evolution*, 17: 319–327.

MacArthur, R.H. and Wilson, E.O. (1967). *The Theory of Island Biogeography.* Princeton University Press, Princeton, N.J.

Mader, H.J. (1988). The significance of paved agricultural roads as barriers to ground dwelling arthropods, In: Schreiber, ed., Connectivity in landscape ecology. *Münstersche geographische arbeiten,* 29: 97–101.

Maehr, D. (1990). The Florida panther and private lands. *Conservation Biology,* 4: 1–4.

Maguire, C.C. (1987). Incorporation of tree corridors for wildlife movement in timber areas: balancing wood production with wildlife habitat management. *Journal of Washington Academy of Science,* 77: 193–199.

Maguran, A.E. (1988). *Ecological Diversity and its Measurement.* Chapman and Hall, London.

Mandelbrot, B. (1982). *The Fractal Geometry of Nature.* W.H. Freeman and Co., New York.

Mandelbrot, B. (1984). *Les Objets Fractals.* Flammarion, Paris.

Mankin, P.C., Brawn, J.D. and Hoover, J.P. (1997). Mammals of Illinois and the Midwest: ecological and conservation issues for human-dominated landscapes. *In:* M.W. Schwartz, *Conservation in Highly Fragmented Landscapes,* Chapman Hall, New York, pp. 135–153.

Marshall, E.J.P. (1989). Distribution patterns of plants associated with arable field edges. *Journal of Applied Ecology,* 26: 247–257.

Marshall, E.J.P. and Arnold, G.M. (1995). Factors affecting field weed and field margins flora on a farm in Essex, UK. *Landscape and Urban Planning,* 31: 205–216.

Marshall E.J.P. and Birnie, J.E. (1985). *Herbicides Effects on Field Margin Flora.* British Crop Protection Conference, Brighton, Thornton Heath, Surrey, pp. 1021–1028.

Marshall, I.B., Dumanski, J., Huffman, E.C. and Lajoie, P.G. (1979). *Soils Capability and Land use in the Ottawa Urban Fringe,* Research Branch, Agriculture Canada and Ontario Ministry of Agriculture and Food, Ottawa, 59 p. + maps.

Marshall, J., Baudry, J., Burel, F., Joenje, W., Gerowitt, B., Paoletti, M.G., Thomas, G., Kleijn, D., Le Coeur, D. and Moonen, A.C. (2002). Field boundary habitats for wildlife, crop, and environmental protection. *In:* L. Ryszkowski, ed., *Landscape Ecology in Agroecosystem Management,* CRC Press, Boca Raton, pp. 219–247.

Martin, M. (2000). *Modélisation de populations en environnement changeant.* UFR Sciences de la Vie et de l'Environnement. Thesis, Université de Rennes 1.

Martin, P., Papy, F., Souchère, V. and Capillon, A. (1998). Maitrise du ruissellement: intérêt d'une modélisation des pratiques de production. *Cahiers Agriculture,* 7: 111–119.

Mascanzoni, D. and Wallin, H. (1986). The harmonic radar: a new method of tracing insects in the field. *Ecological Entomology* 11: 387–390.

Matthiae, P.E. and Stearns, F. (1981). Mammals in forest islands in Southeastern Wisconsin. *In:* R.L. Burgess and D.M. Sharpe, *Forest Island Dynamics in Man-dominated Landscapes,* Springer-Verlag, New York, 41: 55–66.

Mauremooto, J.R., Wratten, S.D., Worner, S.P. and Fry, G.L.A. (1995). Permeability of hedgerows to predatory beetles. *Agriculture, Ecosystems, Environment,* 52: 141–148.

May, R. (1976). Simple mathematical models with very complicated dynamics. *Nature,* 261: 459–467.

May, R.M. (1989). Levels of organization in ecology. *In:* C.J.M. *Ecological Concepts,* Blackwell Scientific Publications, Oxford, pp. 339–363.

McIntyre, N. (1995). Effects of forest patch size on avian diversity. *Landscape Ecology,* 10: 85–99.

McLaughlin, A. and Mineau, P. (1995). The impact of agricultural practices on biodiversity. *Agriculture, Ecosystems and Environment,* 55: 201–212.

Mech, L.D., Fritts, S.H., Raddle, G.L. and Paul, W.J. (1998). Wolf distribution and road density in Minnesota. *Wildlife Society Bulletin,* 16: 85–87.

Meentemeyer, V. and Box, E.O. (1987). Scale effects in landscape studies. *In:* T.M.G. *Landscape Heterogeneity and Disturbance*, Springer-Verlag, New York, pp. 15–34.

Meeus, J., Van der Ploeg, J.D. and Wijermans, M. (1988). *Changing Agricultural Landscapes in Europe: Continuity, Deterioration or Rupture?* IFLA Conference, Rotterdam.

Meeus, J., Wijermans, M. and Vroom, M. (1990). Agricultural landscapes in Europe and their transformation. *Landscape and Urban Planning*, 18: 289–352.

Médail, F., Roche, P. and Tatoni, T. (1998). Functionnal groups in phytoecology: an application to the study of isolated plant communities in Mediterranean France. *Acta Oecologica*, 19: 263–274.

Mérot, P. (1998). Conclusion. *In:* C. Cheverry, *Agriculture Intensive et Qualité des Eaux*, INRA Editions, Paris, pp. 279–282.

Mérot, P. and Bruneau, P. (1993). Sensitivity of bocage landscapes to surface run-off: application of the Kirkby index. *Hydrological Process*, 7: 167–173.

Mérot, P., Gascuel-Odoux, C., Walter, C., Zhang, X. and Molenat, J. (1999). Bocage landscapes and surface water pathways. *Revue des Sciences de l'Eau*, 12(1): 23–44.

Mérot, P. and Ruellan, A. (1980). Pédologie. hydrologie des bocages; caractéristiques et incidences de l'arasement des talus boisés. *Bulletin Technique d'Information du Ministère de l'Agriculture*, 353/355: 657–690.

Merriam, H.G. (1984). Connectivity: a fundamental characteristic of landscape pattern. *In:* J. Brandt and P. Agger, *Methodology in Landscape Ecological Research and Planning*, Roskilde University Centre, Denmark, 1: 5–15.

Merriam, H.G. (1989). Ecological processes in the time and space of farmland mosaic. *In:* I.S. Zonneveld and R.T.T. Forman, *Changing Landscapes: an Ecological Perspective*, Springer-Verlag, pp. 121–133.

Merriam, H.G. (1991). Corridors and connectivity: animals population in heterogeneous environments. *In:* D.A. Saunders and R.J. Hobbs, *Nature Conservation: the Role of Corridors*, Surrey Beatty, Sons, Chipping Norton, pp. 133–142.

Merriam, H.G., Kozakiewicz, M., Tsuchiya, E. and Hawley, K. (1989). Barriers as boundaries for metapopulations and demes of *Peromyscus leucopus* in farm landscape. *Landscape Ecology*, 2: 227–235.

Merriam, H.G. and Lanoue, A. (1990). Corridor use by small mammals: field measurements for three experimental types of *Peromyscus leucopus*. *Landscape Ecology*, 4: 123–131.

Meyer, J. (1972). L'évolution des idées sur le bocage en Bretagne. *In: La Pensée Géographique Française Contemporaine*, Presses universitaires de Bretagne, pp. 453–467.

Meyer, W.B. and Turner II, B.L., eds. (1994). *Changes in Land Use and Land Cover: A Global Perspective.* Cambridge University Press, Cambridge, 537 pp.

Meynier, A. (1966). *La Génèse du Parcellaire Breton.* Norois, pp. 595–609.

Meynier, A. (1970). *Les Paysages Agraires.* Armand Colin, Paris.

Meynier, A. (1976). Typologie et chronologie du bocage. *In:* INRA, ENSA and Université de Rennes. *Les Bocages: Histoire, Écologie, Économie.* ENSA, Rennes, pp. 65–68.

Middleton, J. and Merriam, H.G. (1981). Woodland mice in a farmland mosaic. *Journal of Applied Ecology*, 18: 703–710.

Middleton, J. and Merriam, H.G. (1985). The rationale for conservation: problems from a virgin forest. *Biological Conservation*, 33: 133–145.

Milne, B.T. (1991). Heterogeneity as a multiscale characteristic of landscapes. *In:* J. Kolasa and T.A. Pickett, *Ecological Heterogeneity*, Springer-Verlag, New York, pp. 69–84.

Milne, B.T. (1991). Lessons from applying fractal models to landscape patterns. *In:* M.G. Turner and R.H. Gardner, *Quantitative Methods in Landscape Ecology*, Springer-Verlag, New York, pp. 199–238.

Milne, B.T. (1992). Spatial aggregation and neutral models in fractal landscapes. *The American Naturalist*, 139: 32–57.

Milne, B.T. (1997). Applications of fractal geometry in wildlife biology. *In:* J.A. Bissonette. *Wildlife and Landscape Ecology: Effects of Pattern and Scale*, Springer-Verlag, New York, pp. 32–69.

Minshall, G.W. and Brock, J.T. (1991). Observed and anticipated effects of forest fire on Yellowstone stream ecosystems. *In:* R.B. Keiter and M.S. Boyce, *The Greater Yellowstone Ecosystem: Redifining America's Wilderness Heritage*, Yale University Press, New Haven, pp. 123–135.

Minshall, G.W., Brock, J.T. and Varley, J.D. (1989). Wildfires and Yellowstone stream eco-systems. *BioScience*, 39(10): 707–715.

Moilanen, A. and Hanski, I. (1998). Metapopulation dynamics: effects of habitat quality and landscape structure. *Ecology*, 79(7): 2503–2515.

Molofsky, J. (1994). Population dynamics and pattern formation in theoretical populations. *Ecology*, 75: 30–39.

Monnier, J.L. (1991). *La Préhistoire de Bretagne et d'Armorique*. Jean-Paul Gisserot (Coll. Les universels Gisserot).

Morant, P. (1995). *Système d'Information Géographique et télédétection en pays de bocage: potentialités pour l'inventaire, l'analyse et la gestion des dynamiques paysagères*. Rennes, DEA, Université de Rennes, 2: 167.

Morant, P., Le Henaff, F. and Marchand, J.P. (1995). Les mutations d'un paysage bocager: essai de cartographie dynamique. *Mappemonde*, 1: 5–8.

Morse, D.R., Lawton, J.H., Dodson, M.M. and Williamson, M.H. (1985). Fractal dimension of the vegetation and the distribution of arthropod body lengths. *Nature*, 314: 731–733.

Morvan, N., Delettre, Y.R., Trehen, P., Burel, F. and Baudry, J. (1994). The distribution of Empididae (diptera) in hedgerow network landscapes. Field margins integrating agriculture and conservation, British Crop Protection Council, Ed., British Crop Protection Council. 123–127.

Morvan, N. (1996). *Structure et biodiversité des paysages de bocage: le cas des empidides* (Diptera, Empidoidea). UFR Sciences de la Vie et de l'Environnement, Rennes, Université de Rennes 1.

Morvan, N., Burel, F., Baudry, J., Tréhen, P., Bellido, A., Delettre, Y.R. and Cluzeau, D. (1995). Landscape and fire in Brittany heathlands. *Landscape and Urban Planning*, 31: 81–88.

Morvan, N., Delettre, Y.R., Trehen, P., Burel, F. and Baudry, J. (1994). *The distribution of* Empididae (diptera) *in hedgerow network landscapes*. Field margins integrating agriculture and conservation, pp. 123–127.

Morzadec-Kerfourn, M.T. (1985). Variations du niveau marin à l'Holocène en Bretagne (France). *Eiszeitalter u. Gegenwart*, Hanovre, 35: 15–22.

Mueller-Dombois, D. and Ellenberg, H. (1974). *Aims and Methods of Vegetation Ecology*. Wiley and Sons, New York.

Murphy, D.D., Freas, K.E. and Weiss, S.B. (1990). An environment-metapopulation approach to population analysis (PVA) for a threatened invertebrate. *Conservation Biology*, 4: 41–51.

Naiman, R.J. (1996). Water, society and landscape ecology. *Landscape Ecology*, 11: 193–196.

Naiman, R.J. and Décamps, H., eds. (1990). *The Ecology and Management of Aquatic-Terrestrial Ecotones*. Man and the Biosphere series, UNESCO, Parthenon Publishing, Lancs, U.K., 316 pp.

Naveh, Z. and Lieberman, A.S. (1984). *Landscape Ecology. Theory and Applications*. Springer-Verlag, New York.

Noss, R.F. (1983). A regional landscape approach to maintain diversity. *BioScience*, 33: 700–706.

Noss, R.F. (1987). Corridors in real landscapes. A reply to Simberloff and Cox. *Conservation Biology*, 1: 159–164.

Noss, R.F. (1990). Indicators for monitoring biodiversity: a hierarchical approach. *Conservation Biology*, 4: 355–364.

Noss, R.F. (1991). Landscape connectivity: different functions at different scales. *In:* W.E. Hudson. *Landscape Linkages and Biodiversity*, Island Press, Washington, D.C., pp. 27–39.

Nour, N., Latthhysen, E. and Dhondt, A.A. (1993). Artificial nest predation and habitat fragmentation: different trends in bird and mammal predators. *Ecography*, 16: 111–116.

Nowicki, P., Bennett, G., Middleton, D., Rientjes, S. and Walters, R., eds. (1996). *Perspectives on Ecological Networks*. Man and Nature, European Center for Nature Conservation, Arnhem.

O'Connor, R.J. (1986). *Farming and Birds*. University Press, Cambridge.

Odum, E.P. (1969). The stragegy of ecosystem development. *Science*, 164: 262–270.

Odum, E.P. (1971). *Fundamentals of Ecology*, W.B. Saunders Co., Philadelphia.

Okubo, A. (1980). *Diffusion and Ecological Problems: Mathematical Models*. Springer-Verlag, New York.

Olsson, G. (1988). Nutrient use and productivity for different cropping systems in south Sweden during the 18th century. *In:* H.H. Birks, H.J.B. Birks, P.E. Kaland and D. Moe, eds., *The Cultural Landscape, Past, Present and Future*, Cambridge University Press, Cambridge, pp. 123–138.

O'Neill, R.V. (1988). Hierarchy theory and global change. *In:* T. Rosswall, R.G. Woodmansee and P.G. Risser, eds. *Scales and Global Change*, John Willey & Sons, Chichester, pp. 29–45.

O'Neill, R.V. (1989). Perspectives in hierarchy and scale. *In:* J. Roughgarden, R.M. May and S.A. Levin, *Perspectives in Ecological Theory*, Princeton University Press, Princeton, N.J., pp. 140–156.

O'Neill, R.V., de Angelis, D.L., Walde, J.B. and Allen, T.F.H. (1986). *A Hierarchical Concept of Ecosystems*. Princeton University Press, Princeton, N.J.

O'Neill, R.V., Milne, B.T., Turner, M.G. and Gardner, R.H. (1988). Resources utilization scales and landscape pattern. *Landscape Ecology*, 2: 63–69.

O'Neil, R.V., Gardner, R.H., Turner, M.G. and Romme, W.H. (1992). Epidemiology theory and disturbance spread on landscapes. *Landscape Ecology*, 7: 19–25.

Opdam, P., Rijsdijk, G. and Hustings, F. (1985). Bird communities in small woods in an agricultural landscape: effects of core area and isolation. *Biological Conservation*, 34: 333–352.

Opdam, P. and Schotman, A. (1987). Small woods in rural landscapes as habitat islands for woodland birds. *Acta Oecologica Oecologia Generalis*, 8: 269–274.

Opdam, P., Van Apeldoorn, R., Schotman, A. and Kalkhoven, J. (1993). Population responses to landscape fragmentation. *In:* C.C. Vos and P. Opdam, *Landscape Ecology of a Stressed Environment*, Chapman Hall, Cambridge, pp. 145–171.

OTA (1987). *Technologies to Maintain Biological Biodiversity*, U.S. Government Printing Office, Washington, D.C. (US Congress of Technology Assessment).

Ouborg, N.J. (1993). Isolation, population size and exitinction: the classical and metapopulation approaches applied to vascular plants along the Dutch Rhine System. *Oikos*, 66: 298–315.

Ouin, A. (1997). Perméabilité de la matrice agricole pour le mulot sylvestre (*Apodemus sylvaticus*). Université Paris VII, Paris.

Paillat, G. (1994). *Biodiversité dans un paysage d'agriculture intensive: approche fonctionnelle des populations de petits mammifèrs*. Rennes 1, Tours: 25 pp.

Paillat, G. and Butet, A. (1997). Utilisation par les petits mammifères du réseau de digues bordant les cultures dans un paysage poldérisé d'agriculture intensive. *Ecologia Mediterranea*, 23: 13–26.

Pain, G. (1996). *Contribution à une définition d'une méthodologie d'approche des milieux naturels dans la réalisation des avant-projets autoroutiers*. Scetauroute, INRA, Sad-Armorique, CNRS Ecobio. Rennes, Ecole Nationale Supérieure Agronomique de Rennes, 82 pp. + annexes.

Paoletti, M.G. and Pimentel, D., eds. (1992). *Biotic Diversity in Agroecosystems*. Elsevier. Amsterdam.

Papillon, Y. and Godron, M. (1997). Distribution spatiale du lapin de garenne (*Oryctolague cuniculus*) dans le Puy-de-Dôme: l'apport des analyses de paysages. *Gibier et Faune Sauvage, Game Wildlife*, 14: 303–324.

Papy, F. and Boiffin, J. (1988). Influence des systèmes de culture sur les risques d'érosion par ruissellement concentré. II. Évaluation des possibilités de maîtrise du phénomène dans les exploitations agricoles. *Agronomie*, 8: 745–756.

Papy, F. and Douyer, C. (1991). Influence des états de surface du territoire agricole sur le déclenchement des inondations catastrophiques. *Agronomie*, 11: 201–215.

Papy, F. and Souchere, V. (1993). Control of overland runoff and talweg erosion. A land management approach. *In:* J. Brossier, L. De Bonneval and E. Landais. *Systems Studies in Agriculture and Rural Development*. INRA, Paris, pp. 87–98.

Parr, T.W. and Way, J.M. (1988). Management of roadside vegetation: the long-term effects of cutting. *Journal of Applied Ecology*, 25: 1073–1086.

Pautou, G. and Decamps, H. (1985). Ecological interactions between the alluvial forests and hydrology of the Upper Rhone. *Arch. Hydrobiol.*, 104(1): 13–37.

Pedeviliano, C. and Wright, R.G. (1987). The influence of visitors on mountain goat activities in Glacier National Park. *Biological Conservation*, 39: 1–11.

Peitgen, H.-O., Jürgens, H. and Saupe, D. (1992). *Fractals for the Classroom: Part one, Introduction to Fractals and Chaos*. Springer-Verlag, New York, Berlin, 450 pp.

Peitgen, H.-O. and Richter, H. (1986). *The Beauty of Fractals: Images of Complex Dynamical Systems*. Springer-Verlag, Berlin, 199 pp.

Peitgen, H.-O. and Saupe, D. (1988). *The Science of Fractal Images*. Springer-Verlag, New York.

Perrichon, C. (1994). Une mesure agri-environnementale à l'épreuve de la diversité sur un territoire continu. *Cahiers Agriculture*, 3: 163–169.

Peterjohn, W.T. and Correll, D.L. (1984). Nutrient dynamics in an agricultural watershed: observations on the role of a riparian forest. *Ecology*, 65: 1466–1475.

Peters, R.H. (1983). *The Ecological Implications of Body Size*. Cambridge University Press, Cambridge, 329 pp.

Petit, S. and Burel, F. (1993). Movement of *Abax ater (Col. Carabidae):* do forest species survive in hedgerow networks. *Vie et Milieu*, 43: 119–124.

Petit, S. and Burel, F. (1998). Effects of landscape dynamics on the metapopulation of a ground beetle (*Coleoptera, Carabidae*) in a hedgerow network. *Agriculture Ecosystem and Environment*, 69: 243–252.

Petit, S. and Burel, F. (1998). *Quelle Biodiversité en zone de Grande Culture*. CNRS and Ministry of Environment, Paris.

Petit, S. and Burel, F. (1998). Connectivity in fragmented populations: *Abax parallelepipedus* in a hedgerow network landscape. *Compte rendu Académie des Sciences Paris, Sciences de la vie*, 321: 55–61.

Pfister, C. and Brimblecomte, P. (1990). *The Silent Countdown*. Springer-Verlag, Heidelberg.

Philley, M., McNelly, J., Seidensticker, J., Wickramasinghe, G., Wijewansa, R. and Dissanayake, M. (1985). *Mahaweli Environment Project (N 383-0075) Midterm Evaluation Report*. US Agency for International Development, Colombo, Sri Lanka.

Phipps, M. (1981). Entropy and community pattern analysis. *Journal of Theoretical Biology*, 93: 253–273.

Phipps, M. (1985). Théorie de l'information et problématique du paysage. *In:* V. Berdoulay and M. Phipps, *Paysage et Système de l'Organisation Écologique à l'Organisation Visuelle*. Universite d'Ottawa, pp. 59–74.

Phipps, M. and Berdoulay, V. (1985). Paysage, système, organisation. *In:* M. Berdoulay and M. Phipps, *Paysage et Système de l'Organisation Écologique à l'Organisation Visuelle*. Université d'Ottawa, pp. 3–19.

Phipps, M., Baudry, J. and Burel, F. (1986). Ordre topo écologique dans un espace rural les niches paysagiques. *C.R. Acad. Sc., Paris, vol. 302, series 3*, 20: 691–696.

Pickett, S.T.A., Collins, S.L. and Armesto, J.J. (1987). A hierarchical consideration of causes and mechanisms of succession. *Vegetatio*, 69: 109–114.

Pickett, S.T.A., Kolasa, J. and Jones, C.G. (1994). *Ecological Understanding: the Nature of Theory and the Theory of Nature*. Academic Press, San Diego, 206 pp.

Pickett, S.T.A. and Thompson, J.N. (1978). Patch dynamics and the design of nature reserves. *Biological Conservation*, 13: 27–37.

Pickett, S.T.A. and White, P., eds. (1985). *The Ecology of Natural Disturbance and Patch Dynamics.* Academic Press, New York.

Pickett, S.T.A. and White, P. (1985). Patch dynamics: a synthesis. *In:* S.T.A. Pickett and P. White. *The Ecology of Natural Disturbance and Patch Dynamics.* Academic Press, New York, pp. 371–384.

Pihan, J. (1976). Bocage et érosion hydrique des sols en Bretagne. *In: Les bocages: Histoire, Écologie, Économie.* Rennes, pp. 185–192.

Pinay, G., Décamps, H., Chauvet, E. and Fustec, E. (1990). Functions of ecotones in fluvial systems. *In:* R.J. Naiman and H. Décamps. *The Ecology and Management of Aquatic Terrestrial Ecotones.* UNESCO, Parthenon Publishing, Lancs, U.K., 4: 141–169.

Pinchemel, P. and Pinchemel, G. (1988). *La Face de la Terre. Éléments de Géographie.* Paris.

Pinto Correia, T. (1993). *Landscape monitoring and management in European rural areas: Danish and Portuguese case studies of landscape patterns and dynamics.* Geographica Hafniensia 1 (Inst. Geo. Univi. Copenhagen).

Pitte, J.R. (1983). *Histoire du Paysage Français.* Tallandier, Paris.

Plotnick, R.E., Gardner, R.H. and O'Neill, R.V. (1983). Lacunarity indices as a measure of landscape texture. *Landscape Ecology*, 8: 201–212.

Poiani, K.A., Bedford, B.L. and Merrill, M.D. (1996). A GIS-based index for relating landscape characteristics to potential nitrogen leaching to wetlands. *Landscape Ecology*, 11: 237–255.

Pollard, E., Hooper, M.D. and Moore, N.W. (1974). *Hedges.* W. Collins and Sons, London.

Poly, J. (1976). Preface. *In:* CNRS, INRA, ENSA and Université de Rennes, *Les Bocages: Histoire, Ecologie, Economie*, INRA, Rennes, p. 7.

Pomarède, V. and de Wallens, G. (1986). *Corot. La Mémoire du Paysage.* Gallimard, Evreux.

Ponting, C. (1991). *A Green History of the World: The Environment and the Collapse of Great Civilizations.* Penguin Books USA Inc., New York.

Preobrazensky, V.S. (1984). *Methodology in Landscape Ecological Research and Planning.* First international seminar of the International Association for Landscape Ecology (IALE), Roskilde, Denmark, Roskilde universitetsforlag GeoRuc Ed., pp. 15–27.

Preston, F.W. (1962). The canonical distribution of commonness and rarity: part I. *Ecology*, 43: 185–215.

Primdahl, J. and Hansen, B. (1993). Agriculture in environmentally sensitive areas: implementing the ESA measure in Denmark. *Journal of Environmental Planning and Management*, 36(2): 231–238.

Pulliam, H.R. (1988). Sources, sinks and population regulation. *American Naturalist*, 132: 652–661.

Pullin, A.S., ed. (1995). *Ecology and Conservation of Butterflies.* Chapman Hall, London.

Pungetti, G. (1995). Anthropological approach to agricultural landscape history in Sardinia. *Landscape and Urban Planning*, 31: 47–56.

Rabbinge, R. and van Ittersum, M.K. (1994). Tension between aggregation levels. *In:* L.O.S. Fresco, J. Bouma and H. Van Keulen, *The Future of the Land. Mobilising and Integrating Knowledge for Land Use Options*, John Wiley and Sons, Chichester, pp. 31–40.

Rabbinge, R.V.D., Dijsselbloem, J., De Koning, G.J.H., Van Latesteijn, H.C., Woltjer, E. and Van Zijl, J. (1994). Ground for choices: A scenario study on perspectives for rural areas in the European Community. *In:* L.O.S. Fresco, J. Bouma and H. Van Keulen, *The Future of the Land. Mobilising and Integrating Knowledge for Land Use Options*, John Wiley and Sons, Chichester, pp. 95–121.

Rackham, O. (1986). *The History of the Countryside.* J.M. Dent, Sons Ltd., London, Melbourne.

Ranta, E. (1979). Niche of Daphnia species in rockpools. *Annals of Zoologia Fennici*, 23: 131–140.

Reiners, W.A. and Lang, G.E. (1979). Vegetational patterns and processes in the balsam fire zone, White Mountains. *Ecology*, 60: 403–417.

Remmert, H. (1991). The mosaic-cycle concept of ecosystems: an overview. *In:* H. Remmert, *The Mosaic-Cycle Concept of Ecosystems*, Springer-Verlag, New York, pp. 1–21.

Remmert, H., ed. (1991). *The Mosaic-Cycle Concept of Ecosystems*. Ecological Studies, Springer-Verlag, New York.

Renault-Miskovsky, J. (1991). *L'Environnement au Temps de la Préhistoire*. Masson, Coll. Préhistoire, Paris.

Rex, K.D. and Malanson, G.P. (1990). The fractal shape of riparian forest patches. *Landscape Ecology*, 4: 249–258.

Richard, J.F. (1975). Paysages, écosystèmes, environnement: une approche géographique. *L'espace Géographique*, 2: 81–92.

Richards, J.F. (1990). Land transformation. *In:* B.L. Turner II, W.C. Clark, R.W. Kates, J.F. Richards, J.T. Mathews and W. Meyer, *The Earth as Transformed by Human Action*, Cambridge University Press, Cambridge, pp. 163–178.

Ricou, G. (1978). La prairie permanente du nord-ouest français. *In:* M. Lamotte and F. Bourlière, *Problèmes d'Écologie: Écosystèmes Terrestres*, Masson, Paris, pp. 17–74.

Risser, G., Karr, J.R. and Forman, R.T.T. (1983). *Landscape Ecology Directions and Approaches*. The Illinois Natural History Survey, Champaign, Illinois.

Risser, P. (1989). Landscape pattern and its effect on energy and nutrient distribution. *In:* I.S. Zonneveld and R.T.T. Forman, *Changing Landscapes: An Ecological Perspective*, Springer-Verlag, New York, pp. 45–56.

Roland, J. and Taylor, P.D. (1997). Insect parasitoid species respond to forest structure at different spatial scales. *Nature*, 386: 710–713.

Romme, W.H. (1982). Fire and landscape diversity in Subalpine Forest of Yellowstone National Park. *Ecological Monograph*, 52: 199–221.

Romme, W.H. and Despain, D.G. (1989). The Yellowstone fires. *BioScience*, 39: 695–699.

Romme, W.T. and Knight, D.H. (1982). Landscape diversity: the concept applied to Yellowstone Park. *BioScience*, 32: 664–670.

Rosenzweig, M.L. (1995). *Species Diversity in Space and Time*. Cambridge University Press, Cambridge.

Rost, G.R. and Bailey, J.A. (1979). Distribution of mule deer and elk in relation to roads. *Journal of Wildlife Management*, 43: 634–641.

Roze, F. (1978). *Étude analytique et comparative de la végétation des haies et talus en Bretagne*. Thesis, Université de Rennes 1.

Rudran, R., Jansen, M. and Seidensticker, J. (1980). *Wildlife. Environmental Assessment: Accelerated Mahaweli Development Program*. Ministry of Mahaweli Development, Colombo, Sri Lanka.

Rushton, S.P., Lurz, P.W.W., Fuller, R. and Garson, P.J. (1997). Modelling the distribution of the red and grey squirrel at the landscape scale: a combined GIS and population dynamics approach. *Journal of Applied Ecology*, 34: 1137–154.

Ryszkowski, L. and Kedzoria, A. (1987). Impact of agricultural landscape structure on energy flow and water cycling. *Landscape Ecology*, 1: 85–94.

Saint-Girons, H. and Duguy, R. (1976). Les reptiles du bocage. *In:* INRA, ENSA and Université de Rennes, *Les Bocages: Histoire, Écologie, Économie*, INRA, Rennes, pp. 343–346.

Saint Girons, M.C., Rosoux, R., Philippe, M.A. and A .petit, P. (1987). La typologie des haies et les populations de micromammifères l'exemple du marais Poitevin. *In: Annales de la Société des Sciences Naturelles de la Charente Maritime*. Muséum d'Histoire Naturelle, La Rochelle, 7: 593–608.

Salo, J. (1990). External processes influencing origin and maintenance of inland-water ecotones. *In:* R.J. Naiman and H. Décamps, *The Ecology and Management of Aquatic-Terrestrial Ecotones*, UNESCO, Parthenon Publishing, Lancs, U.K., 4: 37–59.

Sarthou, J.P. (1996). *Contribution à l'étude systématique, biogéographique et agroécocénotique des syrphidae (insecta, diptera) du Sud-Ouest de la France*. Science de la Vie. Toulouse, Thesis, Université de Toulouse.

Sauget, N. and Balent, G. (1993). The diversity of agricultural practices and landscape dynamics: the case of a hill region in the southwest of France. *In:* R.G.H. Bunce, L. Ryskzkowski and M.G. Paoletti, *Landscape Ecology and Agroecosystems*, Lewis, pp. 113–129.

Schippers, P., Verboom, J., Knaapen, J.P. and van Apeldoorn, R.C. (1996). Dispersal and habitat connectivity in complex heterogeneous landscapes: an analysis with GIS-based random walk model. *Ecography*, 19: 97–106.

Schlesinger, W.H. (1989). Discussion: ecosystem structure and function. *In:* J. Roughgarden, R.M. May and S.A. Levin, *Perspectives in Ecological Theory*, Princeton University Press, Princeton, pp. 268–274.

Schoonenboom, I.J. (1995). Overview and state of the art of scenario studies for the rural environment. *In:* J.F. Schoute, P.A. Finke, F.R. Veenklaas and H.P. Wolfert, *Scenario Studies for the Rural Environment*, Kluwer Academic Publisher, Dordrecht, pp. 15–24.

Schoute, J.F.T., Finke, P.A., Veeneklaas, F.R. and Wolfert, H.P., eds. (1995). *Scenario Studies for the Rural Environment*, Environment Policy, Kluwer Academic Publishers, Dordrecht.

Schumaker, N.H. (1996). Using landscape indices to predict habitat connectivity. *Ecology*, 77: 1210–1225.

Senft, R.L., Coughenour, M.B., Bailey, D.W., Rittenhouse, L.R., Sala, O.E. and Swift, D.M. (1987). Large herbivore foraging and ecological hierarchies. *Bioscience*, 37: 789–799.

Shannon, C.E. and Weaver, W. (1949). *Théorie Mathématique de la Communication*. CEL, Paris, 188 pp.

Sheail, J. (1987). *Seventy Five Years in Ecology, the British Ecological Society*. Blackwell Scientific Publications, Oxford.

Shorrocks, B. (1990). Coexistence in a pathchy environment. *In:* B. Shorrocks and I.R. Swingland, *Living in a Patchy Environment*, Oxford Science Publication, pp. 91–106.

Shorrocks, B., Marsters, J., Ward, I. and Evennett, P.J. (1991). The fractal dimension of lichens and the distribution of arthropod body lengths. *Functional Ecology*, 5: 457–460.

Shugart, H.H. (1988). *Terrestrial Ecosystems in Changing Environments*. Cambridge University Press, Cambridge.

Simberloff, D., Farr, A., Cox, J. and Mehlman, D.W. (1992). Movement corridors: conservation bargains or poor investments? *Conservation Biology*, 6: 493–504.

Simpson, J.W., Boerner, R.E.J., DeMers, M.N. and Berns, L.A. (1994). Forty-eight years of landscape change on two continuous Ohio landscapes. *Landscape Ecology*, 9(4): 261–270.

Singer, F.J., Schreier, W., Oppenheim, J. and Garten, E.O. (1989). Drought, fires, and large mammals. *BioScience*, 39: 716–722.

Smith, D.S. (1993). An overview of greenways. *In:* D.S. Smith and C. Hellmund, *Ecology of Greenways: Design and Function of Linear Conservation Areas*. University of Minnesota Press, Minneapolis, London, pp. 1–22.

Smith, T.M. and Urban, D.L. (1988). Scale and resolution of forest structural pattern. *Vegetation*, 74: 143–150.

Solbrig, O.T. (1991). *From Genes to Ecosystems: a Research Agenda for Biodiversity*. IUBS-SCOPE-UNESCO, Paris.

Sotherton, N.W. (1985). The distribution and abundance of predatory arthropods over wintering in field boundaries, *Annals of Applied Biology*, 106: 17–21.

Soulé, M.E. and Gilpin, M.E. (1991). The theory of wildlife corridor capability. *In:* D.A. Saunders and R.J. Hobbs, *Nature Conservation: the Role of Corridors*. Surrey Beatty and Sons, Chipping Norton, NSW, Australia, pp. 3–8.

Stamps, J.A., Buechner, M. and Krishnan, V.V. (1987). The effect of edge permeability and habitat geometry on emigration from patches of habitat. *American Naturalist*, 129: 533–552.

Stangel, P.W., Lennartz, M.R. and Smith, M.H. (1992). Genetic variation and population structure of red-cockaded woodpeckers. *Conservation Biology*, 6: 283–290.

Stearns, R.M. (1985). The evolution of greenways as an adaptive urban landscape form. *Landscape and Urban Planning*, 33: 65–80.

Storm, G.L., Andrews, R.D., Philipps, R.L., Bishop, R.A., Sniff, D.B. and Tester, J.R. (1976). Morphology, reproduction, dispersal and mortality of mid-western red fox populations. *Wildlife Monographs*, 49: 5–82.

Suarez Seoane, S. (1998). *Effectos ecologicos derivados del abandono de tierras de cultivo en la provincia de Leon (Municipio de Chozas de Abajo)*. Departmento de Ecologia, Genetica y Microbiologia, Universidad de Leon (Espagne) Unité SAD-Armorique, Institut National de la Recherche Agronomique (France). Leon, Rennes, 281 pp.

Suarez Seoane, S., Osborne, P.E. and Baudry, J. (2002). Responses of birds of different biogeographic origins and habitat requirements to agricultural land abandonment in northern Spain. *Biological Conservation*, 105(3): 333–344.

Sugihara, G. and May, R.M. (1990). Applications of fractals in ecology. *Tree*, 5(3): 79–86.

Szacki, J. (1987). Ecological corridor as a factor determining the structure and organisation of a bank vole population. *Acta Theriologica*, 32: 113–123.

Szacki, J. and Liro, A. (1991). Movements of small mammals in the heterogeneous landscape. *Landscape Ecology*, 5: 219–244.

Taberlet, P., Fumagalli, L., Wust-Sancy, A.G. and Cosson, J.F. (1998). Comparative phylogeography and postglacial colonization routes in Europe. *Molecular Ecology*, 7: 453–464.

Tamm, C.O. and Troedsson, T. (1955). An example of the amounts of plant nutrients supplied to the ground in road dust. *Oikos*, 6: 61–70.

Tansley, A.G. (1935). The use and abuse of vegetational concepts and terms. *Ecology*, 16: 284–307.

Tatoni, T. (1992). *Évolution post-culturale des agrosystèmes de terrasses en Provence calcaire*. Laboratoire de Biosystèmatique et Ecologie Méditerranéenne. Marseille, Thesis, Université de provence, p. 168.

Tatoni, T. and Roche, P. (1994). Comparison of old-field and forest revegetation dynamics in Provence. *Journal of Vegetation Science*, 5: 295–302.

Taylor, J., Paine, C. and FitzGibbon, J. (1995). From greenbelt to greenways: four Canadian case studies. *Landscape and Urban Planning*, 33: 47–64.

Taylor, P.D. (1997). *Putting the pieces together: fine-scale movements and large-scale distribution of a forest damselfly in agricultural landscapes*. XXth International Congress of Entomology, Florence, Italy, p. 449.

Taylor, P.D., Fahrig, L., Henein, K. and Merriam, H.G. (1993). Connectivity is a vital element of landscape structure. *Oikos*, 68: 571–573.

Taylor, P.D. and Merriam, G. (1996). Habitat fragmentation and parasitism of a forest damselfly. *Landscape Ecology*, 11: 181–189.

Tew, T.E. (1989). The behavioural ecology of the wood mouse (Apodemus sylvaticus) in the cereal field ecosystem. Oxford, University of Oxford. PhD Thesis.

Tew, T.E. and Macdonald, D.W. (1993). The effects of harvest on arable wood mice *Apodemus sylvaticus*. *Biological Conservation* 65: 279–283.

Thenail, C. (1992). *Fonctionnement des exploitations agricoles du pays d'Auge et utilisation des prairies permanentes*. Mémoire de fin d'études, INA-PG, INRA-SAD Normandie.

Thenail, C. (1996). *Exploitations agricoles et territoire(s): contribution à la structuration de la mosaïque paysagère*. Rennes, Thesis Université de Rennes 1.

Thenail, C. (2002). Relationships between farm characteristics and the variation of the density of hedgerows at the level of a micro-region of bocage landscape. Study case in Brittany, France. *Agricultural Systems*, 71: 207–230.

Thenail, C. and Baudry, J. (1996). *Consequences on landscape pattern of within farm mechanisms of land use changes (example in western France)*. Land use changes in Europe and its ecological consequences. European Center for Nature Conservation, Tilburg, pp. 242–258.

Thenail, C., Morvan, N., Le Cœur, D., Burel, F. and Baudry, J. (1997). Le rôle des exploitations agricoles dans l'évolution des paysages: un facteur essential des dynamiques écologiques. *Oecologia Mediterranea*, 23(1–2): 71–90.

Thomas, C.A. and Hanski, I. (1997). Butterfly metapopulations. *In:* I. Hanski and M. Gilpin, *Metapopulation Biology: Ecology, Genetics and Evolution,* Academic Press, San Diego, pp. 359–386.

Thomas, M.B. (1990). The role of man-made grassy habitats in enhancing carabid populations in arable land. *In:* N. Stork, *The Role of Ground Beetles,* Intercept, pp. 77–86.

Tilman, D. and Downing, J.A. (1994). Biodiversity and stability in grassland. *Nature,* 367: 363–365.

Tischendorf, L. and Wissel, C. (1997). Corridors as conduits for small animals: attainable distances depending on movement pattern boundary reaction and corridor width. *Oikos,* 79: 603–611.

Tollens, E. (1993). Agricultural reforms in China and prospects for agricultural development. *In:* P. Frantzen, *China's Economic Evolution,* Vubpress, Brussels, pp. 69–88.

Troll, C. (1939). Luftbildplan und okologische Bodenforschung. *Zeistschraft der gesellschaft fur erdkunde zu Berlin,* pp. 241–298.

Turchin, P. (1991). Translating foraging movements in heterogeneous environments into the spatial distribution of foragers. *Ecology,* 72: 1253–1266.

Turchin, P. (1996). Fractal analyses of animal movement: a critique. *Ecology,* 77: 2086–2090.

Turchin, P.B. (1986). Modelling the effect of host patch size on Mexican bean beetle emigration. *Ecology* 67: 124–132.

Turner, II, B.L., Clark, W.C., Kates, R.W., Richards J.F., Mathews, J.T. and Meyer, W. (1990). *The Earth as Transformed by Human Action.* Cambridge University Press, Cambridge, 713 pp.

Turner II, B.L. and Meyer, W.B. (1994). Global land-use and land-cover change: an overview. *In:* W.B. Meyer and B.L. Turner II, *Changes in Land Use and Land Cover: a Global Perspective,* Cambridge University Press, Cambridge, pp. 3–10.

Turner, M.G., ed. (1987). *Landscape Heterogeneity and Disturbance.* Ecological Studies 64, Springer-Verlag, New York.

Turner, M.G. (1987). Simulation of landscape changes in Georgia: a comparison of 3 transition models. *Landscape Ecology,* 1(1): 29–36.

Turner, M.G. (1990). Spatial and temporal analysis of landscape patterns. *Landscape Ecology,* 4: 21–50.

Turner, M.G., Arthaud, G.J., Engstrom, R.T., Hejl, S.J., Liu, J., Loeb, S. and McKelvey, K. (1995). Usefulness of spatially explicit population models in land management. *Ecological Applications,* 5(1): 12–16.

Turner, M.G. and Gardner, R.H. (1991). *Quantitative Methods in Landscape Ecology.* Springer-Verlag, New York, 536 pp.

Turner, M.G., Gardner, R.H., Dale, V.H. and O'Neill, R.V. (1989). Predicting the spread of disturbance across heterogeneous landscape. *Oikos,* 55: 121–129.

Turner, M.G. and Ruscher, C.L. (1988). Changes in landscape patterns in Georgia, USA. *Landscape Ecology,* 1(4): 227–240.

Turner, M.G., Wear, D.N. and Flamm, R.O. (1996). Land ownership and land-cover change in the southern Appalachian highlands and the Olimpic Peninsula. *Ecological Applications,* 6(4): 1150–1172.

Turner, M.G., Wu, Y., Wallace, L.L., Romme, W.H. and Brenkert, A. (1994). Simulating winter interactions among ungulates, vegetation, and fire in northern Yellowstone Park. *Ecological Applications,* 4: 472–496.

Turner, S.J., O'Neill, R.V., Conley, W., Conley, M.R. and Humphries, H.C. (1991). Pattern and scale: statistics for landscape ecology. *In:* M.G. Turner and R.H. Gardner, *Quantitative Methods in Landscape Ecology,* Springer-Verlag, New York, pp. 17–50.

Urban, D.L., O'Neill, R.V. and Shugart, H.H. (1987). Landscape ecology. *BioScience,* 37: 119–127.

Uusi-Kämppä, J., Turtola, E., Hartikainen, H. and Yläranta, T. (1996). The interactions of buffer zones and phosphorus runoff. *In:* N.E. Haycock, T.P. Burt, K.W.T. Goulding and G. Pinay, *Buffer Zones: Their Processes and Potential in Water Protection,* Quest Environmental, Harpenden, pp. 43–53.

van Apeldoorn, R.C., Oostenbrink, W.T., Van Winden, A. and Van der Zee, F.F. (1992). Effects of habitat fragmentation on the bank vole, *Clethrionomys glareolus*, in an agricultural landscape. *Oikos*, 65: 265–274.

van der Zande, A.N. ter Eurs, W.J. and van der Weijden, W.J. (1980). The impact of roads on the densities of four bird species in an open field habitat: evidence of a long distance effect. *Biological Conservation*, 18: 299–321.

van Dorp, D. and Opdam, P.F.M. (1987). Effects of patch size, isolation and regional abundance on forest bird communities. *Landscape Ecology*, 1: 59–73.

van Hees, W.W.S. (1994). A fractal model of vegetation complexity in Alaska. *Landscape Ecology* 9(4): 271–278.

Varley, J.D. and Schullery, P. (1991). Reality and opportunity in the Yellowstone fires of 1988. *In:* R.B. Keiter and M.S. Boyce, *The Greater Yellowstone Ecosystem: Redefining America's Wilderness Heritage.* Yale University Press, New Haven, pp. 105–121.

Vaudour, J., Bonin, G. and Tatoni, T. (1991). Terrasses de culture: leur évolution après abandon et mode de gestion minimum. Ministry of the Environment, Comité EGPN, Paris.

Verboom, B. and Van Apeldoorn, R. (1990). Effects of habitat fragmentation on the red squirrel, *Sciurus vulgaris* L. *Landscape Ecology*, 4: 171–176.

Verboom, J. (1995). Dispersal of animals and infrastructure. A model study: summary. The Netherlands, Road and hydraulic engineering division.

Verboom, J., Schotman, A., Opdam, P. and Metz, J.A.J. (1991). European muthatch metapopulations in a fragmented agricultural landscape. *Oikos*, 61: 149–156.

Vermeulen, R. (1993). The effects of different vegetation structures on the dispersal of carabid beetles from poor sandy heath and grasslands. *In:* K. Desender, M. Dufrêne, M. Loreau, M. Luff and J.P. Maelfait. *Carabid Beetles: Ecology and Evolution.* Kluwer, Dordrecht, pp. 289–394.

Vermeulen, R. and Opdam, P.F.M. (1995). Effectiveness of roadside verges as dispersal corridors for small ground-dwelling animals: a simulation study. *Landscape and Urban Planning*, 31: 233–248.

Vernet, J.L. (1997). *L'Homme et la Forêt Méditerranéenne de la Préhistoire à Nos Jours.* Errance, Coll. des Hespérides, Paris.

Vernier, L.A. and Fahrig, L. (1996). Habitat availability causes the species abundance-distribution relationship. *Oikos*, 76: 564–570.

Vink, A.P.A. (1983). *Landscape Ecology and Land Use.* Longman, London and New York.

Vos, C.C. and Opdam, P. eds. (1993). *Landscape Ecology of a Stressed Environment.* IALE Studies in Landscape Ecology. Chapman Hall, Cambridge.

Voss, R.F. (1998). Fractals in nature: from characterization to simulation. *In:* H.-O. Peitgen and D. Saupe, *The Science of Fractal Images*, Springer-Verlag, New York, pp. 21–70.

Wace, N.M. (1977). Assessment of dispersal of plant species—the carborne flora in Canberra. *Proceedings of the Ecological Society of Australia*, 10: 167–186.

Waliczky, Z. (1991). Guild structure of beetle communities in three stages of vegetational succession. *Acta Zoologica Hungarica*, 37: 313–324.

Walmsley, A. (1995). Greenways and the making of urban form. *Landscape and Urban Planning*, 33: 81–127.

Watt, A.S. (1947). Pattern and process in the plant community. *Journal of Ecology*, 35: 1–22.

Way, J.M. and Greig-Smith, P.W., eds. (1987). *Field Margins.* BCPC Monograph., British Crop Protection Council.

Wear, D.N., Turner, M.G. and Naiman, R.J. (1998). Land cover along an urban-rural gradient: implications for water quality. *Ecological Applications*, 8(3): 619–630.

Weber, E. (1983). *La fin des terroirs. La modernisation de la France rurale* (1870–1914), Fayard, Paris.

Welsh, H. (1990). Relictual amphibians and old growth forest. *Conservation Biology*, 3: 309–319.

Wetzel, J.F., Wambaugh, J.R. and Peek, J.M. (1975). Appraisal of white-tailed deer winter habitats in northeastern Minnesota. *Journal of Wildlife Management*, 38: 59–66.

Whitcomb, R.F., Robbins, C.S., Lynch, J.F., Whitcomb, B.L., Klimkiewicz, M.K. and Bystrak, D. (1981). Effects of forest fragmentation on avifauna of eastern deciduous forest. *In:* R.L. Burgess and D.M. Sharpe, *Forest Island Dynamics in Man-dominated Landscapes.* Springer-Verlag, New York, 41: 125–206.

Wieber, J.C. (1985). Le paysage visible, un concept nécessaire. *In:* V. Berdoulay and M. Phipps, *Paysage et Système,* Université d'Ottawa, pp. 167–177.

Wiens, J.A. (1989). Spatial scaling in ecology. *Functional Ecology,* 3: 385–397.

Wiens, J.A. (1992). Ecological flows across landscape boundaries: a conceptual overview. *In:* A.J. Hansen and F. Di Castri, *Landscape Boundaries: Consequences for Biotic Diversity and Ecological Flows,* Springer-Verlag, New York, pp. 217–235.

Wiens, J.A. (1995). Landscape mosaics and ecological theory. *In:* L. Hansson, L. Fahrig and G. Merriam, *Mosaic Landscapes and Ecological Processes,* Chapman Hall, London, pp. 1–26.

Wiens, J.A. (1997). Metapopulation dynamics and landscape ecology. *In:* I. Hanski and M. Gilpin, *Metapopulation Biology: Ecology, Genetics and Evolution,* Academic Press, San Diego, pp. 43–62.

Wiens, J.A., Crawford, C.S. and Gosz, J.R. (1985). Boundary dynamics: a conceptual framework for studying landscape ecosystems. *Oikos,* 45: 421–427.

Wiens, J.A. and Milne, B.T. (1989). Scaling of landscapes in landscape ecology, or landscape ecology from a beetles' perspective. *Landscape Ecology,* 3: 87–96.

Wiens, J.A., Schooley, R.L. and Weeks, R.D.J. (1997). Patchy landscapes and animal movements: do beetles percolate? *Oikos,* 78: 257–264.

Wiens, J.A., Stenseth, N.C., Van Horne, B. and Ims, R.A. (1993). Ecological mechanisms and landscape ecology. *Oikos,* 66: 369–380.

Wilcove, D.S. (1985). Nest predation in forest tracks and the decline of migratory songbirds. *Ecology,* 66: 1211–1214.

Wilcove, D.S. and May, R.M. (1986). National park boundaries and ecological realities. *Nature,* 324: 206–207.

Wilson, E.O. (1988). *Biodiversity.* National Academic Press, Washington, D.C.

With, K.A. and Crist, T.O. (1995). Critical thresholds in species' responses to landscape structure. *Ecology,* 76(8): 2446–2459.

With, K.A. and Crist, T.O. (1996). Translating across scales: simulating species distribution as the aggregate response of individuals to heterogeneity. *Ecological Modelling,* 93: 125–137.

With, K.A., Gardner, R.H. and Turner, M.G. (1997). Landscape connectivity and population distribution in heterogeneous environments. *Oikos,* 78: 151–169.

Yahner, R.H. (1993). Seasonal dynamics, habitat relationships, and management of avifauna in farmstead shelterbelts. *J. Wild. Management,* 47: 85–104.

Zhang, Z.B. and Usher, M.B. (1991). Dispersal of wood mice and bank voles in an agricultural landscape. *Acta Theriologica,* 36: 239–245.

Zonneveld, I.S. (1995). *Land Ecology.* SPB Publishing, Amsterdam.

Zonneveld, I.S. and Forman, R.T.T. (1989). *Changing Landscapes: an Ecological Perspective.* Springer-Verlag, New York.

Glossary

Biodiversity: The variety and variability of living organisms and the ecosystems in which they grow. Diversity can be defined as the number and relative abundance of the elements considered. The components of biological diversity are organized at several levels, from ecosystems to the chemical structures that are the molecular bases of heredity. This term thus encompasses ecosystems, species, genes, and their relative abundance.

Bocage: Agrarian landscape characterized by the presence of living hedgerows that surround cultivated fields and grasslands. These hedgerows form networks connected to woods, moors, or other uncultivated areas.

Buffer zone: Protective zone surrounding a sensitive zone, for example, buffer zones protecting water courses and perimeters harnessing water, or buffer zones surrounding nature reserves.

Community (or **species assemblage**): The set of populations of species belonging to a single taxonomic group and presenting a certain number of common traits in their ecology.

Connectedness: Measurement of the spatial arrangement of landscape elements that takes into account the contiguity of elements of the same nature. It is a cartographic measurement.

Connectivity: Measurement of the possibilities of movement of organisms between the patches of the landscape mosaic. It is a function of the landscape composition, its configuration (spatial arrangement of the landscape elements), and the adaptation of the behaviour of organisms to these two variables.

Corridor: Linear element of the landscape that differs in physiognomy from the adjacent environment. Corridors play several roles. They serve as conduits that favour movement or as barriers that limit movement.

Denitrification: Bacterial process of transformation of nitrates into a gaseous form of nitrogen.

Dispersal: Process of displacement of an individual in the course of its life cycle: from one type of habitat to another for multi-habitat species, and from one patch to another of the same kind for specialist species. Depending on the organism considered, the term "dispersal" signifies movement (wide sense) or a particular type of movement. For example, for vertebrates, it means movements outside the home range, excepting short exploratory incursions (Lidicker, 1975: 497).

Dispersion: Modality by which individuals of a population are spatially distributed.

Disturbance: A disturbance is any relatively discrete event in time that disrupts ecosystem, community, or population structure and changes resources, substrate availability, or the physical environment (Pickett and White, 1985).

Diversity: The measurement of diversity takes into account the number of elements (landscape elements, species) and their relative abundance. The most often used diversity index in ecology is the Shannon-Weaver index:

$$H = \Sigma \; p_i \; log \; p_i$$

where p_i is the relative frequency of element i.

Ecosystem: Tansley defines ecosystem as an element in the hierarchy of physical systems ranging from the universe to the atom, as the basic system of ecology, and as the combination of all living organisms and the physical environment. The ecosystem has for a long time been defined as a homogeneous and aspatial entity. Duvigneaud (1980: 20) defines it as a homogeneous biocoenosis that develops in a homogeneous environment.

Fragmentation: Forman (1995) defines fragmentation as the breaking up of an object. When a saucer falls and breaks into pieces, it is fragmented. A more generally accepted definition in ecology is the dynamic process of reduction in the area of a habitat and its separation into several fragments.

Generalist: A species is said to be generalist when it indiscriminately uses a wide variety of resources. A species is generalist in relation to the landscape when it indiscriminately uses the elements of the mosaic.

Gradient: Continuous variation of an ecological factor in space or in time. A thermal gradient, for example, is a continuous variation of the temperature along a geographic transept.

Habitat: The place in which a given species lives. In the narrow sense, a habitat contains all the landscape elements, of whatever kind, used by the species. By extension, one of the types of elements used by a species is often called a "habitat". Empidids can be said to use several habitats: open water, the shore, and soil that is not ploughed.

Heterogeneity: Character of land having elements that differ in form, size, or nature. In landscape ecology, heterogeneity integrates the diversity of elements and their spatial arrangement.

Hierarchy theory: A theory that organizes observations at different scales of space and time and allows us to break up complex systems into levels of organization.

Interior zone: Central portion of a landscape element, located within the margin. Some species are characteristic of interior zones.

Landscape: Portion of space corresponding to a scale appropriate to human activities. It is defined by its spatial and temporal heterogeneity, the human activities that occur in it, and its environment.

Matrix: Dominant element of the landscape. In agrarian landscapes, the term "agricultural matrix" signifies all the parcels that are reserved for agricultural production.

Metapopulation: A population made up of populations that die out and are recolonized locally.

Mosaic: An assemblage of elements of different kinds. The average size of these elements defines the grain of the mosaic.

Movement: Any type of displacement by an organism. Movements vary in intensity and nature (direction, frequency) as a function of mobility and associated biological processes.

Network: Set of interconnected linear elements. The hydrographic network, for example, is the set of water courses associated with a watershed.

Organization: Quality of a system of being structured, endowed with a determined constitution and a mode of functioning.

Patch: Landscape element defined by its size, form, and nature.

Percolation: Percolation is a problem of communication that arises in an extended environment including a large number of "sites" susceptible of relaying information locally. These communicate among themselves by links that are randomly efficient. Depending on whether the proportion of links is greater or lesser than a threshold value, there may or may not be a possibility of transmitting the information over a long distance.

Population: Set of individuals belonging to a single species and forming a functional demographic unit.

Scale: Scale in ecology is defined as the degree of resolution, the scale of geographers, and the extent of the study area. The time scale is a function of the study time frame and of the interval between two measurements.

Specialist: A species that has particular ecological needs is said to be specialist. It can only survive in a well-defined type of environment and/ or can only use a particular type of resource. Its presence is therefore directly determined by the presence of the type of environment with which it is associated.

Species: The fundamental taxonomic unit of classification of the living world. A species is made up of the set of individuals belonging to inter-fertile populations that freely exchange their gene pool.

Species richness: The number of species present in an environment.

Sub-surface flow: Oblique flow that occurs in the surface horizons of soil.

Watershed: Land that receives the rainfall that feeds a watercourse.

Index